Noxious Range Weeds

Noxious Range Weeds

EDITED BY

Lynn F. James John
O. Evans Michael
H. Ralphs
R. Dennis Child

ASSISTANT EDITORS

Terrie L. Wierenga
Joyce A. Johnson
Betty J. Sigler

ed 1991 by Westview Press

2021 by Routledge
venue, New York, NY 10017
e, Milton Park, Abingdon, Oxon OX14 4RN

an imprint of the Taylor & Francis Group, an informa business

1991 Taylor & Francis

ongress Cataloging-in-Publication Data
ge weeds / edited by Lynn F. James ... [et al.].
special studies in agriculture science and policy)
hy: p.
idex.
33-8395-1
ck poisoning plants—West (U.S.). 2. Rangelands—Weed
est (U.S.). 3. Livestock poisoning plants—Control—West
eeds—West (U.S.). I. James, Lynn F. II. Series.
47 1991
ic20   91-31637
                                                            CIP

8-0-3670-1662-3 (hbk)
8-0-3671-6649-6 (pbk)

Contents

Preface

Rangelands, private and public, are of considerable importance to the general populace as well as those directly associated with them because of vocational pursuits. Businesses such as livestock, lumber, mining, and their supporting industries and communities rely directly on these lands for their existence. Other people enjoy and use these same resources for recreational pursuits, and everyone associates high aesthetic values with them.

Recently there has been great concern about the invasion of rangeland and forests by noxious weeds, most of which are not native to North America and pose peculiar threats for most uses and values of these lands. The rapid invasion of the ranges and pasture areas by these plants has not been correlated with any particular use or land management strategy to this point. Plants such as leafy spurge, knapweeds, thistles, and certain woody plants seem to have strong competitive advantages in native plant communities.

To address these concerns, a group of interested scientists, range managers, and weed maintenance personnel representing agencies such as the USDA Forest Service, Bureau of Land Management, USDA Agricultural Research Service, USDA Soil Conservation Service, Montana Department of Agriculture, Montana Weed Control Association, Intermountain Noxious Weed Ad Hoc Committee (INWAC), and interested ranchers and individuals met in February 1989 in Billings, Montana, to discuss undesirable plants and investigate avenues of weed management to realize a sustained cooperative effort between federal, state, and local agencies and with Canada to deal with noxious weeds. At that meeting it was concluded that a national conference would be held to exchange information regarding noxious range weeds and explore means for their control and management. The U.S. Department of Agriculture, Agricultural Research Service (USDA-ARS) was asked to serve as the lead agency to develop the conference. The USDA-ARS Poisonous Plant Research Laboratory had simultaneously prepared for a similar but smaller conference concerning noxious and poisonous range plants.   The Poisonous Plant Research Laboratory was subsequently requested to host a "National Noxious Range Weed Conference" on behalf of those represented at the Billings meeting. The proceedings of this national conference are reported herein.

At this conference, the Billings group, whose name was formalized to the "Greater Northwest Noxious Weed Management Council," held a one-day meeting on "A Forum for Continuing Cooperation." A report of the activities of this meeting is included in the proceedings.

The program committee made every attempt to ensure that the leaders in research participated in presenting the most recent and up-to-date information on the problem of noxious range weeds. We believe that we succeeded in accomplishing this task.

Lynn F. James
John O. Evans
Michael H. Ralphs
R. Dennis Child

Acknowledgments

We express appreciation to all who contributed to the success of this conference and the publication of this proceedings. Several people helped make the conference a pleasant and enjoyable experience. These include Larry Riley of Utah State University Conferences and Institute Division, who ensured congenial meeting facilities, good housing, and arrangements for banquets, etc.; Betty Sigler for her help in organizing, preparing and sending out announcements, and attending to many details essential for a successful and smooth conference; Phil Kechele and Terrie Wierenga for their efficient operation of lights and projectors, etc.; the Utah Section of the Society for Range Management and Bob Hamner, USFS, for a very successful cookout in Logan Canyon; and Steve Dewey, Jack Evans, Rodney Lym, Peter Fay, Lloyd Haderlie, Mike Ralphs, and Kirk McDaniel who set up tours and served as tour guides.

Thanks to Joe Antognini, National Program Leader for Weeds, USDA-ARS, and George Hittle, Weed Supervisor, Wyoming State Department of Agriculture, for acting as observers and for summarizing the accomplishments at the conclusion of the meeting.

The editors express their appreciation to assistant editors Terrie Wierenga, Joyce Johnson, and Betty Sigler for their invaluable assistance in editing and preparing this proceedings for publication.

Program Committee:

Lynn F. James, Chairman
John O. Evans
Michael H. Ralphs
Steven A. Dewey
Curtis M. Johnson
Larry H. Maxfield
R. Dennis Child
Deen Boe

List of Contributors

Ansley, R. James. MS, PhD. Texas Agricultural Experiment Station, Texas A&M University, PO Box 1658, Vernon, TX 76384

Asay, Kay H. USDA-ARS Forage and Range Research Laboratory, Utah State University, Logan, UT 84322-6300

Beck, K. George. BSc, MS, PhD. Colorado State University, 115 Weed Research Lab, Fort Collins, CO 80523

Bedell, Thomas E. PhD. Department of Rangeland Resources, Oregon State University, Corvallis, OR 97331

Boe, Deen. BSc, MS. Deputy Director, Range Management, USDA Forest Service, PO Box 96090, Washington, DC 20090-6090

Bovey, Rodney W. BSc, MS, PhD. USDA-ARS, Dept. of Range Science, Texas A&M University, College Station, TX 77843

Briede, Jan W. PhD. Dept. of Animal and Range Sciences, New Mexico State University, Las Cruces, NM

Bruckart, W. L. PhD. USDA-ARS Foreign Disease Weed Science Research, Building 1301, Room 218, Fort Detrick, Frederick, MD 21702

Call, Christopher A. BSc, MS, PhD. Range Science Department, Utah State University, Logan, UT 84322-5230

Callihan, Robert H. PhD. Department of Plant Science, University of Idaho 83843

Campobasso, Gaetano. BA. USDA-ARS Biological Control of Weeds Laboratory–Europe, American Embassy, AGRIC-ARS, APO New York 09794-0007

Chicoine, Timothy K. MS. Dupont Agrichemicals, 1825 71st Avenue, Vero Beach, FL 32966

Coombs, Eric M. MS. Biological Control of Weeds, Oregon Department of Agriculture, 635 Capitol Street NE, Salem, OR 97310

Dahl, Bill E. BSc, MS, PhD. Texas Tech University, Range and Wildlife Management Department, Lubbock, TX 79409

Davis, Edward S. MS. Plant and Soil Science Department, Montana State University, Bozeman, MT 59717

DeLoach, C. Jack. BSc, MS, PhD. USDA-ARS Grassland, Soil, and Water Research Laboratory, 808 E. Blackland Road, Temple, TX 76502

Dewey, Steven A. PhD. Plants, Soils, and Biometeorology Department, Utah State University, Logan, UT 84322-4820

Duncan, Keith W. BSc, MA, PhD. Brush & Weed Specialist, New Mexico State University, 67 East Four Dinkus Road, Ag Science Center, Artesia, NM 88210

Edrington, Thomas S. BSc, MS. Department of Animal and Range Sciences, New Mexico State University, Las Cruces, NM 88003-0003

Evans, John O. PhD. Plants, Soils, and Biometeorology Department, Utah State University, Logan, UT 84322-4820

Fay, Peter K. MS, PhD. Plant and Soil Science Department, Montana State University, Bozeman, MT 59717

Flores-Rodriguez, Gonzalo I. MVZ, MC, PhD. Department of Animal and Range Sciences, New Mexico State University, Las Cruces, NM 88003-0003

Fornasari, Luca. Perfez. (Pisa). USDA-ARS Biological Control of Weeds Laboratory-Europe, American Embassy, AGRIC-ARS, APO New York 09794-0007

Frandsen, Ed. BSc, MS. Range Management Staff, USDA Forest Service, Economics and Natural Resource Specialist, PO Box 96090, Washington, DC 20090-6090

Frasier, Gary W. BSc, MS. USDA-ARS Aridland Rangeland Management Unit, 2000 East Allen Road, Tucson, AZ 85719

Haderlie, Lloyd C. BSc, PhD. Agra-Serv, Inc., 3243 W. JoAnna Ct., American Falls, ID 83211

Hoefer, Ray H. PhD. DowElanco, 3905 Vincennes Road, Indianapolis, IN 46268

Horton, William H. BSc. USDA-ARS Forage and Range Research Laboratory, Utah State University, Logan, UT 84322-6300

Jacoby, Pete W., Jr. PhD. Texas Agricultural Experiment Station, Texas A&M University Agricultural Research Center, PO Box 1658, Vernon, TX 76384

James, Lynn F. PhD. USDA-ARS Poisonous Plant Research Laboratory, 1150 East 1400 North, Logan, UT 84321

Johnsen, Thomas N., Jr. PhD. USDA-ARS Aridland Rangeland Management Unit, 2000 East Allen Road, Tucson, AZ 85719

Johnson, Hyrum B. BSc, MS, PhD. USDA Agricultural Research Service Grassland, Soil, and Water Research Laboratory, 808 East Blackland Road, Temple, TX 76502

Jones, Thomas A. BSc, MS, PhD. USDA-ARS Forage and Range Research Laboratory, Utah State University, Logan, UT 84322-6300

Knutson, Lloyd. PhD. USDA-ARS Biological Control of Weeds Laboratory - Europe, American Embassy - Agriculture, APO NY 09794

Lacey, Celestine A. MS. Consultant, Weed Management Services, PO Box 9055, Helena, MT

Lacey, John R. PhD. Extension Range Management Specialist, Montana State University, Bozeman, MT 59717

Leino, Phil W. PhD. Consultant, Iowa Natural Heritage Foundation, 2016 74th Street, Des Moines, IA 50322

Lym, Rodney G. PhD. Crop and Weed Sciences Department, North Dakota State University, Fargo, ND 58105

McAllister, Ray S. BSc, MS, MEd, PhD. National Agricultural Chemicals Association, 1155 - 15th St. NW, Suite 900, Washington, DC 20005

McDaniel, Kirk C. BSc, MS, PhD. Department of Animal and Range Sciences, New Mexico State University, Las Cruces, NM 88001

McEvoy, Peter B. BA, PhD. Entomology Department, Oregon State University, Corvallis, OR 97331-2907

Mayeux, Herman S., Jr. BSc, MS, PhD. USDA Agricultural Research Service Grassland, Soil, and Water Research Laboratory, 808 Blackland Road, Temple, TX 76502

Morishita, Don W. Plant, Soils and Entomology Department, University of Idaho, Twin Falls, ID 83301

Nelson, C. Riley. PhD. Department of Zoology, University of Texas, Austin, TX 78712-1064

Oetting, Bryan C. BSc, MS. Department of Animal and Range Sciences, New Mexico State University, Las Cruces, NM 88003-0003

Olson, Bret E. PhD. Assistant Professor, Department of Animal and Range Sciences, Montana State University, Bozeman, MT 59717

Panter, Kip E. PhD. USDA-ARS Poisonous Plant Research Laboratory, 1150 East 1400 North, Logan, UT 84321

Polley, H. Wayne. BSc, MS, PhD. USDA Agricultural Research Service Grassland, Soil, and Water Research Laboratory, 808 Blackland Road, Temple, TX 76502

Quimby, P. C., Jr. PhD. USDA-ARS Biological Control of Weeds Research Unit, Rangeland Weeds Lab, Montana State University, Culbertson Hall, Bozeman, MT 59717-0056

Ralphs, Michael H. PhD. USDA-ARS Poisonous Plant Research Laboratory, 1150 East 1400 North, Logan, UT 84321

Rasmussen, G. Allen. PhD. Extension Range Specialist, Assistant Professor, Range Science Department, Utah State University, Logan, UT 84322-5230

Rees, Norman E. MS. USDA-ARS Rangeland Weeds Lab, Culbertson Hall, Montana State University, Bozeman, MT 59717

Roché, Ben F., Jr. BSc, MS, PhD. Extension Range Management, Department of Natural Resource Sciences, Washington State University Cooperative Extension, Pullman, WA 99164-6410

Roché, Cindy Talbott. BSc, MS. Department of Natural Resource Sciences, Washington State University Cooperative Extension, Pullman, WA 99164-6410

Rosenthal, Sara S. PhD. USDA-ARS Rangeland Weeds Laboratory, Culbertson Hall, Montana State University, Bozeman, MT 59717-0056

Ross, Timothy T. BSc, MS, PhD. Department of Animal and Range Sciences, New Mexico State University, Las Cruces, NM 88003-0003

Rumbaugh, Melvin D. PhD. USDA-ARS Forage and Range Research Laboratory, Utah State University, Logan, UT 84322-6300

Smith, G. Stanley. BSc, MS, PhD. Department of Animal and Range Sciences, New Mexico State University, Las Cruces, NM 88003-0003

Sosebee, Ronald E. BSc, MS, PhD. Department of Range and Wildlife Management, Texas Tech University, Lubbock, TX 79409

Sobhian, Rouhollah. PhD. USDA-ARS Biological Control of Weeds Laboratory-Europe, Rome, Italy, American Consulate General, APO New York 09693

Spencer, Neal R. BSA. USDA-ARS Northern Plains Soil and Water Research Center, PO Box 1109, Sidney, MT 59270

Stevens, Kenneth L. BSc, MA, PhD. USDA-ARS Western Regional Research Laboratory, 800 Buchanan Street, Albany, CA 94596

Thorne, Kaye Hugie. MS. Herbarium, Room 375 MLBM, Brigham Young University, Provo, UT 84602

Torell, James M. MSc. Research Associate - Crop Management, SW Idaho Research and Extension Center, 29603 University of Idaho Lane, Parma, ID 83660-6701

Torell, L. Allen. PhD. Department of Agricultural Economics and Agricultural Business, New Mexico State University, Las Cruces, NM 88003

Turner, Charles E. PhD. USDA-ARS-PBA, Western Region Research Center, 800 Buchanan Street, Albany, CA 94710

Ueckert, D. N. PhD. Texas A&M University Agricultural Research and Extension Center, San Angelo, TX 76901-9782

Welsh, Stanley L. PhD. Life Science Museum and Department of Botany and Range Science; Room 375 MLBM, Brigham Young University, Provo, UT 84602

West, Neil E. BSc, PhD. Department of Range Science, Utah State University, Logan, UT 84322-5230

Whitson, Thomas D. BSc, MS, PhD. Extension Weed Specialist, Box 3354, University of Wyoming, Plant, Soil, and Insect Sciences, Laramie, WY 82071

Young, James A. PhD. Landscape Ecology of Rangelands, USDA-ARS, 920 Valley Road, Reno, NV 89512

1

What Is a Noxious Weed?

Steven A. Dewey and James M. Torell

Abstract

The term "weed" has been defined many different ways, but can be considered a very general term referring to any plant that interferes with the management objectives for a given area of land at a given point in time. "Noxious" weed is a more selective term, referring either to weeds requiring management or control because of legislative action or to weeds that are extremely prolific, invasive, competitive, harmful, or difficult to control.

Discussion

During this conference, nearly every speaker will be using the terms "weed" or "noxious weed". Our assignment is to review definitions of these two terms, hopefully answering the question: "What is the difference between a weed and a noxious weed?"

The first, and perhaps most difficult, term to define is "weed". Finding a single definition that all can agree upon is probably impossible, based on the fact that there are so many weed definitions in the literature (1-8). Common definitions include:

- A plant out of place.
- Unwanted plants.
- A plant growing where it is not desired.
- A plant whose virtues have not yet been discovered.

- A plant that is more detrimental than beneficial.
- A plant that grows spontaneously in a habitat greatly modified by man.
- Any plant other than the crop sown.
- A plant is a weed if, in any specified geographical area, its populations grow entirely or predominantly in situations markedly disturbed by man.
- A plant interfering with man's intended use of the land.
- A plant that originated under a natural environment and, in response to imposed and natural environments, evolved, and continues to do so, as an interfering associate with our crops and activities.
- Plants that are competitive, persistent, and pernicious. They interfere with human activities and, as a result, are undesirable.
- Any herbaceous or woody plant that is objectionable or interferes with the activities or welfare of man.
- A plant that has little value or grows to the disadvantage of other more desirable plants.
- Unwanted and undesirable plants which interfere with the utilization of land and water resources and thus adversely affect human welfare.

In our opinion, each of these definitions has one or more weaknesses. Some are too vague. Others are too restrictive, addressing only cultivated cropping situations and not recognizing unwanted or harmful plants in pastures, rangeland, forests, recreation sites, or waste areas. Most definitions fail to recognize aquatic weeds. Some definitions erroneously imply that all weeds require significant man-related soil disturbance (tillage, erosion, or livestock overgrazing) to become established and thrive; and that weeds cannot or do not infest nondisturbed land or "native" plant communities. Some definitions recognize volunteer crops or other "conditionally" undesirable plants as weeds, while others limit the term "weeds" to plants that could be considered undesirable by almost all people in all situations.

Nearly all weed definitions are based on man's interpretation of what are "desirable" and "undesirable" plants. Opinions in this matter will obviously differ, depending on each individual's point of view. Whether a plant is considered a weed usually depends on intended land use. For example, big sagebrush (*Artemisia tridentata*) growing on big game winter range is almost always regarded as a desirable species by wildlife managers. Yet, in a recent survey of weed scientists in the 12 Western states, big sagebrush was ranked as the most important woody *weed* species on rangeland managed for cattle production. Volunteer crop plants, such as potatoes or rye, can be difficult "weed" problems in subsequent rotational crops, even though otherwise they are important crops.

A universal weed definition would have to be broad and flexible, allowing for differences in human perspectives as well as diversity and change in land (or water) use. The following definition (9) fits those criteria, making it appropriate for nearly all situations:

A weed is any plant that interferes with the management objectives for a given area of land (or body of water) at a given point in time.

The term "noxious" applies to a more specific group of weeds. However, the fact that there are at least two distinct usages of the word suggests some degree of confusion. Correctly applied to weeds, the term "noxious" refers to undesirable plant species that are regulated in some way by law. For example, federal and state pure seed laws have been enacted which prohibit or limit the sale of crop seeds containing noxious weed seeds. The Utah Seed Act, which typifies most state seed purity laws, states that "no person in this state shall offer or expose any agricultural seed for sale or sowing unless it is free of *noxious* weed seed, subject to any tolerance allowed for restricted *noxious* weeds." State and federal weed control laws also exist, which require the control of designated noxious weed species growing on private and public lands. The Utah Noxious Weed Act, representative of many state weed control laws, requires that "*noxious* weeds standing, being, or growing shall be controlled and the spread prevented by cutting, tillage, cropping, pasturing, chemicals, or other effective methods, as often as required to prevent the weed from blooming, maturing seeds, or spreading by root or other means."

In both types of laws, noxious weed species are declared and listed as part of accompanying regulations. Lists will vary in length and content from state to state and law to law, with some weeds considered noxious in one state but not in another. As an example, 168 plant species are considered noxious in California. In Washington, there are only 87 species designated as noxious, while in Idaho, Utah, and Colorado there are 31, 17, and 4 noxious species declared, respectively. "Noxious", in a weed control sense, is a regulatory designation rather than a biological or agronomic term (10). According to this usage, a *noxious weed* might be defined simply as:

A plant species requiring management or control because of legislative action.

However, the term "noxious" is frequently used in a broader sense by the general public to describe weeds that are especially troublesome, difficult to control, or prolific. To some people the word "noxious" indicates a perennial weed. To others the term implies poisonous, spiny, or otherwise injurious weeds. Still others reserve the category for highly competitive invading species introduced from other countries. In the dictionary "noxious" is defined as "harmful, deadly, or pernicious". Pernicious is defined as "harmful in the extreme, destructive, or deadly". Combining and modifying these definitions, to reflect a broad agricultural usage of the term, a *noxious weed* might be defined as:

A plant that is extremely prolific, invasive, competitive, harmful, destructive, or difficult to control.

The broad and narrow definitions of "noxious weed" are often used interchangeably in weed control discussions. It would be less confusing if weeds in the broader category were called "pernicious" and the term "noxious" were reserved for regulated species only. It is unlikely that such a change will ever be adopted; therefore, at least two interpretations of the term "noxious" weed must be recognized.

References

1. Aldrich RJ. *Weed-Crop Ecology: Principles in Weed Management.* Breton Publishers, North Scituate, MA, 1984.
2. Anderson WP. *Weed Science: Principles,* 2nd ed. West Publishing Co., St. Paul, MN, 1983.
3. Baker HG. Ann. Rev. Ecol. Syst. 5:1-24 (1974).
4. Bohmont BL. *The New Pesticide User's Guide.* Reston Publishing Co., Reston, VA, 1983.
5. Crafts AS. *Modern Weed Control.* California Press, Berkeley, CA, 1975.
6. Harper JL. *Population Biology of Plants.* Academic Press, New York, NY, 1977.
7. Klingman GC and FM Ashton. *Weed Science: Principles and Practices,* 2nd ed. John Wiley & Sons, New York, NY, 1982.
8. Ross MA and CA Lembi. *Applied Weed Science.* Burgess Publishing Co., Minneapolis, MN, 1985.
9. Whitson TD, SA Dewey, MA Ferrell, JO Evans, SD Miller, and RJ Shaw. *Weeds and Poisonous Plants of Wyoming and Utah* (TD Whitson, Ed.). Cooperative Extension Service, Univ. of Wyoming, Laramie, 1987.
10. Lorenz RJ and SA Dewey. In: *The Ecology and Economic Impact of Poisonous Plants on Livestock Production* (LF James, MH Ralphs, and DB Nielsen, Eds.), pp. 309-336. Westview Press, Boulder, CO, 1988.

2

Environmental and Economic Impacts of Noxious Range Weeds

John R. Lacey and Bret E. Olson

Abstract

Noxious range weeds conflict, restrict, or otherwise interfere with range management objectives. Published literature was reviewed to assess the effects of these weeds on the environment and the economy. The displacement of many native range plant communities by noxious range weeds has reduced rangeland biodiversity. Noxious weeds, such as spotted knapweed, are detrimental to elk and soil conservation. The appraised value of rangeland is often reduced by the presence of noxious weeds. Similarly, economic returns from rangeland are reduced by poisonous noxious weeds and by weeds with low forage value that displace desirable forage plants. Additional data are needed to fully assess the environmental and economic impact of noxious range weeds.

Introduction

Many plant species growing on range and pasture lands conflict, restrict, or otherwise interfere with land management objectives. These are commonly defined as "weeds". Once a plant has been classified as a weed, it only attains a "noxious" status by legislation. More than 500 weeds in the United States and Canada are designated as noxious by either weed or seed laws (1).

Most noxious weeds are not native to the area in which they are a problem. They were either introduced intentionally for their perceived

TABLE 2.1 Distribution of six noxious weeds.

Plant Species	Hectares
Spotted knapweed (*Centaurea maculosa*)	2,938,221[a]
Diffuse knapweed (*Centaurea diffusa*)	1,301,739[a]
Russian knapweed (*Centaurea repens*)	568,441[a]
Yellow starthistle (*Centaurea solstitialis*)	3,822,464[b]
Cheatgrass (*Bromus tectorum*)	41,000,000[c]
Leafy spurge (*Euphorbia esula*)	729,167[d]

[a] Reported infestation in 9 states and 2 Canadian provinces (2).
[b] Reported infestation in 10 states and 2 Canadian provinces (2,3).
[c] Reported as "dominant plant" in the Intermountain West (4).
[d] Reported infestation in Montana (5) and North Dakota (6).

value to man, or unintentionally as contaminants in seed grain or along transportation corridors. Noxious weeds are strong competitors for moisture and nutrients. Many of the natural diseases and predators that kept them controlled in their native countries were not introduced. Without these "natural" controls, weeds have displaced native range plants on millions of hectares in Western North America (Table 2.1). Of greater concern, many of the noxious weeds are continuing to spread, raising questions about their ecological amplitude (2,3,7-9).

Effects of weed infestations on environment and economy need to be understood. This information would be useful to evaluate current management alternatives and control strategies and to allocate resources for research, education, and control. Thus, environmental and economic impacts of noxious range weeds are reviewed in this report.

Methods

We used the AGRICOLA Literature Search to locate environmental and economic reports on 13 noxious weeds: cheatgrass, dyers woad (*Isatis tinctoria*), rush skeletonweed (*Chondrilla juncea*), houndstongue (*Cynoglossum officinale*), St. Johnswort (*Hypericum perforatum*), scotch thistle (*Onopordum acanthium*), musk thistle (*Carduus nutans*), Canadian thistle (*Cirsium arvense*), spotted knapweed, diffuse knapweed, Russian knapweed, yellow starthistle, and leafy spurge. We felt that these were the species in which environmental and economic information were most likely to be available. Poisonous range plants, such as snakeweed (*Gutierrezia sarothrae*), and other weeds without the noxious declaration were not evaluated in this report. However, their economic impact should not be overlooked (10).

Scientific and common names, environment, and economics were the key words used in the AGRICOLA search. The 1979-1990 period was included in the literature review.

Results and Discussion

Environmental Impact of Noxious Weeds

Many weeds became widespread and locally abundant during the early part of the twentieth century. When plotted in terms of area occupied through time, weed invasions are modeled as sigmoidal, or S-shaped patterns (Figure 2.1). For example, cheatgrass spread in the western United States was initially limited by small populations (1). Once a viable population was established, subsequent expansion was rapid. Although Figure 2.1 suggests that cheatgrass may have exploited the suitable habitat by 1930, the plant continues to expand its range (4).

Displacement of native range plants by noxious weeds has reduced rangeland biodiversity. Introduced annuals in California and cheatgrass in the Intermountain West have replaced native plant species, drastically changing these plant communities and affecting human activities (11). Although cheatgrass communities are comprised of fewer species and life forms than the *Artemisia tridentata/Agropyron spicatum* association, they are capable of occupying a site for a long time (12). The conservation of

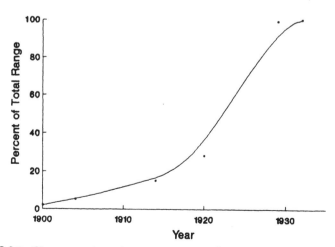

FIGURE 2.1 Cheatgrass invasion as a percent of total range occupied by 1900, 1904, 1914, 1920, 1929, and 1932. Taken with permission from Mack (4).

8

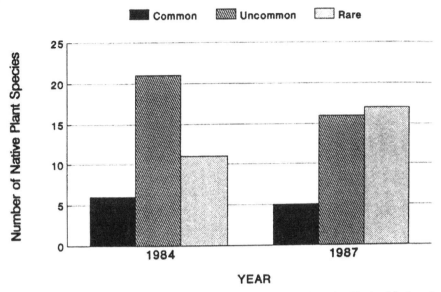

FIGURE 2.2 Spotted knapweed invasion affected native flora in Glacier National Park. Adapted from Tyser and Key (13).

climax communities should recognize that climax communities, rather than being inherently stable, are vulnerable to man's activities (12). Some native plant communities could disappear if they are not protected (12). Spotted knapweed is replacing native flora in Glacier National Park where frequency of seven plant species changed from a higher to a lower category (13). Six of the 21 species in the uncommon category (5-50% of the plots) in 1984 were in the rare category (less than 5% of the plots) by 1987 (Figure 2.2). Another species which was classed as "common" (in more than 50% of the plots) in 1984 was "uncommon" in 1987. Similarly, cover values of 10 and frequency of five common native species of the mixed grass prairie were negatively correlated with leafy spurge (14). Native species are also being replaced in western Montana by spotted knapweed (15), in Utah by dyers woad (16), and in California by yellow starthistle (5).

The presence of a noxious plant such as cheatgrass increases the hazard of fire (17). Dry cheatgrass is highly flammable, and permits fires to kindle and spread rapidly. It thus increases the number and size of fires thereby increasing fire suppression costs (18). The burned lands may be more susceptible to erosion (19).

The relationship of noxious range weeds to rare plants is not known. A rare species is one that is limited to a restricted geographic range, or one that occurs sparsely over a wider area. Aggressive introduced plants could displace "rare" native plants. However, we did not find any documentation of a threatened or endangered species being replaced by a noxious weed.

The impact of noxious weeds on wildlife appears to be detrimental. In the *Sand County Almanac*, Aldo Leopold (20) lamented how cheatgrass prevented newly hatched ducklings from reaching water. In Western Montana, elk (*Cervus canadensis*) use, estimated by pellet groups/ha, varied from 84 on spotted knapweed to 3,780 on bunchgrass sites, indicating a drastic reduction of use on spotted knapweed-infested sites (15). A Forest Service assessment on the Lolo National Forest concluded that forage loss due to spotted knapweed invasion on big game winter ranges could result in a loss of 220 elk annually by 1998 (21). Although spotted knapweed was common on a mule deer (*Odocoileus hemionus*) winter range in Montana, the plant was not detected in the deer's diet (22). The negative impact of noxious weeds on wildlife could adversely affect Montana's hunting and tourism industries by degrading habitat, restricting use and reducing wildlife populations (21).

Noxious weeds are not entirely harmful to wildlife. They provide cover, habitat diversity, and a source of feed for many game and nongame birds (23). In contrast to Montana, knapweed rosettes were important components of deer and elk diets during winter and early spring in the Kootenay Region of British Columbia (24). Unfortunately, Miller did not discuss vegetative composition. Whether big game selectively graze knapweed, or graze it because it is the only available forage, has not been evaluated.

The influence of spotted knapweed on surface runoff and sediment yield was determined under simulated rainfall conditions near Garrison, Montana (25). These responses were compared on 12 paired plots; one of each pair was dominated by bunchgrasses whereas the other was dominated by spotted knapweed. Runoff and sediment yield during the initial 30-min simulated rainfall period were 56% and 192% higher, respectively, for the spotted knapweed compared with the bunchgrass-dominated plots (Table 2.2). This indicates that spotted knapweed invasion onto bunchgrass rangelands of western Montana is detrimental to the protection of soil and water resources (25). Similar soil and water losses have probably occurred, and are continuing to occur, on millions of hectares where grass communities have been replaced by tap-rooted forb communities.

The most alarming aspect of the relationship between weeds and the environment is the lack of data. The impact of noxious weeds on threatened and endangered plants and animals, wildlife species, and environmental cycles (hydrologic, energy, and carbon) needs to be

TABLE 2.2 Surface runoff and sediment yield on 12 grass-dominated and 12 spotted knapweed-dominated plots subjected to two consecutive 30-min simulated rainfall periods in November 1987. The initial 30-min period was on the unaltered site to measure the effect of vegetative cover plus soil surface characteristics. Vegetative cover was removed before the second 30-min period to measure the effects of surface characteristics alone. Taken with permission from *Weed Technology*, Lacey et al. (25).

	First 30-min.		Second 30-min.	
Site Characteristics	Grass	Spotted Knapweed	Grass	Spotted Knapweed
Surface runoff (%)	23	36*	39	45
Sediment yield (kg/ha)	44	128*	95	161

* An asterisk indicates significance, $P<0.10$, as compared with paired T-test.

identified and quantified. Their impact on sustainable agriculture needs to be understood.

General Principles for Evaluating Economic Impact of Noxious Weeds

Land Values. The effect of noxious weeds on land value can be measured by either the sales approach or the income approach (26). The sales approach relies on market sales to estimate value. The amount that buyers are willing to pay for comparable lands is used to estimate the value of lands that are to be sold. A condition such as the presence of noxious weeds may make it inappropriate to directly compare two pieces of property. When this occurs, the sales value is adjusted to make the two pieces of property comparable. As long as the buyer and seller are aware of the existence and impact of noxious weeds, property with noxious weeds is expected to sell for less than property without noxious weeds. Thus, the sales approach reflects what a purchaser is "willing to pay".

The income approach calculates the projected income to be received and uses it as a basis to calculate what someone can "afford to pay" for the land. Principles associated with capital budgeting are used:

$$\sum_{j=1}^{N} NR_j/(1+i)^j$$

where NR_j = net returns in period j, i = discount rate, and N = length of the period over which the returns are to be received (26). For example, if a landowner leases his pasture for 10 years with a payment of $5 per hectare and a discount rate of 5% is used, the value of the income stream would be $38.60. However, land values are usually based over longer planning periods. By using an infinite planning period, the income stream increases to $100 (or $5/0.05). This also is the capitalized value of the land. "Capitalization" is a technique used to calculate the present value of a future perpetual flow of annual net income.

Once the general value of grazing land has been estimated, the impact of noxious weeds on asset value can be estimated (26). The productive value of the land is decreased by weeds that detract from or limit its productivity, or increase operating and management costs. Any factor that reduces potential net returns from the land will diminish its value. These factors can be estimated by using a "with" versus "without" approach in the basic discounting procedure:

$$\sum_{j=1}^{N} NR(WO)_j/(1+i)^j - \sum_{j=1}^{N} NR(W)_j/(1+i)^j$$

where $NR(WO)_j$ = net returns without noxious weeds, $NR(W)_j$ = net returns with noxious weeds, i = the discount rate, and N = the planning horizon.

This analysis reveals the impact of noxious weeds on land value. For example, if the presence of spotted knapweed reduces carrying capacity by 60%, then net returns decrease from $5 per hectare to $2 per hectare. Thus, land values drop from $100 to $40 per hectare, assuming a 5% discount rate and an infinite planning period. In theory, the negative impact of noxious weeds on net returns should be reflected in reduced land values.

Knapweed infestations have reduced the appraised value of land in Oregon (27). In western Montana, however, the effect of noxious weeds on income is frequently not considered in land sales (personal communication Jim Wiley, Senior Appraiser, Farm Credit Services, Bozeman, MT). Recent buyers are more interested in mountains, streams, and other amenities than in productive agriculture. In contrast, the presence of weeds is usually considered when land sells in the more production-oriented central and eastern Montana (personal communication Chuck Egan, Extension Agent, Stillwater County, Columbus, MT).

Economic Returns. Net returns are calculated by subtracting total expenses from total returns. Noxious weeds usually reduce net returns by either increasing operating expenses, decreasing total returns, or both. Operating expenses increase when landowners implement strategies to limit weed invasion or eradicate current infestations. Total returns are directly affected when animals are poisoned by noxious weeds. More than 125 species appearing on state or provincial noxious weed lists either impair livestock performance or result in death (1). Fifteen of the 125 are so common that they often force operators to change either the kind of grazing animal or the season of grazing (Table 2.3). Thus, operating expenses increase.

Net economic loss from poisonous noxious range weeds is probably dwarfed by the loss from nonpoisonous noxious weeds. Because of nutritional quality, season of growth, and palatability, most noxious range weeds have less forage value than the native plants they displace. However, the economic importance of cheatgrass as spring forage (17,29) and leafy spurge as forage for sheep (30,31) should not be overlooked. The relationship between noxious range weeds and economics is not always straightforward.

TABLE 2.3 Some commonly-occurring noxious range weeds which have been reported as poisonous. Adapted with permission from Lorenz and Dewey (1).

Plant Name	Poisonous Principle	Primary Kind of Livestock Affected[a]
Common tansy (*Tanacetum vulgare*)	unknown	unknown
Halogeton (*Halogeton glomeratus*)	oxalate	sheep, cattle
Henbane (*Hyoscyamus niger*)	alkaloid	all
Houndstongue (*Cynoglossum officinale*)	alkaloids	cattle, horses
Kochia (*Kochia scoparia*)	alkaloids, oxalates, nitrates	all
Larkspur (*Delphinium* spp.)	alkaloids	cattle
Leafy spurge (*Euphorbia esula*)	unknown	sheep, cattle
Lupine (*Lupinus* spp.)	alkaloid	sheep, cattle
Poison hemlock (*Conium maculatum*)	alkaloid	all
Russian knapweed (*Centaurea repens*)	unknown	horses
Russian thistle (*Salsola kali*)	nitrate	all
St. Johnswort (*Hypericum perforatum*)	hypericin	all
Tansy mustard (*Descurainia pinnata*)	unknown	cattle
Water hemlock (*Cicuta* spp.)	cicutoxin	cattle
Yellow starthistle (*Centaurea solstitialis*)	unknown	horses

[a] Adapted from Leininger et al. (28).

Grazing capacities of leafy spurge-infested land in North Dakota and knapweed-infested land in Montana have been reduced by 75% and 63%, respectively (6,32). Similar reductions probably result from monocultures of other unpalatable, noxious weeds. Livestock grazing would decrease proportionately with the loss of forage. For example, when an AUM is valued at $10, the annual grazing value of a hectare of land rated at 1 AUM/ha is $10. A 75% reduction in forage would reduce the stocking rate to 0.25 AUM/ha and the value to $2.50. Thus, the economic impact of a weed on livestock forage could be estimated by placing a value on forage and then multiplying forage reduction due to displacement of desirable forage by the number of affected hectares.

Only a few economic models have been developed at the microeconomic or ranch level to evaluate the feasibility of controlling range weeds. Six principles (diminishing returns, fixed and variable costs, optimality, marginality, opportunity costs, and time comparisons) were used to evaluate Scotch thistle (33). Scotch thistle infestations in northeastern California resulted in annual losses to ranchers of $24.50/ha for wet meadows, $16.00/ha for wheatgrass stands, and $8.16/ha on cheatgrass rangelands. Annual application of 0.05 Kg/ha of picloram provided the most economical ($8.57/ha) means of suppressing Scotch thistle (33). Economic returns from controlling leafy spurge with four herbicides were evaluated in east-central South Dakota (34). Forage yields were similar among treatments controlling 90% or more leafy spurge. Marginal net return over marginal cost from herbicide treatments ranged from $35 to $63/ha (34). The most cost-effective treatment in North Dakota was Tordon plus 2,4-D (35). Annual spring applications provided a net return of $284 and $109/ha in eastern and western North Dakota, respectively (35).

The threat of spotted knapweed spread, infesting additional acreage and reducing forage, was considered in an economic evaluation of controlling the weed in western Montana (36). An economic loss in current dollars of $2.38/ha was incurred during a 20-year period under the no treatment strategy when 25% of the management unit was initially infested with spotted knapweed. After-tax present value of added AUMs in the eradication strategy was greater than the after-tax present value of added costs, $3.41/ha and $1.99/ha, respectively. Economic returns increased relative to treatment costs as productivity, value of an AUM, and rate of knapweed spread to new acres increased (36).

Most economic studies of range weeds at the macroeconomic level have been cursory. Bucher (32) reported that spotted knapweed could potentially invade 13.4 million ha of rangeland in Montana (9). This would reduce annual gross revenue of the state's livestock industry by $155 million.

14

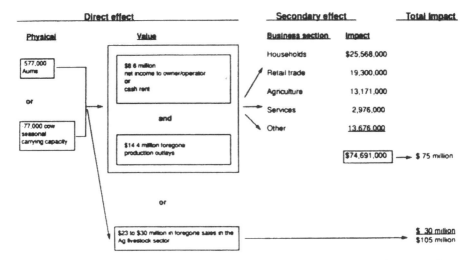

FIGURE 2.3 Economic impacts of leafy spurge infestation in North Dakota, 1989. Taken with permission from Thompson et al. (6).

The relationship between leafy spurge infestations and carrying capacity reductions was modeled by agronomists and range scientists in North Dakota (Figure 2.3) (6). A reduction of 580,000 AUMs, or enough forage to graze 77,000 cows for 7.5 months, was attributed to leafy spurge. Production foregone by ranchers approached $9 million. Value of grazing land was reduced by $137 million. By using an Input-Output model, the total business activity for all sectors was reduced by nearly $75 million.

The North Dakota study should serve as a model for additional evaluation of noxious weeds. However, further analyses will not be possible until more biological data are collected. Reliable acreage estimates, forage values, and economic relationships are needed. Therefore, studies conducted at state or local levels are recommended over regional studies.

Conclusion

The impacts of noxious range weeds on the environment and economy were reviewed. Several major points need to be reiterated. Environmentally, weeds: 1) have a detrimental effect on soil and water resources; 2) reduce forage production for wildlife and livestock; and 3) reduce biodiversity.

Economically, noxious range weeds: 1) reduce value of land; 2) reduce net returns by increasing operating costs, decreasing total returns, or both; 3) can be economically controlled at the microeconomic level; and 4) negatively impact regional economies.

The most disturbing conclusion regards weed data. There are little data documenting the impact of noxious weeds on our environment and economy. Until these impacts are fully known, future allocation of funding and legislation will continue to be made with inadequate information.

References

1. Lorenz RJ and SA Dewey. In: *The Ecology and Economic Impact of Poisonous Plants on Livestock Production* (LF James, MH Ralphs, and DB Nielsen, Eds.), pp. 309-336. Westview Press, Boulder, CO, 1988.
2. Lacey C. In: *Proc. 1989 Knapweed Symp.* (PK Fay and JR Lacey, Eds.), pp. 1-6. Montana State Univ., Bozeman, MT, 1989.
3. Maddox DM and A Mayfield. California Agric. Nov.-Dec.:10-12 (1985).
4. Mack RN. Agro-Ecosystems 7:145-165 (1981).
5. Lacey CA, PK Fay, RG Lym, CG Messersmith, B Maxwell, and HP Alley. Montana State Univ. Circ. 309 (1985).
6. Thompson F, FL Leistritz, and JA Leitch. North Dakota State Univ., Agric. Econ. Rep. 257 (1990).
7. Messersmith CG and RG Lym. North Dakota Farm Res. 40:8-13 (1983).
8. Roché BF Jr, GL Piper, and CJ Talbott. Washington State Univ. Bull. EB1393 (1986).
9. Chicoine TK, PK Fay, and GA Nielsen. Weed Sci. 34:57-61 (1988).
10. Torell LA, HN Gordon, KC McDaniel, and A McGinty. In: *The Ecology and Economic Impact of Poisonous Plants on Livestock Production* (LF James, MH Ralphs, and DB Nielsen, Eds.), pp. 57-59. Westview Press, Boulder, CO, 1988.
11. Mack RN. In: *Range Weeds Revisited* (BF Roché Jr and CT Roché, Eds.), pp. 1-4. Washington State Univ., Pullman, WA, 1990.
12. Richard WH and JF Cline. Northwest Sci. 54:216-221 (1980).
13. Tyser RW and CH Key. Northwest Sci. 62:151-160 (1988).
14. Belcher JW and SD Wilson. J. Range Manage. 42:172-175 (1989).
15. Bedunah D and J Carpenter. In: *Proc. 1989 Knapweed Symp.* (PK Fay and JR Lacey, Eds.), pp. 205-212. Montana State Univ., Bozeman, MT, 1989.
16. West NE and KD Farah. J. Range Management. 42:5-10 (1989).
17. Upadhyaya MK, R Turkington, and D McIlvride. Can. J. Plant Sci. 6:689-709 (1986).
18. Stewart G and AC Hull. Ecology 30:58-74 (1949).
19. Morrow LA and PW Stahlman. Weed Sci. 32:2-6(1984).

20. Leopold A. In: *A Sand County Almanac with Essays on Conservation from Round River*. Ballantine Books, NY, 1970.

21. Spoon CW, HR Bowles, and A Kulla. *Noxious weeds on the Lolo National Forest. A situation analysis staff paper*. USDA Forest Serv. North. Reg. (1983).

22. Guenther GE. MS Thesis, Montana State Univ., Bozeman, MT, 1989.

23. Weigand J. Montana Outdoors, Jan-Feb, p. 6 (1977).

24. Miller VA. In: *Range Weeds Revisited* (BF Roché Jr and CT Roché, Eds.), pp. 35-37. Washington State Univ., Pullman, WA, 1990.

25. Lacey JR, CB Marlow, and JR Lane. Weed Tech. 3:627-631 (1989).

26. Godfrey EB, DB Nielsen, and NR Rimbey. In: *The Ecology and Economic Impact of Poisonous Plants on Livestock Production* (LF James, MH Ralphs, and DB Nielsen, Eds.), pp. 17-25. Westview Press, Boulder, CO, 1988.

27. Maddox DM. Rangelands 1:139-140 (1979).

28. Leininger WC, JE Taylor, and CL Wambolt. Montana State Univ. Bull. 348 (1977).

29. DeFlon JG. Rangelands 8:14-17 (1986).

30. Landgraf BK, PK Fay, and KM Havstad. Weed Sci. 32:348-352 (1984).

31. Lacey CA, RW Kott, and PK Fay. Rangelands 6:202-204 (1984).

32. Bucher RF. Montana State Univ., Bozeman, MT. Montguide 8423, 1984.

33. Hooper JF, JA Young, and RA Evans. Weed Sci. 18:583-586 (1970).

34. Gylling S and WE Arnold. Weed Sci. 33:381-385 (1985).

35. Lym RG and CG Messersmith. Weed Tech. 4:635-641 (1990).

36. Griffith D and JR Lacey. J. Range Manage. 44:43-47 (1991).

3

Naturalization of Plant Species in Utah — 1842 to Present

S. L. Welsh, C. R. Nelson, and K. H. Thorne

Abstract

Introduced plant species that have become naturalized in the state of Utah are listed. The list contains some 310 species in 47 plant families. Somewhat more than half of them, primarily from Europe, were introduced before the beginning of this century. New ones, some of major consequence, have continued to arrive to the present time. Many more await introduction, especially in Europe.

Introduction

It seems probable that at least some species of plants were introduced to Utah by Indians, who carried them from place to place as food plants. The presence of *Lycium pallidum*, the tomatilla, on trash mounds of prehistoric Indian ruins in southern Utah is one such example. The plant persists there today. The advent of European peoples, their agrarian way of life and their crop plants, marks the beginning of an influx of plant species, both cultivated and weedy species, that has continued to the present time (Figure 3.1). It is with the cultivated plants and weeds that this paper is concerned.

Data for this work are derived from *Spring Flora of the Wasatch* (1,2), *Flora of Utah and Nevada* (3), *Weeds of Utah* (4), *A Utah Flora* (5), specimens in the herbarium at Brigham Young University, and a lifetime of observation of naturalization of plants in Utah. Basis for the work is in

18

large part, however, from the list of adventive species in the state compiled specifically for this treatment. All will be included in a statistical accounting of the introduced flora, but only those that are definitely established in the flora will be regarded as naturalized. The list presented below contains those species of plants introduced mainly from foreign countries that are growing and reproducing naturally in Utah. While the list is not comprehensive, it does contain a large proportion of the naturalized flora of the state. Naturalized plants in the list include those classified as weeds, poisonous plants, crop plants, and ornamental and shade trees. Crop plants often occur as weedy components in other crop areas or along roadsides or in other areas where native vegetation has been removed or modified. Included in the list of naturalized crop plants and trees are alfalfa, sweet clover, white and red clover, asparagus, peaches, apricots, various plums and cherries, catalpa, Russian olive, Siberian elm, and tamarix. Excluded are wheat, barley, corn, and other small grains that occur occasionally following cultivation but do not persist. Also excluded from the list are most of the weedy species native to the United States, some of which were likely introduced also. Many of those are so ingrained within the native flora as to be considered as native here also. Common and petiolate sunflowers are examples of such native weedy species.

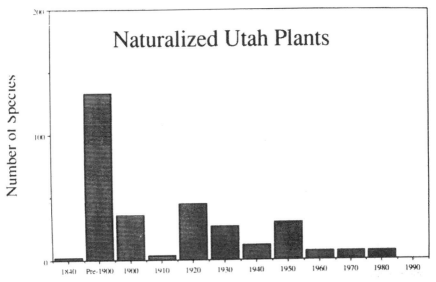

FIGURE 3.1 Approximate schedule of introduction of Utah's naturalized plant species.

Origins of Utah's Naturalized Plants

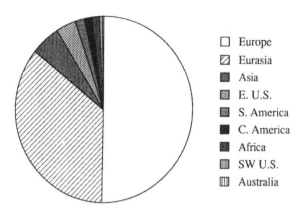

FIGURE 3.2 Probable origins of Utah's naturalized plant species.

Places of origin are not clearly understood. Plants are designated as having originated in Europe, Eurasia, Asia, South America, Australia, eastern U.S., and others (Figure 3.2). Many of them undoubtedly arrived by stages from some initial focus of introduction, only reaching Utah following the primary establishment elsewhere. Some plants evidently spread rapidly. Others apparently build to a certain population size and then spread almost explosively. Estimates of time of introduction are taken, where possible, from published literature. Such dates are not to be taken too literally. Publication of occurrences is often made years or decades following the initial introduction. Negative data is often as important as the actual publication of an occurrence. If the citation is missing the author either left it out on purpose or did not know of the presence of the particular plant. Authors also disagree on whether some of the plants are introduced or not. It seems probable, for instance, that the common bluegrass of our lawns belongs to one or more of a series of introductions from the Old World. The species, *Poa pratensis*, is certainly native to portions of North America and possibly even to Utah. It is not included in the list of species cited below, but it is certainly naturalized throughout our area.

The naturalized species treated are in 47 families of flowering plants and include 310 species. This number includes 184 species not treated by Holmgren (4) in his publication on the weeds of Utah. There is little doubt that the list of species treated by Holmgren was not exhaustive and many of the inclusions in the present work are of little consequence as weeds. Nevertheless, there are numerous examples of additional weedy introductions since the 1958 publication by Holmgren.

Several weedy species have moved rapidly over the state during the past five decades. Included in this category are halogeton, nodding thistle, yellow starthistle, Scots thistle, broadleaf pepperweed, *Bryonia alba*, and Dalmatian toadflax. First collected in Nevada in the mid-1930s, halogeton spread quickly through the desert ranges of western Utah, and in the 1950s continued its spread into eastern Utah. The potential habitat for this poisonous plant was essentially occupied completely by the 1960s. Pattern for movement across the state was probably similar to that of Russian thistle, which reached the state prior to 1900. The nodding thistle was first noted by me near Alpine in Utah County in 1961. Since then, it has been recorded for most of the counties in the state. In 1963 I first saw the Scots thistle near Fillmore and in Washington County. It has since spread into the canyons along the Wasatch and to the east. Yellow starthistle was first observed by me in Washington County about 1968. It has spread rapidly over much of Utah. A treatment of the mustard family by Welsh and Reveal (6) did not include the broadleaf peppergrass as there were no available records of its occurrence in the state. Since then, it has become common in the Uinta Basin, Utah Valley, and at scattered localities elsewhere in Utah. The cucurbit *Bryonia* has become established in several counties. Dalmatian toadflax is reputed to have been introduced in a wildflower mixture near Aspen Grove in Utah County in the 1940s. It has spread through most of the northern and some southern counties. The discovery of medusahead ryegrass in Northern Utah gives indication of still another weedy species with capacity for spreading across the state.

The entire representation of the 11 *Centaurea* species in Utah is introduced (5). Some of them were probably introduced early in contaminated seed. Others were grown initially as ornamentals. All are more or less naturalized in the state. More than a hundred additional species are known from Europe and Asia. Many of these will likely be introduced in the future. Indeed, *Centaurea* might be one of the largest genera included in the sunflower family in Utah a century from now.

The rate of naturalization of tamarix in Utah was reviewed by Christensen (7). Christensen discussed historical data in journals of various explorers to the state and could find no record, except for a possible misidentification by Escalante, of the species for Utah prior to 1925, when it was recorded by Tidestrom (3) on the Virgin River in southern Nevada. Christensen (7) concludes that "the histories of invasion of tamarix at Utah Lake, Great Salt Lake, Colorado River, and Green River are similar. The period from approximately 1925 to 1960 was one of rapid spread and increase in importance of tamarix. The greatest degree of invasion occurred during the twenty-year period from 1935 to 1955." Christensen (8) also studied the naturalization of the

Russian olive in Utah. He concluded that Russian olive was not one of the species planted in Utah prior to 1900. Increment borings of large trees at Powell's Slough near Utah Lake indicated that they began growth between 1924 and 1929. The trees have spread in moist sites, especially in soils that are moderately saline, throughout northern Utah. The species is very abundant in portions of the Uinta Basin. A similar pattern was described for the Siberian elm (*Ulmus pumila*). Christensen (9) notes that the earliest establishment in nature in Utah is 1935; it was extensively planted during the 1930s and 1940s.

In large part, naturalized plants are aggressive colonizers of disturbed sites. Many of them have been intimately associated with the history of agriculture and have evolved parallel to cultivated plants. The most successful examples of naturalized plants are those that fit within a recurring niche of some breadth and often of large areal extent. Plants with great reproductive potential, either through massive seed production or through vegetative offsets, or both, seem to have been the most successful. Included in these categories are cheatgrass, Russian thistle, halogeton, Dyer's woad, field bindweed, Canada thistle, Russian knapweed, Russian olive, Siberian elm, and tamarix.

The following list includes scientific name, common name, place of suspected origin, estimate of time of introduction to Utah, and earliest date and collector of specimens in the herbarium at Brigham Young University.

MAGNOLIOPSIDS -- THE DICOTS
AIZOACEAE
Trianthema portulacastrum Horse purslane Cent. Amer.1980s (1985 Higgins)
AMARANTHACEAE
Amaranthus albus Pale amaranth. Trop. Amer. pre-1900. (1925 Diehl)
Amaranthus retroflexus Redroot pigweed. Cent. Amer. pre-1900. (1895 Diehl)
BIGNONIACEAE
Catalpa speciosa Catalpa. E. U.S. pre-1900. (1930 Reimschussel)
BORAGINACEAE
Anchusa azurea Alkanet. Europe. ca 1950. (1950 Woodbury)
Anchusa officinalis Bugloss. Eurasia. ca 1920. (1933 Harrison)
Asperugo procumbens Catchweed. Eurasia. ca 1920. (1937 Snell)
Borago officinalis Common borage. Europe. ca 1970. (1979 Neese)
Cynoglossum officinale Houndstongue. Eurasia. pre-1900. (1927 Harrison)
Lappula squarrosa European stickseed. Eurasia. ca 1920. (1934 Harrison)
CAMPANULACEAE
Campanula rapunculoides Rover bellflower. Eurasia. ca 1940. (1953 Burkey)
Triodanis perfoliata Venus looking-glass. Europe. pre-1950. (1979 Clark)
CANNABACEAE
Cannabis sativa Hemp; marijuana. Eurasia. pre-1900.

CARYOPHYLLACEAE

Cerastium fontanum Mouse-ear chickweed. Eurasia. pre-1900. (1927 Harrison
Gypsophila paniculata Babysbreath. Eurasia. ca 1950. (1973 Atwood & Higgins)
Gypsophila scorzonerifolia Perennial babysbreath. Europe. ca 1970. (1979
Goodrich)
Herniaria glabra Herniaria. Europe. ca 1970. (1982 Arnow)
Holosteum umbellatum Holosteum. Europe. ca 1930. (1942 Maguire)
Lychnis alba White cockle. Europe. ca 1900. (1944 Chamberlin)
Sagina procumbens Sagina. Eurasia. ca 1950. (1980 Arnow)
Saponaria officinalis Bounding Bet. Europe. pre-1900. (1925 Matley)
Silene armeria Silene. Europe. ca 1970. (1981 Neese)
Silene noctiflora Night-flowering catchfly. Asia. ca 1920.
Stellaria media Common chickweed. Eurasia. pre-1900. (1933 Harrison)
Vaccaria pyramidata Cowcockle. Europe. pre-1900. (1896 Diehl)

CHENOPODIACEAE

Atriplex heterosperma Two-seed orach. Eurasia. ca 1950. (1965 Collotzi)
Atriplex hortensis Garden orach. Eurasia. ca 1930. (1944 Harrison)
Atriplex rosea Tumbling orach. Eurasia. ca 1900. (1913 Diehl)
Atriplex semibaccata Australia saltbush. Australia. ca 1900. (1918 Hall)
Bassia hyssopifolia Bassia. Eurasia. ca 1930. (1933 Harrison)
Chenopodium album Lambsquarter. Eurasia. pre-1900. (1895 Diehl)
Chenopodium ambrosioides Mexican-tea. Mexico. ca 1920. (1917 Hall)
Chenopodium botrys Jerusalem-oak. Eurasia. pre-1900. (1896 Diehl)
Chenopodium hybridum Maple-leaf goosefoot. Europe. pre-1900. (1939 Ream)
Chenopodium murale Nettleleaf goosefoot. Eurasia. pre-1900. (1921 Hall)
Halogeton glomeratus Halogeton. Eurasia. ca 1930. (1934 Stahman)
Kochia prostrata Prostrate kochia. Eurasia. 1980. (1981 Goodrich)
Kochia scoparia Summer-cypress. Eurasia. ca 1900. (1927 WSF)
Salsola iberica Russian-thistle. Eurasia. pre-1900. (1913 Diehl)
Salsola paulsenii Barbwire Russian-thistle. Eurasia. ca 1920. (1930 Stanton)

COMPOSITAE

Ambrosia confertiflora Cutleaf ragweed. SW U.S. ca 1980. (1988 Baird)
Anthemis cotula Mayweed. Eurasia. pre-1900. (1880 Howard)
Anthemis tinctoria Yellow chamomile. Eurasia. pre-1900. (1945 Garrett)
Arctium lappa Great burdock. Eurasia. ca 1950. (nd)
Arctium minus Burdock. Eurasia. pre-1900. (1925 Cottam)
Artemisia abronatum Garden sagebrush. Eurasia. ca 1950. (1981 Welsh)
Artemisia absinthium Absinthe. Eurasia. ca 1920. (1925 Cottam)
Artemisia annua Sweet wormwood. Eurasia. ca 1920. (1935 Blake)
Bellis perennis European daisy. Europe. ca 1900. (1981 Neese)
Carduus nutans Nodding thistle. Europe. ca 1950s. (1967 Welsh)
Centaurea calcitrapa Star-thistle. Eurasia. ca 1940. (1984 Chase)
Centaurea cyanus Bachelors button. Europe. pre-1900. (1967 Stubbs)
Centaurea diffusa Diffuse centaurea. Europe. 1980. (1985 Atwood)
Centaurea jacea Brownscale centaurea. Europe. ca 1950. (nd)
Centaurea melitensis Centaurea. Europe. ca 1950. (1986 Warrick)
Centaurea moschata Centaurea. Europe. pre-1900. (1899 nc)

Centaurea repens Russian knapweed. Eurasia. pre-1900. (1964 Fairbourn)
Centaurea scabiosa Centaurea. Europe. ca 1950. (1974 Arnow)
Centaurea solstitialis Yellow starthistle. Europe. ca 1950. (1972 Clark)
Centaurea virgata Squarrose knapweed. Eurasia. ca 1940. (1964 Moore)
Chamomilla rescuita Chamomile. Europe. ca 1900. (1964 Foster)
Chamomilla suaveolens Pineapple weed. Europe. pre-1900. (1896 Diehl)
Chrysanthemum balsamita Costmary. Europe. ca 1900. (1928 Blake)
Chrysanthemum leucanthemum Oxeye-daisy. Eurasia. ca 1930. (1935 Cottam)
Chrysanthemum parthenium Perennial matricary. Europe. ca 1930. (1971 Arnow)
Cichorium intybus Chickory. Europe. pre-1900. (1935 Harrison)
Cirsium arvense Canada or creeping thistle. Europe. pre-1900. (1927 Biddulph)
Cirsium vulgare Bull thistle. Eurasia. pre-1900. (1964 Foster)
Conyza bonariensis Conyza. S Amer. ca 1980. (1985 Baird)
Crepis capillaris Thread hawksbeard. Europe. ca 1950. (1981 Neese)
Gnaphalium luteo-album Yellow-white cudweed. Europe. ca 1940. (1983 Neese)
Gnaphalium uliginosum Marsh cudweed. Europe? pre-1900. (1977 Higgins)
Hypochaeris radicata Cats-ears. Europe. ca 1950. (1981 Arnow)
Inula helenium Elecampane. Eurasia. ca 1920. (1944 Galway)
Lactuca serriola Prickly lettuce. Europe. pre-1900. (1925 Matley)
Lapsana communis Nipplewort. Eurasia. ca 1950. (1975 Arnow)
Matricaria maritima Chamomile. Europe. ca 1920. (1977 Higgins)
Onopordum acanthium Scotch thistle. Europe. ca 1950. (1964 Welsh)
Senecio vulgaris Common groundsel. Europe. ca 1930. (1938 Harrison)
Sonchus arvensis Field sowthistle. Europe. pre-1900. (1933 Garrett)
Sonchus asper Spiny sowthistle. Europe. pre-1900. (1925 Cottam)
Sonchus oleraceus Common sowthistle. Europe. pre-1900. (1900 sc)
Sonchus uliginosus Meadow sowthistle. Europe. pre-1900. (1955 Welsh)
Tanacetum vulgare Tansy. Europe. pre-1900. (1925 Cottam)
Taraxacum officinale Dandelion. Eurasia. pre-1900. (1927 Biddulph)
Tragopogon dubius Goatsbeard. Europe. pre-1900. (1932 Garrett)
Tragopogon porrifolius Oysterplant. Europe. pre-1900. (1932 Harrison)
Tragopogon pratensis Field goatsbeard. Europe. ca 1930. (1981 Thorne)
Xanthium strumarium Cocklebur. E. U.S. pre-1900. (1927 Biddulph)
CONVOLVULACEAE
Calystegia sepium Hedge bindweed. Europe?. ca 1920. (1925 Cottam)
Convolvulus arvensis Bindweed. Eurasia. pre-1900. (1914 Diehl)
Ipomoea purpurea Morning-glory. Europe. pre-1900. (1934 Galway)
CRUCIFERAE
Alliaria officinalis Garlic mustard. Eurasia. ca 1960. (1983 Thorne)
Alyssum alyssioides Alyssum. Europe. ca 1900. (1930 Cottam)
Alyssum desertorum Desert alyssum. Eurasia. ca 1920. (1966 Welsh)
Alyssum minus Small alyssum. Eurasia. ca 1950. (1969 Rose)
Arabidopsis thaliana Mouse-ear cress. Europe. ca 1930. (1944 Dunn)
Barbarea vulgaris Common wintercress. Eurasia. ca 1950. (1978 Welsh)
Berteroa incana Berteroa. Europe. ca 1960. (1985 Atwood)
Brassica campestris Rape. Europe. pre-1900. (1929 Harrison)
Brassica juncea Indian mustard. Asia. pre-1900. (1966 Higgins)

Brassica kaber Charlock. Europe. pre-1900. (1924 Decker)
Brassica nigra Black mustard. Europe. pre-1900. (1913 Diehl)
Camelina microcarpa Falseflax. Europe. pre-1900. (1913 Diehl)
Capsella bursa-pastoris Shepherds purse. Europe. pre-1900. (1929 Harrison)
Cardaria chalepensis Coarse whitetop. Europe. ca 1920. (1933 Maguire)
Cardaria draba Whitetop. Europe. pre-1920. (1939 Farley)
Cardaria pubescens Hairy whitetop. Europe. ca 1950. (1976 Arnow)
Chorispora tenella Musk-mustard. Asia. ca 1920. (1937 Garrett)
Conringia orientalis Hares-ear mustard. Eurasia ca 1900. (1935 Mason)
Descurainia sophia tansy-mustard. Europe. pre-1900. (1913 Diehl)
Diplotaxis muralis Diplotaxis. Europe. ca 1950. (1980 Arnow)
Erysimum cheiranthoides Treacle. Eurasia. ca 1920. (1939 Maguire)
Erysimum repandum Spreading wallflower. Europe. pre-1900. (1906 Garrett)
Euclidium syriacum Syria-weed. Europe. ca 1930. (1940 Maguire)
Isatis tinctoria Dyer's woad. Europe. ca. 1910. (1947 Garrett)
Lepidium latifolium Broadleaf peppergrass. Europe. ca 1960. (1969 Stenquist)
Lepidium perfoliatum Peppergrass. Europe. ca 1900. (1924 Decker)
Malcolmia africana African mustard. Africa. ca 1920. (1933 Decker)
Nasturtium officinale Watercress. Eurasia. ca 1900. (1933 Harrison)
Rorippa austriaca Austrian fieldcress. Europe. ca 1920. (nr)
Rorippa sylvestris Wild fieldcress. Europe. ca 1970. (1981 Goodrich)
Sisymbrium altissimum Jim Hill mustard. Europe. pre-1900s. (1925 Cottam)
Sisymbrium irio Southern cress. Europe. ca 1920. (1942 Gould)
Sisymbrium officinale Hedge mustard. Europe. ca 1920. (1920?)
Thlaspi arvense Fanweed. Europe. pre-1900. (1927 Cottam)
CUCURBITACEAE
Bryonia alba Bryonia. Europe. ca 1960. (1974 Arnow)
CUSCUTACEAE
Cuscuta approximata Slender dodder. Europe. pre-1900. (1940 Garrett)
DIPSACACEAE
Dipsacus sylvestris Teasel. Europe. pre-1900. (1940 Harrison)
EBENACEAE
Diospyros virginiana Common persimmon. E U.S. pre-1900. (1941 Reimschussel)
ELAEAGNACEAE
Elaeagnus angustifolia Russian olive. Europe. ca 1920. (1929 Stanton) (8).
EUPHORBIACEAE
Euphorbia cyparissias Cypress spurge. Europe. ca 1940. (1981 Neese)
Euphorbia esula Leafy spurge. Eurasia. ca 1930. (1936 Garrett)
Euphorbia marginata Snow-on-the-mountain. Cent. U.S. ca 1900. (1983 Higgins)
Euphorbia peplus Petty spurge. Eurasia. ca 1940. (1974 Arnow)
Euphorbia prostrata Prostrate spurge. E U.S. ca 1940. (1982 Thorne)
FRANKENIACEAE
Frankenia pulverulenta Alkali-heath. Europe. ca 1950. (1972 Arnow)
FUMARIACEAE
Fumaria officinalis Fumitory. Europe. ca 1900. (1906 Garrett)
GERANIACEAE
Erodium cicutarium Storksbill. Europe. ca 1840. (1913 Diehl)

Erodium texanum Gray Texas storksbill. Europe. ca 1840. (1935 Galway) Note: This undoubtedly has an earlier name in Europe.

Geranium pusillum Slender cranesbill. Europe. pre-1900. (1906 Garrett)

JUGLANDACEAE

Juglans major Arizona walnut. SW U.S. pre-1900. (1944 Galway)

LABIATAE

Dracocephalum thymiflorum Dragonhead. Europe. ca 1960. (1976 Welsh)

Glecoma hederacea Ground-ivy. Europe. ca 1920. (1929 Harrison)

Lamium amplexicaule Deadnettle. Europe. ca 1900. (1929 Garrett)

Lamium purpureum Red henbit. Europe. ca 1900. (nr)

Leonurus cardiaca Motherwort. Europe. pre-1900. (1927 Cottam)

Marrubium vulgare Horehound. Europe. pre-1900. (1925 Decker)

Mentha citrata Bergamot mint. Europe. pre-1900. (1927 Cottam)

Mentha piperita Peppermint. Europe. pre-1900. (1896 Diehl)

Mentha spicata Spearmint. Europe. pre-1900. (1925 Cottam)

Molucella laevis Shellflower. Asia. ca 1920. (1925 Woodbury)

Nepeta cataria Catnip. Europe. pre-1900. (1896 Diehl)

Stachys byzantina Woolly betony. Asia. ca 1970. (1981 Neese)

LEGUMINOSAE

Anthyllis vulneraria Kidney-vetch. Europe. ca 1970. (1971 Stevens)

Astragalus cicer Chickpea milkvetch. Europe. ca 1950. (1965 Stevens)

Astragalus falcatus Russian milkvetch. Europe. ca 1950. (1977 Welsh)

Coronilla varia Crown-vetch. Europe. ca 1950. (1970 Welsh)

Cytisus scoparius Scots-broom. Europe. ca 1950. (1989 Welsh)

Galega officinalis Goatsrue. Europe. ca 1900. (1932 Maguire)

Lathyrus latifolius Perennial sweetpea. Europe. pre-1900. (1953 Burkey)

Lathyrus sylvestris Scots sweetpea. Europe. pre-1900. (1950 Flatley)

Lotus corniculatus Birds-foot trefoil. Europe. ca 1900. (1971 Stevens)

Lotus tenuis Slender trefoil. Europe. ca 1920. (1947 Galway)

Medicago falcata Yellow alfalfa. Europe. ca 1940. (1966 Letham)

Medicago lupulina Black medick. Europe. pre-1900. (1900 Diehl)

Medicago sativa Alfalfa; lucerne. Europe. pre-1900. (1931 Rowe)

Melilotus albus White sweet-clover. Europe. pre-1900. (1895 Diehl)

Melilotus indicus India sweet-clover. Eurasia. pre-1900. (1935 Hall)

Melilotus officinalis Yellow sweet-clover. Europe. pre-1900. (1925 Matley)

Onobrychis viciifolia Sainfoin. Europe. ca 1940. (1943 Harrison)

Robinia hispida Rose-acacia. E U.S. ca 1900. (1944 Galway)

Robinia pseudoacacia Black locust. E U.S. pre-1900. (1933 Harrison)

Sphaerophysa salsula Sphaerophysa. Europe. ca 1930. (10).

Trifolium fragiferum Strawberry clover. Europe. ca 1940. (1948 Galway)

Trifolium hybridum Alsike clover. Europe. pre-1900. (1925 Harrison)

Trifolium pratense Red clover. Europe. pre-1900. (1935 Hall)

Trifolium repens White clover. Europe. pre-1900. (1924 Decker)

Vicia villosa Hairy vetch. Europe. ca 1930. (1940 Chamberlain)

LYTHRACEAE

Lythrum salicaria Loosestrife. Europe. ca 1940. (1943 Harrison)

MALVACEAE

Abutilon theophrasti Velvet-leaf. Europe. ca 1910. (1917 Hall)
Althaea rosea Hollyhock. Europe. pre-1900. (1943 Burkey)
Hibiscus trionum Flower-of-an-hour. Africa. ca 1900. (1929 Cottam)
Malva neglecta Cheeses; mallow. Eurasia. pre-1900. (1935 Beck)
Malva sylvestris High mallow. Europe. ca 1920. (1933 Nelson)

MELIACEAE

Melia azedarach Chinaberry. Asia. ca 1900. (1927 Hafen)

OXALIDACEAE

Oxalis corniculata Creeping woodsorrel. Europe. pre-1900. (1929 Stanton)
Oxalis dillenii Erect woodsorrel. Europe. ca 1900. (1964 Hogan)

PAPAVERACEAE

Roemeria refracta Roemer poppy. Europe. ca 1930. (1936 Maguire)

PLANTAGINACEAE

Plantago lanceolata Narrow-leaf plantain. Europe. pre-1900. (1933 Harrison)
Plantago major Broad-leaf plantain. Europe. pre-1900. (1895 Diehl)

POLYGONACEAE

Polygonum aviculare Knotweed. Europe? pre-1900. (1895 Diehl)
Polygonum convolvulus Black bindweed. Europe. pre-1900. (1906 Garrett)
Polygonum cuspidatum Fleece-flower. Asia. ca 1930. (1956 Cottam)
Polygonum hydropiper Waterpepper. Europe. pre-1900. (1895 Diehl)
Polygonum lapathifolium Willow-weed. Europe? pre-1900. (1925 Harrison)
Polygonum persicaria Ladysthumb. Europe. pre-1900. (1895 Diehl)
Rumex acetosella Sheep sorrel. Eurasia. pre-1900. (1928 Cottam)
Rumex crispus Curled dock. Eurasia. pre-1900. (1895 Diehl)
Rumex dentatus Toothed dock. Asia. ca 1920. (nr)
Rumex obtusifolius Bitter dock. Eurasia. ca 1900. (1938 Stanton)
Rumex patentia Patent dock. Eurasia. ca 1910. (1949 Galway)
Rumex stenophyllus Ledeb. Slender-leaf dock. Eurasia? ca 1930. (1970 Atwood)

PORTULACACEAE

Portulaca oleracea Purslane. Europe. pre-1900. (1937 Snell)

PRIMULACEAE

Anagallis arvensis Scarlet pimpernel. Europe. ca 1920. (nr)

RANUNCULACEAE

Adonis annua Pheasanteye. Eurasia. ca 1920. (nr)
Clematis orientalis Oriental clematis. Asia. ca 1900. (1930 Fitsroy)
Ranunculus acris Tall buttercup. Eurasia. pre-1900. (1941 Harrison)
Ranunculus arvensis Field buttercup. Europe. ca 1900. (1943 Harrison)
Ranunculus repens Creeping buttercup. Europe. pre-1900. (1929 Harrison)
Ranunculus testiculatus Bur buttercup. Eurasia. ca 1930. (1932 Garrett) (11).

ROSACEAE

Malus pumila Apple. Eurasia. pre-1900s (1938 Morris)
Prunus americana American plum. Eastern U.S. pre-1900. (1976 Coombs)
Prunus avium Sweet cherry. Eurasia. pre-1900. (1902 Blackwood)
Prunus mahaleb Mahaleb, St. Lucie cherry. Eurasia. pre-1900. (1963 Treshow)
Sanguisorba minor Burnet. Europe. ca 1950. (1958 Anderson)

RUBIACEAE
Rubia tinctoria Madder. Europe. pre-1900. (1927 Cottam)
SALICACEAE
Populus alba White poplar. Europe. pre-1900. (1944 Galway)
Salix fragilis Crack willow. Eurasia. pre-1900. (1944 McKnight)
SCROPHULARIACEAE
Linaria dalmatica Dalmatian toadflax. Europe. ca 1930. (1939 Ream)
Linaria vulgaris Butter-and-eggs. Eurasia. 1900. (1964 Welsh)
Verbascum blattaria Moth mullein. Europe. ca 1920. (1927 Cottam)
Verbascum thapsus Woolly mullein. Europe. pre-1900. (1925 Decker)
Verbascum virgatum Stokes Wand mullein. Europe. ca 1950. (1977 Holmgren)
Veronica anagallis-aquatica Water speedwell. Europe. pre-1900. (1928 Cottam)
Veronica arvensis Corn speedwell. Eurasia. ca 1900. (1974 Allred)
Veronica biloba Bilobed speedwell. Asia. ca 1900. (1967 Arnow)
Veronica hederifolia Ivy-leaved speedwell. Europe. ca 1920. (1944 Garrett)
Veronica persica Persian speedwell. Eurasia. ca 1900. (1896 Diehl)
SIMAROUBACEAE
Ailanthus altissima Tree-of-heaven. Asia. pre-1900. (1944 Galway)
SOLANACEAE
Datura stramonium Jimson weed. Eastern U.S. pre-1900. (1934 Galway)
Hyoscyamus niger Henbane. Europe. ca 1920. (1928 Vanbor)
Lycium barbarum Matrimony vine. Europe. pre-1900. (1925 Cottam)
Solanum carolinense Carolina horsenettle. E U.S. pre-1900. (nr)
Solanum dulcamara European bittersweet. Europe. pre-1900. (1933 Larsen)
Solanum elaeagnifolium Silverleaf nightshade. E U.S. ca 1900. (1953 Harrison)
Solanum nigrum Black nightshade. Europe. pre-1900. (1880 sc)
Solanum rostratum Buffalobur. E U.S. pre-1900. (1927 Cottam)
Solanum sarrachoides Nightshade. S Amer. pre-1900. (1972 Clark)
TAMARICACEAE
Tamarix parviflora Small-flower tamarix. Europe. ca 1920. (1926 Galway)
Tamarix ramosissima Tamarix. Europe. ca 1920. (1925 Cottam) (7)
ULMACEAE
Ulmus pumila Siberian elm. Asia. 1920. (1935 Galway) (9).
UMBELLIFERAE
Carum carvi Caraway. Europe. pre-1900. (1947 Allred)
Conium maculatum Poison hemlock. Europe. pre-1900. (1944 Dunn)
Daucus carota Queen Ann's lace. Europe. pre-1900. (1932 Stanton)
Pastinaca sativa Parsnip. Europe. pre-1900. (1925 Harrison & Cottam)
Torilis arvensis Hedge parsley. Europe. ca 1960. (1964 Barnum)
VALERIANACEAE
Valerianella locusta Valerianella. Europe. ca 1920. (nr)
ZYGOPHYLLACEAE
Tribulus terrestris Puncture vine. Eurasia. ca 1910. (1920 Garrett)

LILIOPSIDS -- THE MONOCOTS
GRAMINEAE
Aegilops cylindrica Jointed goatgrass. Eurasia. ca 1920. (1932 Bush)

Agropyron cristatum Crested wheatgrass. Eurasia. ca 1930. (1939 Hull)
Agrostis capillaris Colonial bentgrass. Europe. ca 1920. (1977 Neese)
Agrostis stolonifera Redtop. Europe. pre-1900. (1930 Stanton)
Alopecurus pratensis Meadow foxtail. Europe. pre-1900. (1964 Matthews)
Avena fatua Wild oats. Europe. pre-1900. (1929 Larsen)
Bromus briziformis Rattlesnake chess. Europe. pre-1900. (1927 Saling)
Bromus catharticus Rescue grass. S Amer. ca 1900. (1932 Harrison)
Bromus inermis Smooth brome. Eurasia. pre-1900. (1913 (Diehl)
Bromus secalinus Rye chess. Eurasia. ca 1900. (nr)
Bromus rubens Red brome. Eurasia. ca 1930. (1932 Harrison)
Bromus sterilis Poverty brome. Eurasia. ca 1920. (1932 Harrison)
Bromus tectorum Cheatgrass. Eurasia. ca 1890. (1920 Harrison) (5)
Crypsis alopecuroides Pricklegrass. Eurasia. ca 1940. (1976 Welsh)
Crypsis schoenoides Common pricklegrass. Eurasia. ca 1950. (1973 McDonald)
Cynodon dactylon Bermuda grass. S Amer.? pre-1900. (1935 Galway)
Dactylis glomerata Orchard grass. Europe. pre-1900. (1928 Cottam)
Digitaria ischaemum Smooth crabgrass. Eurasia. pre-1900. (1938 Harrison)
Digitaria sanguinalis Hairy crabgrass. Eurasia. pre-1900. (1927 Biddulph)
Echinochloa crus-galli Barnyard grass. Europe. pre-1900. (1933 Harrison)
Eleusine indica Goosegrass. Asia. pre-1900. (1934 Galway)
Elymus caput-medusae Medusa-head. Eurasia. 1980s. (1988 Dewey)
Elymus elongatus Tall wheatgrass. Eurasia. ca 1950. (1951 Harrison) (12)
Elymus hispidus Intermediate wheatgrass. Eurasia. ca 1930. (1954 Plummer)
Elymus junceus Russian wildrye. Eurasia. ca 1950. (1964 Blackburn)
Elymus repens Quackgrass. Eurasia. pre-1900. (1933 Larsen)
Eragrostis barrelieri Mediterranean lovegrass. Europe. ca 1920. (1971 Meyer)
Eragrostis cilianensis Stinkgrass. Eurasia. ca 1900. (1921 Hall)
Eragrostis curvula Weeping lovegrass. S Africa. ca 1930. (1966 Higgins)
Eragrostis lehmanniana Lehmann lovegrass. S Africa. ca 1930. (1986 Higgins)
Eragrostis minor lovegrass. Eurasia. ca 1930. (1981 Neese)
Eremopyrum triticeum Annual wheatgrass. Asia. ca 1950. (19??)
Festuca arundinacea Tall fescue. Europe. ca 1900. (1970 Moore)
Festuca bromoides Brome sixweeks-fescue. Europe. ca 1920. (1935 Cottam)
Festuca myuros Myur fescue. Eurasia. ca 1930. (1954 Harrison)
Festuca pratensis Meadow fescue. Eurasia. ca 1920. (1932 Weight)
Holcus lanatus Velvetgrass. Eurasia. pre-1900. (1939 Harrison)
Hordeum marinum Mediterranean barley. Eurasia. pre-1900. (1932 Cottam)
Hordeum murinum Rabbit barley. Europe. pre-1900. (1935 Galway)
Lolium perenne Ryegrass. Eurasia. pre-1900. (1925 Harrison)
Panicum antidotale Blue panic-grass. India. ca 1980. (1985 Higgins)
Panicum miliaceum Broomcorn panicum. Asia. pre-1900. (1935 Menzies)
Phleum pratense Timothy. Eurasia. pre-1900. (1925 Harrison)
Poa annua Annual bluegrass. Eurasia. ca 1900. (1925 Harrison)
Poa bulbosa Bulbous bluegrass. Eurasia. ca 1920. (1932 Stanton)
Polypogon monspeliensis Rabbitfoot grass. Eurasia. pre-1900. (1925 Harrison)
Polypogon semiverticillatus Water polypogon. Eurasia. pre-1900. (1935 Galway)
Puccinellia distans Weeping alkaligrass. Eurasia. pre-1900. (1880 Howard?)

Puccinellia fasciculata Torrey alkaligrass. Eurasia. pre-1900. (1938 Stanton)
Saccharum ravennae Ravenna grass. Eurasia. pre-1930. (1935 Galway)
Schismus arabicus Arabian grass. Eurasia. ca 1930. (1966 Higgins)
Schismus barbatus Mediterranean grass. Eurasia. ca 1930. (1941 Harrison)
Sclerochloa dura Hardgrass. Eurasia. ca 1920. (1935 Harrison)
Secale cereale Rye. Eurasia. pre-1900. (1970 Moore)
Setaria glauca Yellow bristlegrass. Eurasia. pre-1900. (1925 Garrett)
Setaria verticillata Bur bristlegrass. Eurasia. pre-1900. (1933 Larsen)
Setaria viridis Green bristlegrass. Eurasia. pre-1900. (1927 Biddulph)
Sorghum halepense Johnson grass. Eurasia. pre-1900. (1940 Snow)
IRIDACEAE
Iris germanica Flag. Europe. pre-1900. (1974 Allred)
LILIACEAE
Asparagus officinalis Asparagus. Eurasia. pre-1900. (1929 Cottam)
Hemerocallis fulva Day-lily. Eurasia. pre-1900. (1964 Hogan)
Ornithogalum umbellatum Star-lily. Europe. pre-1900. (1944 Skousen)

References

1. Garrett AO. *Spring Flora of the Wasatch Region*, 1st Ed. Skelton Publishing Co., Salt Lake City, UT, 1911.
2. Garrett AO. *Spring Flora of the Wasatch Region*, 6th Ed. Stevens & Wallis, Inc., Salt Lake City, UT, 1936.
3. Tidestrom I. Contr. U. S. Nat. Herb. 25:1-665 (1925).
4. Holmgren AH. Utah State Univ. Agric. Expt. Sta. Spec. Rep. 12 (1958).
5. Welsh SL, Atwood ND, Goodrich S, and Higgins LC. Mem. Great Basin Naturalist 9:1-894 (1987).
6. Welsh SL and JL Reveal. Great Basin Naturalist 37:279-365 (1977).
7. Christensen EM. Amer. Midl. Naturalist 68:51-57 (1962).
8. Christensen EM. Amer. Midl. Naturalist 70:133-137 (1963).
9. Christensen EM. Great Basin Naturalist 24:103-106 (1964).
10. Christensen EM. Rhodora 65:764 (1963).
11. Barkworth ME. Utah Science Spring 1982:6-9 (1982).
12. Graham E. Ann. Carnegie Museum 26:1-432 (1937).

4

Benchmarks for Rangeland Management and Environmental Quality

Neil E. West

Abstract

In the past, benchmarks for range management have been chosen based on classical concepts of plant succession. Many assumptions of how succession operates have since been overturned by basic ecologists. Range scientists and managers have been slow in modernizing their applications of successional theory and adopting an ecosystem view. They have also been slow in recognizing the connections of approaches for range condition and trend to changing human value systems. A few individuals and institutions are now proposing fresh approaches to choosing benchmarks and monitoring and interpreting change. New ways of examining both old and new vegetation data and including soil characteristics are being tested. Our view of rangeland weeds and how to deal with them should be modified to fit within these new perspectives.

Introduction

Evaluation and redirection of any human activity involves establishment of standards. A benchmark is a standard by which something can be measured and judged. While establishment of standards can be fairly easily arrived at in the short-term, we need to be cognizant that the social and economic contexts of evaluation are constantly changing. Devising benchmarks for land management and environmental quality is a daunting task that we are increasingly being forced to address. After a long period of moving on to new lands as older ones were settled and degraded, we turned to technology to enhance the short-term productivity of lands previously altered. It is now, however, becoming obvious that many technological inputs are having some unintended and cumulative consequences. There are currently many emerging populist movements that intend to limit the kinds of technological fixes we will be allowed to apply in the future.

Rather than using only measures of short-term economic return, many are demanding that we also monitor environmental quality or ecosystem health and focus on sustainability (1). These issues are important because ecosystem health eventually determines the temporal and spatial extent of economic liability or opportunity. But, while it is easy for the public to desire such values and employ them in political slogans, it falls to the scientists they employ to determine how this may be accomplished.

Environmental quality, ecosystem health and sustainability are strongly related concepts. The concept of sustainable development has been defined as "development that meets the needs of the present without compromising the ability of future generations to meet theirs" (2). While this definition recognizing intergenerational equity is now widely accepted, it needs to be made more operational in terms of the criteria by which technical, social, economic, and political interventions can be designed to accomplish these ends. Is a quality environment, a healthy ecosystem, and a sustainable society like good art--different to every beholder--or do these things have at least a partial scientific basis? I can easily visualize many of us spending the rest of our careers just starting to make substantial progress toward those ideals and training the next generation to build upon our initial efforts.

Although environmental quality, health and sustainability are principles that apply to the entire biosphere, we have only space here to focus on a part--the western rangelands of the U.S. Rangelands, contrary to popular perception, are a kind of land, not a kind of use. Rangelands are desert, grassland, tundra, and some kinds of forest ecosystems that are managed extensively on ecological principles. These wildland ecosystems are too hot, cold, steep, rocky, salty, dry, or otherwise environmentally stressed such that intensive management based on agronomic principles never has and never will make much economic sense (3). Without subsidization of energy, materials, and information, we have little control over such "marginal" systems. Surely we can improve them, but the changes are subtle compared to the urban, industrial, and intensive agricultural fractions of our landscape.

We, in the range management profession, have already had a long history of trying to devise low input and sustainable management schemes. First native ungulates and then livestock were used as harvesting machines of the coarse, widely dispersed, low value (directly to humans) vegetation. This has been the only way we have, so far, been able to consistently first subsist and then, occasionally, accumulate economic wealth from such lands.

We now recognize that wildlands also serve society as repositories of minerals, biotic diversity, and ground water. These ecosystems also do much service in cleaning the biosphere's air and water. These lands also serve as a place to recreate and/or to expand all kinds of human activities. We formerly disregarded many of these values and services or considered them "free" because of the style of economics we practiced. We are beginning to devise a more inclusive economic theory (1). No longer will the short-term economic health of a few human users be the only way we monitor the value of wildlands. But how are we to judge the quality, health, or sustainability of the wildlands known as rangelands? Let's first briefly review the history of our attempts and then make suggestions for better approaches.

Earlier Approaches

The western hemisphere is unique in that people are a relatively recent biotic component compared to elsewhere on the globe. Although plant gatherers may have been here longer, the hunting style of life didn't develop here until about 12,000 years ago. Aboriginal farming was much more recent and geographically restricted. Europeans began their much more influential invasion about 500 years ago. It is significant that this latter invasion coincided with the rise of science and technology as a cultural underpinning.

Because Europeans had a scientific and mechanistic view, they tended to see the natural world as something apart from mankind. Europeans and their descendants felt that it was their task to subdue and subjugate the world for their personal benefit. Despite this emphasis on the importance of the individual human, they also held the romantic notion that the pre-Columbian systems were perfect and stable. So, even as the majority of the European colonists went about the task of reshaping the land in their commercial image, artists, writers, and naturalists extolled the beauties and wonders of the noble savages and primeval landscapes. With this cultural background, it is no wonder that the earliest ecologists developed very idealistic notions of what climates, landscapes, soils, vegetation, and animal communities once were. Without the understanding of ecosystem and culture dynamics that we possess today, the earliest scientists studying these lands assumed that prior to Columbus, everything was at a pinnacle of perfection and balance. Vegetation and soils were viewed as developing toward an optimum in balance with an unvarying climate.

Vegetational climaxes (4) and zonal soils (5) were natural consequences of the culture-wide assumptions of the day. The ideals did not accommodate much climatic or genetic change or include humans as part of nature. The embracement of these idealistic notions has been so complete that the operation of whole federal agencies has been designed around the assumptions that pre-Columbian systems were in an unchanging balance. We have only recently begun to realize the folly of this backward look for benchmarks (6).

First, we don't and can't ever know in any detail what the makeup and functioning of pre-Columbian systems were. While paleo-ecological evidence gives us good lists of species, we rarely can gauge their relative abundances and functional interactions with each other. Furthermore, paleo-ecological research usually shows different combinations of species than we are familiar with today (7).

The aboriginals didn't write or take photographs. Furthermore, the earliest European explorers rarely left us with other than a few words on what the land looked like and a few of the more obvious plants and animals. Even if the earliest European explorers had taken data, it would have been suspect because the European influence moved more rapidly than the Europeans did. For instance, the reintroduction of the horse by the Spanish explorers caused such a rapid and profound reorganization of the native culture that full characterizations of the pre-horse systems cannot be easily reconstructed now. Furthermore, European diseases spread so rapidly and virulently that by the time northern European explorers first observed and documented the West, the aboriginal populations may have been reduced to a tenth of their former numbers (8). The timing between the pandemics and northern European writers is removed enough such that a case can be made for a recovery of the plants and animals from previously higher levels of utilization.

If we can't rely very heavily on historical evidence, then the case is often made that we can substitute space for time (9) to determine what the original system was like and the successional steps toward and away from this ideal. This approach usually involves the search for relicts-- areas that have escaped direct human impact. We then collect data to use as reference points for the types of biotic communities that could eventually develop on other disturbed areas with similar environments. The Soil Conservation Service (SCS) and other users of its range condition and trend guides have relied heavily on this technique.

There are several problems with using relicts as benchmarks on rangelands. The first is their availability. They simply don't remain in any abundance in areas of gentle topography, e.g. the Great Basin or Great Plains. Even in areas where they are common, e.g. Colorado Plateau, they are more abundant on the most rugged and less frequently used parts that have inherently less productive range sites than the fraction with gentler topography. An additional problem is that the relicts may be so topographically isolated and/or poorly watered that native ungulates and Indians were not and are not part of the system. Such places thus represent a less complex and complete system than the areas under typical multiple use.

At this point, I need to make clear that ecosystems are open-ended by definition. Thus, there are no fully complete systems anywhere and using the complete ecosystem idea is specious, e.g. the Greater Yellowstone Ecosystem (10). Ecosystems having more participants are, however, more typical of the majority of rangelands.

In order for the relict method to be scientifically supportable, the relict and the disturbed areas it is compared with must be alike in all ways except time since disturbance occurred (9). This is a tall order. Furthermore, recent work has shown that timing of disturbances can influence successional trajectories, particularly in semiarid regions (11). Use of relicts frequently leads to employment of a linear, deterministic, and reversible model of succession to a single stable point, a model that is possible, but probably exceptional, on most rangeland sites (6,12). These simplistic assumptions are imbedded in nearly all of the range condition and trend guides produced by the SCS and habitat-type work of the U.S. Forest Service. Only one relict may exist for a given range site and it may not be modal for the conditions being addressed. If relicts are lacking, climax conditions for given range sites and habitat types are often hypothecated in order to complete range condition and trend guides for a given area.

It may not be possible for an ecosystem like that on the relict area to recover on the related disturbed areas. On the disturbed areas there may already be insurmountable alterations of the soil, loss of species or ecotypes of particular species, loss of soil seed reserves or seed rain of the later seral species, and the addition of new species that now control future trajectories of plant succession, soil development, animal communities, and utility to people. This is where the introduced weed issue enters.

Earlier ecologists believed that equilibrium or steady state conditions were possible. Some still do (13). The pristine systems were assumed to be complete and stable except for very obvious places where natural (landslides, river flooding, volcanoes, lava flows, wildfires, etc.) and human-induced (fields, roads, cities etc.) disturbances intervened. Biotic influences on microclimate (what Clements called "reaction") was once thought to be the only major mechanism of succession (14). Evolution was thought to have produced the most efficient collection of organisms to utilize the resources of a given site. Pristine communities were therefore closed to invaders by definition. Merely designating a park or wilderness area and keeping man's heavy hand off was considered adequate because of faith in the system's strong self-regenerative ability to come to equilibrium.

When it became more obvious that we didn't exactly know what pristine systems were like and we doubted that we could maintain or restore them anyway, some ecologists shifted to a forward looking ideal. These ideals were called "potential natural communities" (PNCs) (15). Küchler's map (15), the only one with a reasonable number of units nationwide, was adopted by the Forest Service in their need to meet the requirements of the Resources Planning Act. However, PNCs still assume removal of humans from the system. A firm time frame to reach stability is not stated. Lamar Smith (12) recently concluded that the details of PNCs are as ill-defined as those of the backward-looking climaxes. Both exclude climatic and genetic change, altered fire regimes, and the role of naturalized invaders. For instance, both Clements and Küchler lead us to believe that perennial bunchgrasses will eventually dominate stable grasslands of the Great Central Valley of California, denying the realities that now apply to most of that area. What little remains as rangeland is dominated by about 300 species of introduced annual grasses and forbs. Despite long concerted efforts, no one has consistently been able to replant or otherwise manage toward a return to the *Stipa* (needlegrass) dominance supposedly there 350 years ago.

Newer Approaches

It is, of course, much easier to criticize long extant approaches than to devise better new ones. The proof of what is better does not emerge instantly. Accordingly, my brief review of new, promising approaches is only offered as a list of ideas for you to consider in your own circumstances.

Approaches are now available that recognize there is no separate ecological ideal, that men are both components and manipulators of ecosystems. Society is no longer ecological in the older sense because the world is not ruled solely by ecological laws, but primarily by ethical and economic ones (13).

Desired Plant Communities

If the idealistic notions discussed previously don't work well, how about adopting more pragmatic benchmarks? The BLM (16) has recently proposed to adopt "desired plant communities" (DPCs) as benchmarks. They recommended the following procedure for selecting the DPC for a given range site (3). "Identify from the land use plan the multiple use values that are affected or dependent on vegetation. Then, identify the vegetation characteristics or attributes that provide for the desired uses and values. Next, analyze the seral plant communities occurring on the site and assess their capability for providing the desired attributes. Finally, select a community or communities that best represents the desired attributes. This community becomes the DPC and the vegetation management objective."

The BLM's DPC approach has some potential pitfalls. Who does the judging of what is desirable? Can the criteria for desirability be quantified? Is the desired plant community possible on a given site through ordinary means of management in a reasonable amount of time?

Because both vegetation and management objectives change through time, there will be temptations to move the targets (17). Nevertheless, I encourage the BLM and others to press forward with the testing of this fresh approach. Although other pitfalls will undoubtedly be discovered, at least this new approach potentially avoids some of the major, mostly mental, obstacles of the past. For instance, assumptions that the climax and near climax communities are the most productive, most diverse, and least erosive are avoided. We now know that the shape of these relationships are unique for each type of disturbance on each range site.

Whereas ranchers have focused on the availability of forage, largely irrespective of plant species composition, management and advisory agencies have emphasized native species composition and seral status and ignored the forage value of introduced species. Forage availability and seral status are usually only related in complex, nonlinear ways. No wonder confusion and disputes have developed about assessment of condition and interpretation of trend. The optimal quantity and quality of vegetation produced surely must depend on desired rangeland use (3). The optimum obviously should be determined by the intended uses of the range and whether they are compatible and sustainable.

Bioindicators

Most efforts to develop benchmarks and monitor change have involved vegetation analysis. Vegetation is abundant and is relatively easy to measure via a variety of techniques. Choice of vegetational (and any other) variables and how to measure them inevitably reflects individual and institutional biases and value systems as well as technical advances (18). Because of connectivity of soil protection, forage and wildlife habitat to vegetation, vegetational characteristics are often used as a proxies for overall ecosystem health (13). The public is, however, often interested in particular species (e.g. desert tortoise), products (e.g. water), or qualities (such as wilderness) that are only indirectly related to vegetation in complex ways. Monitoring of such things may have to be done more directly.

Because of the taxonomic difficulties faced in assessing total community composition and expressing diversity, there have been calls to use the principle of bioindicators or key species. When the chosen species is relatively abundant, easy to monitor, and noncharismatic, little disputation arises. Population levels of these organisms is thought to be a "litmus test" of environmental quality. Recently proposed groups in mesic environments are earthworms and amphibians. They are, however, naturally infrequent in the drier rangelands. If rare, charismatic vertebrates are chosen as bioindicators of environmental health, serious debate can erupt. Witness the use of the northern spotted owl as an indicator of environmental quality in old-growth forests of the Pacific Northwest. The analogue on some rangelands is the desert tortoise.

Quantification of significant changes in vertebrate animal population levels (trends) is especially difficult. The diverse, much more abundant, and supposedly more stable populations of invertebrates and soil microorganisms may provide indices of ecosystem health. We shouldn't let the lack of completeness of taxonomic references thwart us. All organisms can be sorted into broader groups based on size, life cycles, mouth parts, and thus their probable roles in food webs. New ways of obtaining functional groups of soil microflora now make assessment of their diversity much more feasible. Microfloral diversity is often greater in moderately grazed situations than within livestock exclosures or relicts (19). Thus, simple linear trends of increasing productivity, diversity, and stability with decreasing levels of use will probably not be found. Some of us are currently working on this topic as yet another angle on range conditions and trends.

Ecosystem-level Phenomena

Some have proposed that ecosystem-level phenomena be monitored and certain levels used as benchmarks. This supposedly avoids the questions of differing human value systems that are more obvious with single component approaches. Ecosystem structural and functional criteria that are often proposed are productivity, food webs, decomposition rates, and nutrient cycling. Whereas these characteristics are known to be important, there is often no consensus of opinion on which features should be studied, let alone how to do so. The ecosystem concept provides one with a solid basis for ordering information, but it does not, in itself, determine how great a rate of change is acceptable, nor which components of a system should be altered and which should not. Rather than being solely a technical decision, selecting which variables to monitor and how to do so is very value-laden (20).

Ethical Concerns

The fundamental problem with all previous approaches is that no objective, quantitative standards of naturalness are possible (20). Human values dictate what attributes to emphasize. Managers often set ecological standards based on the ecosystem characteristics they value most. Each definition of naturalness and each ecosystem characteristic to monitor can lead to different ethical, as well as ecological, implications.

The subjectivity in deciding which variables to measure obviously opens the door for potential abuse and is probably a major reason why there have been so many efforts to find an objective ideal. If presettlement ecosystems are undecipherable, then by this reasoning what exists today is natural or the only quantifiable reality, even if modified by users and managers. The lack of solely objective quantitative standards for ecosystem management means that it is impossible to strictly measure how effectively a rancher or land management agency is adhering to the idealistic legislative and administrative mandates applying to the land (20). Furthermore, each management action or inaction (e.g. failure to try and control invading species like pinyons and junipers) moves us further from the old ideal.

Our values for land management are thus not held in isolation but are components in a larger value system, as a hierarchy or ordered set (20). We have continuous reordering of values in our society, but perhaps more reordering is now going on for rangelands than at any other time in this century. Many fail to realize that it is impossible to maximize more than one value at a time. The compromises that inevitably must follow are what creates dissension.

The solution to the task at hand will take us far beyond technical and scientific concerns. It is the ethical and political considerations that now dominate the debate. Scientists and technicians can participate as only one of many groups interested in forging new ways of managing rangeland.

Is it possible to devise some ethically-based rules for managing rangelands in our present society? What principles should we aspire to?

I will try. I have to do so before I can suggest new ways to set standards.

1. Acknowledge that use inevitably leads to change of rangeland ecosystems. Society has designated certain areas as parks and wilderness where attempts are being made to preserve and/or restore species and other characteristics. Society, however, has to allow some land to produce the goods and services it requires to maintain the desired standard of living.

2. For those lands chosen for multiple use, strive for sustainability. Heady's (21) concept of the "management equilibrium" applies. This will probably result in reduction or displacement of some species from part of their range. Moderate management action will be required to sustain major functional processes and components. A more holistic view (22) is required to more successfully tinker with ecosystems. The institutional and disciplinary balkanization we have allowed to develop hinders our abilities to deal holistically (23). Public debate, focusing on the desired overall results rather than the means to achieve them, will cut down on the acrimony (22).

3. Monitor critical characteristics of rangelands and make the results readily available to all interested.

4. Publically discuss the changes on rangelands, especially those on federal lands. Discuss the alternatives of action and inaction and how they are likely to influence the character of the land.

5. Don't expect all land to serve all purposes. Zone conflicting uses to lands best able to tolerate those uses without permanent damage, e.g. off-road vehicles, feral horses, and burros.

6. Keep options open for the future. Minimize irreversible actions. Stress renewability so as to provide for intergenerational equity.

Given these guidelines, what can be done to develop benchmarks for rangeland management in the future?

Sustainability

For sustainability of ecosystem structure and function, I consider the soil to be the most essential component. Soils are highly buffered systems preserving the 'memory' of former systems. Furthermore, maintaining soils is essential for capturing, storing, and releasing water and nutrients, especially in semiarid environments (24). Some range types (e.g. "mature" pinyon-juniper woodlands) exhibit undetectably slow changes in vegetation structure while undergoing accelerated soil erosion. Unless soil profile depths and other features (e.g. soil organic matter and aggregate stability) are retained, desertification ensues (25). With unknown consequences of global climatic change (26), it behooves us to maintain soils to allow options for future agricultural needs. Accordingly, I feel we should put more emphasis on soils in describing benchmarks and monitoring change. We have to recognize that the uses that export goods from the system will inevitably involve a long-term run-down of nutrient status (24). These nutrients will have to eventually be replaced. We can, however, preserve profile depths by maintaining certain threshold plant and litter cover on most, but not all, sites.

Soil Erosion and Microrelief

Monitoring soil erosion directly and separating natural and human-induced acceleration is exceedingly difficult (27). Much different natural rates of erosion may be expected from various landscape positions (28, 29). Geoff Pickup and colleagues (29) have recently shown in Australia that changes in the proportions of different landscape components can be correlated with vegetation change over large areas via satellite imagery. I believe this approach has considerable potential in other rangeland contexts.

Smaller scale patterns in soil microrelief may also offer ways to judge health of ecosystems in arid and semiarid environments. Eckert et al. (30) have shown some interesting trends in shrub steppes of the Great Basin. Tongway and Smith (31) have done similar work in semiarid woodlands of Australia. Lower level remote sensing (e.g. via ultralight aircraft) could make such approaches more feasible.

Newer Ways to Interpret Vegetation

Just because vegetation has been the most studied element of wildlands does not mean we have run out of possibilities for using it in describing benchmarks and monitoring change. We have largely over-looked whole facets of the total vegetation [(e.g. small scale aggregation and population dynamics (32) and microphytic crusts (33)] and different ways to organize the vascular data e.g. functional groups (34) and interpret meaning from them [e.g. ordination (35)].

Ordination allows us to reduce mutivariate phenomena to a few axes. Trends in both space and time may be interpreted from large, complicated data sets. Furthermore, I see some possibilities for untangling the influences of climatic fluctuations from disturbances due to grazing, fire, off-road vehicles, etc. via ordination. This confoundment has been a major problem in interpreting vegetation data in the past. Conventional benchmarks have shown naturally high amounts of variability due to climate, natural waxing and waning of populations of plant species, insect and vertebrate herbivores, and pathogens. Not all observed change has been due to livestock. Livestock use just happens to be one of the few things that can be changed easily.

New Benchmarks

If idealistic and hypothetical notions of climax, PNC, and relicts no longer adequately serve as benchmarks, what can be used? I propose that we instead choose areas that have, in the judgement of people with long experience in a given region (both users and technicians), received what is perceived to be sustainable management (an expert system approach). These areas (multiple examples preferred) should then be monitored along with some others that are obviously abused, as well as some that are unused. We can then monitor vegetation, microtopography, soil microbiota, or whatever else we can afford. Next we can ordinate the *differences* to see if there is divergence. Provided these samples are all on the same range site, the major differences should be due to the local management manipulations we are making. The more global influences of climatic change, air pollution, etc. should cause all samples to drift. These cumulative effects should be detectable in temporal ordinations of the *absolute* data. If regional trends are undesirable, political changes at much higher levels will be required. Introduced species and their desirability will be considered along with everything else. Introduced species are only serious concerns if they become functionally important and/or alter major aspects of structure at the ecosystem level (36). Complete eradication of invaders is rarely practical or possible. We can only afford to concentrate on the major players.

Stewardship of Records

Retention of all land related data is very important in order to allow future scientists and managers means to define and understand change. As local extinctions of old species and introduction of new organisms takes place, there is a tendency to forget how things once were. After introduced weeds have been around awhile, they tend to be taken for granted. A new norm is accepted and management goes on. While that is pragmatic, it may cover up the possibility of greater production from the land. Sustainability may be fully judged only from long-term observations. Small increments of change are difficult to observe. Retaining records, especially photographs or video imagery that anyone can re-interpret, is important in convincing all those concerned that enormous change is the norm for rangelands.

The short term views of modern society and its institutions discourage preservation of earlier collected information. While most kinds of experimental science do improve rapidly, making publications of just a few years ago obsolete, this is not the case for issues of environmental change. Any data or information collected earlier can be marshalled together to build up a record of change. This could be such unconventional sources as surveyor's records (37). New means of image analysis and statistical procedures have been devised to allow detection of vegetation change (38) and soil erosion (39) from photos. Computer-assisted summarizations and comparisons of all types of imagery are expected to increase enormously in the near future. Therefore, don't throw away, but securely archive all photographs and data files involving rangelands. Some of us are working on better ways of analyzing and interpreting them.

New Views

Some range extensions and contractions of species are natural. Some invasion and extinction of species is natural. The difficulty is defining the acceleration due to human interference and interpreting what the consequences are. Native species are not always the most efficient users of the resources of a given environment (36). Native species are in large measure the product of the phylogenetic possibilities--simply what stocks accidentally became available first. There may well be other biota on other continents that are better pre-adapted to our conditions. When they move to a new locale, they may also escape predators and pathogens, at least for a while. If the newcomers play a major role in capturing energy, cycling water and nutrients, changing microclimates and fire regimes, they may well cause a cascade of changes throughout the entire ecosystem (36).

Our notions of what is normal, what is possible, and what is sustainable--our benchmarks--determine our approaches to land management. Unrealistic benchmarks will thwart our ability to cope with change. More realistic benchmarks will guide us toward sustainability. Let's get on with the search for better ways to reach sustainable levels of development.

References

1. Holden C. Science 249:18 (1990).
2. Anon. *World Conservation Strategy: Living Resource Conservation for Sustainable Development*. Int. Union Conserv. Nature, Geneva, Switzerland (1980).

3. Gardner BD. Discussion Paper ENR 90-04. *Resources for the Future,* Washington, DC. 1990.
4. Weaver JE and FE Clements. *Plant Ecology,* McGraw Hill, NY, 1938.
5. Baldwin MC, CE Kellogg, and J Thorp. *USDA Yearbook of Agric.* pp. 979-1001. 1938.
6. Smith EL. In: *Secondary Succession and the Evaluation of Rangeland Condition* (WK Lauenroth and WA Laycock, Eds.), pp. 103-141. Westview Press, Boulder, CO, 1989.
7. Davis MC. In: *Community Ecology* (J Diamond and TJ Case, Eds.), pp. 269-284. Harper and Row, New York, NY, 1986.
8. Roberts L. Science 246:1245 (1989).
9. Pickett STA. In: *Long-term Studies in Ecology* (GE Likens, Ed.), pp. 110-135. Springer-Verlag, New York, 1989.
10. Case A. *Playing God in Yellowstone: The Destruction of America's First National Park.* Harcourt, Brace and Jovanovich, New York, 1987.
11. Westoby M, BH Walker, and I Noy-Meir. J. Range Mgt. 42:266 (1989).
12. Smith EL. In: *Vegetation Science Applications for Rangeland Analysis and Management* (PT Tueller, Ed.), pp. 111-133. Kluwer Sci. Publ., Dordrecht, The Neth., 1988.
13. Looman J. Folia Geobot. Phytotax. 11:1 (1976).
14. Pickett STA and MJ McDonnell. Trends Res. Ecol. Evol. 4:241 (1989).
15. Küchler AW. In: *The National Atlas of the USA,* pp. 90-91. USDI Geol. Surv., Washington, DC, 1970.
16. Anon. *State of the Public Rangelands, 1990,* USDI Bur. Land Manage., Washington, DC (1990).
17. West NE. In: *Vegetation Science Applications for Rangeland Analysis and Management* (PT Tueller, Ed.), pp. 11-27. Kluwer Acad. Publ., Dordrecht, The Netherlands, 1988.
18. West NE. In: *Proc. Int. Conf. on Renewable Resource Inventories for Monitoring Changes and Trends* (JT Bell and T Atterbury, Eds.), pp 636-639. Oregon State Univ., Corvallis, OR. 1983.
19. Seasedt TR, RA Ramundo, and DC Hayes. Oikos 51:243 (1988).
20. Lemons J. Environ. Conserv. 14:329 (1987).
21. Heady HF. *Rangeland Management.* McGraw-Hill, New York, 1975.
22. Savory A. *Holistic Resource Management.* Island Press, Covelo, CA, 1988.
23. Wood GW. J. Forestry 88:8 (1990).
24. West NE. In: *Semiarid Lands and Deserts* (J Skujins, Ed.), pp 295-332. Marcel Dekker, New York, 1990.
25. Schlesinger WH, JF Reynolds, GL Cunningham, LF Huenneke, WM Jarrell, RA Virginia, and WG Whitford. Science 247:1043 (1990).
26. Neilson RP, GA King, RL DeVelice, J Lenihan, D Marks, J Dolph, B Campbell, and G Glick. *Sensitivity of Ecological Landscapes and Regions to Global Climatic Change.* US Environmental Protection Agency, Corvallis, OR, 1989.
27. Blackburn WH, TL Thurow, and CA Taylor. In: *Using Cover, Soil and Weather Data in Rangeland Monitoring* (EL Smith, SS Coleman, CE Lewis, and

GW Tanner, compilers), pp. 31-40. Society for Range Management, Kissimmee, FL, 1986.

28. Wondzell SM, GL Cunningham, and D Bachelet. In: *Symposium on Strategies for Classification and Management of Natural Vegetation for Food Production in Arid Zones* (EF Aldon, CFG Vincent, and WH Moir, Eds.), pp. 15-23. Rocky Mtn Forest & Range Expt. Sta., Forest Service, U.S. Dept. Agric., Ft. Collins, CO, 1988.

29. Pickup G and VH Chewings. Int. J. Remote Sens. 9:69 (1989).

30. Eckert RE, FF Peterson, MK Wood, WH Blackburn, and JL Stephens. Nevada Agric. Expt. Sta. Bull. 89-01 (1989).

31. Tongway DJ and EL Smith. Aust. Rangel. J. 11:15 (1989).

32. West NE. In: *Biology and Utilization of Shrubs* (CM McKell, Ed.), pp. 283-305. Academic Press, Orlando, FL, 1989.

33. West NE. Adv. Ecol. Res. 20:179 (1990).

34. Friedel MH, GN Bastin, and GF Griffin. J. Environ. Manage. 27:85 (1988).

35. Uresk D. J. Range Manage. 43:282 (1990).

36. Vitousek PM. In: *Ecology of Biological Invasions of North America and Hawaii* (HA Mooney and JA Drake, Eds.), pp. 163-173. Springer-Verlag, New York, 1986.

37. Sparks SR, NE West, and EB Allen. USDA For. Serv. Gen. Tech. Rep. INT-276 (1990).

38. Noble IR. Aust. J. Bot. 25:639 (1977).

39. Graf WL. Geol. Soc. Amer. Bull. 99:261 (1987).

5

Seed Dynamics

James A. Young

Abstract

Seedbank is the collective term used to describe dispersal, accumulation, germination, predation, and death of seeds in seedbeds. In terms of the suppression of noxious weeds, knowledge of seedbanks in terms of species composition, germination, density, and duration is essential in designing weed control practices. Seedbanks are studied by either physically recovering the seeds from samples or by germinating the seeds from a soil sample in place, termed bioassaying of seedbanks. The knowledge gained from studies of seedbanks can be used to construct predictive models of subsequent seedling emergence. The development of innovative methodologies to conduct meaningful, appropriate studies of seedbanks and especially seedbanks containing seeds of noxious range weeds is a very challenging field for science.

Introduction

Major points of interest when dealing with plant species that are considered noxious weeds are: (a) how many seeds exist under the plants in the litter and soil; (b) are these seeds viable; (c) are the viable seeds capable of germinating; and (d) how rapidly do they accumulate? Collectively, these factors fall under the category of seedbanks.

Halogeton provides a good example of how knowledge of seedbanks aids in planning strategies for dealing with noxious weeds. An annual herbaceous species, halogeton must reestablish from seed each year in order to persist on an infested site. It was obvious that halogeton produced abundant seeds; a single plant was capable of producing as many as 250,000 seeds. The seeds were small and black, once they were freed of their papery coverings. Germination was nearly instantaneous when the seeds were moistened, and the black seeds showed no evidence of dormancy (1). Due to the lack of dormant seeds, it was assumed that a single year's control of halogeton plants (so the seedbank could not renew) would eradicate halogeton from a given site. A second form of halogeton seeds was found in some seed collections. These brown seeds were assumed to be immature black seeds and since they could not be germinated, they were assumed to be nonviable. M.C. Williams first showed that the brown seeds were viable and secondly demonstrated that the brown seeds were produced first. Therefore, they could not be immature seeds (2). He then demonstrated that the production of brown and black seeds was controlled by photoperiod as expressed in nature by day length. After a prolonged study of halogeton seed burial and recovery, Robocker et al. (3) were able to demonstrate that brown seeds of halogeton were capable of germination and they represented a means of prolonging the species in seedbanks. The significance of these studies was that eradication of halogeton was practically impossible.

Nature of Seedbanks

Harper (4) visualized the soil as a seedbank in which both deposits and withdrawals are made. Deposits occur by seed production and dispersal, whereas withdrawals occur by germination, senescence and death, and predation. Most research conducted on seedbanks has concerned cultivated soils where the tillage process results in the distribution of seeds throughout the surface horizon. Brenchley and Warington (5), in the classical study of seeds in variable soils in the United Kingdom, estimated the number of seeds in the soil at 34,000 to 75,000 per m^2.

There are huge differences in the estimates of viable seeds in seedbanks depending on the ecosystem being sampled. Radosevich and Holt (6) provide the following estimates they compiled on a m^2 basis from numerous sources for different ecosystems: pasture, 2,000-17,000; early-old field succession, 1,200-13,200; tropical fields, 7,600; tropical secondary forest, 1,900-3,900; tropical rain forest, 170-900; prairie, 300-800; and forest 200-3,300. In degraded big sagebrush (*Artemisia tridentata*) plant communities in the Great Basin, seedbanks ranged from 5,200 to 12,600 per m^2 (7). The seedbank in these communities consisted almost entirely of caryopses of cheatgrass (*Bromus tectorum*). In degraded big sagebrush/bunchgrass communities in the Great Basin, the soil seedbank is usually nearly devoid of seeds of the native perennial bunchgrasses (7). One exception was the pluvial lake sand deposits of the Lahontan Basin where numerous seeds of Indian ricegrass (*Oryzopsis hymenoides*) were found distributed to considerable depth in the sands of dunes (8).

The very low number of seeds found in the seedbanks of tropical rain forests, despite the huge number of plant species found in such environments, apparently reflects the rapid rate of decay of organic matter in such environments and the occurrence of recalcitrant seeds with short half-lives. The seeds of several species of conifers found in temperate forests are not adapted to germination or subsequent seedling growth under the canopy of mature trees of the same species. Such species rely on dispersal to suitable ash or mineral seedbeds after stand renewal is initiated by wildfires.

In the Great Basin, seeds of the landscape-characterizing species, big sagebrush, have a very short half-life in soil seedbanks (9). In western Nevada, big sagebrush seeds are mature about December 1 and the superabundant seed rain is gone from seedbanks by the next June.

It is obvious that the only way annuals can persist on a given site is through seedbanks. The annual grass-dominated ranges of Mediterranean California provide an example of the extreme dynamics of the seedbanks of such ecosystems. Plant densities, early in the fall after the first effective rain following the summer drought, may reach 200,000 plants per m^2 (10). Estimates, made by various researchers, of the number of viable seeds that compose the seedbank immediately before the first effective rain range from 300 to 250,000 per m^2 (11). This seemingly impossible range in variability is apparently reflective of the extreme dynamics of such seedbanks and of different methodologies used to estimate the size of the seedbanks. By six weeks after the initial rain that supports germination on the California annual-type ranges, the seedbanks are down to a few hundred per m^2 (11). This reflects the very rapid decay of organic matter in such Mediterranean-type environments with the occurrence of warm, moist conditions after the summer drought and the highly germinable condition of the majority of the seeds in the seedbank. The seeds that remain in the seedbank after six weeks are largely those of species of legumes with extremely hard seed coats.

Sampling Seedbanks

There are two basic approaches to estimating the number and viability of seeds in seedbanks. The first involves the classical bioassay approach developed as described by Brenchley and Warington (5). In this procedure, numerous samples are collected from a random grid of the area to be sampled. The samples are placed in containers that are large enough to bring all the seeds in the sample close to the surface. A moisture-holding substrate and/or covering is applied to the samples and they are incubated at a suitable temperature. Emerging seedlings are identified and counted. Samples can be stirred periodically to enhance germination. The advantage of this system is that a large number of samples can be processed with a minimum amount of labor. Continuous or interrupted incubation can continue for years. A second advantage is the seeds are germinated in the presence of their natural soil-borne microflora.

The second approach to estimating seedbanks involves actual recovery of seeds from soil samples. The difficulties that this procedure presents depend on the parent material and texture of the soil and the size and characteristics of the seeds being recovered. Indian ricegrass seeds can be recovered from pluvial lake sands by simply allowing the dry sand to flow through an appropriate sized screen (8). In this situation, the recovery is simple because the soil lacks structure and the seeds are definitely larger than the soil particles.

Fine-textured soils that have well-developed structures present a much more difficult situation for the recovery of seeds. Soil structure can influence seed emergence and the acquisition of dormancy by soil-borne seeds (12). In order to separate seeds from hard clods, it is necessary to deflocculate the soil structure. This is accomplished by soaking the soil samples, often with agitation. Sodium hexametaphosphate and/or sodium bicarbonate may be added to the solution to aid in deflocculation (13).

The viable seeds in soil-borne seedbanks are living, biological entities in a state of rest or dormancy through the action of desiccation. Once you have wet soils, or wet the soils with aqueous solutions containing one or more chemicals to aid in deflocculation, the physical and biochemical status of the seeds in the sample is influenced. The problem is how to interpret the influence of the soaking among the seeds of different species that may be represented in the sample. Obviously, if your interest is only in the number of seeds in the seedbank and not in their germination characteristics, the soaking is not a problem.

Solvents other than water can be used in separating seed from soil. Solvents of a specific gravity can be used to separate seeds from soil particles in a process called floatation. Magnesium sulphate solutions have been used as a floatation medium for recovery of seeds from soils (13). Again, the solvent may very well influence the germination characteristics of the seeds being recovered.

Almost all seed recovery systems involve the wetting of soils. The samples are dried and the seeds recovered from other organic matter through air screening. Air drying is preferred to forced drying with elevated temperatures to avoid adding another variable to those influencing the germination characteristics of the seeds. If a lot of organic matter is recovered along with the seeds, care must be taken to spread the material sufficiently so that the seeds will not imbibe moisture during the drying process.

A common problem with the recovery of seeds from soil samples is recognizing seeds once they have spent some time in the soil. This is especially true with grass caryopsis where the lemma and palea may decompose and the naked seed will be considerably different in appearance, shape, and specific gravity than the fresh caryopsis.

The recovery of seeds provides material for use in experiments for determining the viability of seeds and/or germination characteristics. If viability is the only consideration, seeds can be sectioned and tested by the tetrazolium method (14). A positive reaction to tetrazolium indicates the seed contains living material but does not necessarily indicate the seed will germinate. Seeds exposed in the soil can accumulate a considerable microflora. These microorganisms may interfere with subsequent germination testing. Incubation at low temperatures or surface seed sterilization may be required. Dipping the seeds in solutions of hydrogen peroxide or sodium hypochlorite has been used to surface sterilize seeds, but both materials may change the germination characteristics of certain seeds.

Despite the difficulties with recovering seeds from soils for estimates of seedbanks, excellent studies have been conducted using variations of this basic methodology [an example for Great Basin rangelands would be Ralphs and Cronin (15)].

The variability inherent in the spacial distribution of seedbanks usually necessitates the collection of large numbers of samples. Forcella (16) provides a procedure for estimating the number of samples required. Collecting and surface caching of seeds by rodents or birds may greatly increase the spatial variability in the distribution of seeds in seedbanks (17,18).

Comparison of Recovery Techniques

A recent study by Ball and Miller (19) compared the germination in place or bioassay method of estimating seedbanks with actual recovery of seeds from paired samples. Both techniques were suitable for determining changes in seedbanks brought on by weed control practices.

The two techniques provided effective detection of the species that composed the seedbanks and provided comparable estimates of the relative density of seeds in the seedbank. However, counts of seeds produced poor estimates of subsequent seedling emergence, with seedling emergence representing less than 10% of the potential estimated seedbank.

Most of the research conducted with seedbanks as estimates of subsequent seedling emergence has been conducted in amiable land ecosystems with the aim of predicting weed populations. H.A. Roberts (20) has been one of the most innovative investigators in this field. In the intensively cropped areas of the United States, considerable progress has been made toward using estimates of weed seeds in soilbanks to predict future weed problems (21). Models have been developed using such data to help make economic decisions on application of weed control measures in crop systems (22). A matrix model has been developed for the vertical distribution of weed seeds within the soil, which was then inserted in a life cycle model of *Alopecurus myosuroides* in order to predict seed population dynamics (23). In simulations, stability of vertical distribution was reached sooner under plowing than under rigid time cultivation. Stable distributions were very different for the two cultivation methods, one being reached by damped oscillation in the case of plowing and the other asymptotically in the case of rigid time cultivation.

Manipulating Bioassay Samples

One obvious problem with using estimates of seedbanks to predict subsequent seedling emergence is determining the nature of the safe sites for germination in the field seedbed. Will the seedbed have adequate moisture for germination, and how long and how continuous will that period be? Will seedbed moisture and optimum temperatures for germination coincide, slightly overlap, or occur at different times? Another obvious question is what is the inherent potential of the physiological systems of the various plant species whose seeds form the seedbank? The physiological potentials for germination and for persistence are both vital. As previously mentioned for seeds of big sagebrush, the half-life of seeds can be so short in seedbanks that there is no annual carryover of seeds for the species. Ideally, the most precise prediction of subsequent emergence would be obtained by bioassaying the samples collected to estimate the seedbank species composition and density under conditions that closely approximate the subsequent field seedbed. This is a seemingly impossible goal because despite the widespread acceptance of the hypothesis of safe sites (as originally proposed by J.L. Harper), no one has adequately described the physical and biological dimensions of a safe site for germination for any species. On the other hand, the best predictions of subsequent seedling emergence based on samples of seedbanks have occurred when one dominant species composed the seedbank and the biology of the species was well understood (24).

The seeds (caryopses) of the annual cheatgrass provide a model of how this complex process of defining safe sites for germination can be approached experimentally. At maturity the seeds of cheatgrass are usually highly viable and usually lack dormancy. Germination over a fairly wide range of incubation temperatures is commonly greater than 80% (25). Seeds that find favorable microtopography and/or litter cover through dispersal actions will germinate simultaneously with the first effective moisture event in the fall or early spring (26,27). Seeds that are dispersed to unfavorable sites in the seedbed (nonsafe sites) acquire a dormancy over winter (7). This dormancy breaks down slowly over the next two to three years (28). The dormancy can be overcome by adding nitrate or gibberellin to the germination substrate (29).

The detailed knowledge of the biology and acquired dormancy of cheatgrass seeds can be used to manipulate bioassay procedures so germination results will reflect potential seedling emergence. Malone (13) objected to bioassay procedures to estimate seedbanks because of the time required to complete such tests. Time of germination from bioassay test can be used as an experimental variable. The results of prolonged or repeated bioassays can be used to adjust predictive models of seedling emergence. For example, a single week of soil moisture will allow 25% of the seeds in the seedbank to germinate, while 6 weeks of seedbed moisture will allow 60% of the seedbank to come out of dormancy and have the chance to emerge, or three interrupted periods of one week each of soil moisture will produce 30% emergence (29). Nitrate enrichment can be entered as a second variable, interacting with time (30). The number of variables entered in the bioassay is limited only by the ingenuity of the investigator.

Seedbed Litter

Most studies of seedbanks, as previously mentioned, have been conducted on tilled soils. Litter has been ignored in these studies. In wildland situations, most of the seedbank may be located in the scant and discontinuous to extensive accumulations of organic matter on the surface of soils. The development of minimum tillage or conservation tillage procedures for amiable lands may be increasing the apparently little-appreciated importance of litter for these crop sites as a physical component of the seedbank. Litter can limit the potential of the seedbed to support germination, as is the case for seeds of several species of conifers that germinate best in contact with mineral soil, or it can modify extremes of temperature to permit germination for some species (26). The seeds of some species are capable of germinating in litter without contact with the mineral soil (31).

The important point concerning litter in reference to seedbanks is to be sure to include it in both sampling and bioassay procedures. If the seedbed for which you are predicting emergence is going to be devoid of litter, do not cover the bioassay sample with a moisture-protecting material such as vermiculite. Bioassay containers located on a mist bench will have much higher seedling emergence compared to a normal greenhouse bench.

Seed Longevity

Obviously, one of the first places to start in studies of seedbanks is to examine published information on the longevity of seeds of the species in which you are interested. Simple longevity figures for storage under laboratory or some form of defined conditions, such as sealed containers at a certain moisture percentage and storage temperature, provide leads to the life of seeds in seedbanks. For agronomic crops and weeds, great volumes of data have been developed on seed longevity (6). The longevity of seeds stored in the soil ranges from a few weeks to thousands of years. For secondary successional species in the Great Basin, information is much more limited. Information for species such as Russian thistle (*Salsola australis*) largely comes from the work of W.S. Chepil conducted during the 1930s (32).

Long Distance Transportation of Seeds

A major means of spread of noxious weeds on rangelands is through contaminated hay. This source of weed seeds is more important than seed for revegetation because of the relatively small amount of expensive seed used for seeding on rangelands. Many old hay fields are infested with noxious weeds. Long distance transportation of livestock is a secondary source of noxious weed seeds. Tracklaying equipment used for construction and maintenance of stock water facilities is a major dispersal mechanism for specific weed seeds. Granivorous birds may be long distance dispersal agents for noxious weed seeds either through hard seeds that pass through the digestive system or through mucilaginous seeds that stick to feathers. In areas of irrigated agriculture, there is often concern about the water transportation of seeds from rangeland watersheds to crop land.

In terms of the interest of this conference, the central issue is seedbanks for the range weeds that are classed as noxious weeds. Throughout the literature, there are references to the seeds of species such as rush skeletonweed (*Chondrilla juncea*), which remain viable for about 18 months or less (33). However, seedbank studies require a lot of effort, time, and money; they are difficult to interpret if the seed and seedbed ecology of the species is not understood. Obviously, the information produced by seedbank studies is essential in planning suppression technologies for noxious weeds.

References

1. Dye WB. Weeds 4:55-60 (1956).
2. Williams MC. Weeds 8:452-461 (1960).
3. Robocker WA, MC Williams, RA Evans, and PJ Torrel. Weed Sci. 17:63-65 (1969).
4. Harper JL. *The Population Biology of Plants*. Academic Press, London, 1956.
5. Brenchley WE and K Warington. J. Ecol. 18:235-272 (1930).
6. Radosevich SR and JS Holt. *Weed Ecology*. John Wiley & Sons, New York, 1984.
7. Young JA, RA Evans, and RE Eckert Jr. Weed Sci. 17:20-26 (1964).
8. Young JA, RA Evans, and BA Roundy. J. Range Manage. 36:82-86 (1983).
9. Young JA and RA Evans. Weed Sci. 37:201-206 (1989).
10. Biswell HH and CH Graham. J. Range Manage. 9:116-118 (1957).
11. Young JA, RA Evans, CA Raguse, and JR Larsen. Hilgardia 49(2):1-37 (1981).
12. Terpstra R. Weed Sci. 34:889-895 (1986).
13. Malone CR. Weeds 15:381-382 (1967).
14. Grabe DF (ed.). *Handbook on Seed Testing*, Agric. Official Seed Analysis Contrib. 29:1-62 (1970).
15. Ralphs MH and EH Cronin. Weed Sci. 35:792-795 (1987).
16. Forcella F. Aust. J. Agric. Res. 35:645-652 (1984).
17. Latourette JE, JA Young, and RA Evans. J. Range Manage. 24:118-120 (1971).
18. McAdoo JK, CC Evans, BA Roundy, JA Young, and RA Evans. J. Range Manage. 36:61-64 (1983).
19. Ball DA and SD Miller. Weed Res. 29:365-373 (1989).
20. Roberts HA. Adv. Appl. Biol. 6:1-55 (1981).
21. Wilson RG, ED Kerr, and LA Nelson. Weed Sci. 33:171-175 (1985).

22. King RP, DW Lybecker, EE Schweizer, and RL Zimdahl. Weed Sci. 34:972-979 (1986).
23. Cousens R and SR Moss. Weed Res. 30:61-70 (1910).
24. Naylor REL. Weed Res. 10:296-299 (1970).
25. Young JA and RA Evans. ARR-W27 Agric. Res. Ser., U.S. Dept. Agric. Berkeley, CA, 1987.
26. Evans RA and JA Young. Weed Sci. 18:697-703 (1970).
27. Evans RA and JA Young. Weed Sci. 20:350-356 (1972).
28. Young JA, RA Evans, and RE Eckert Jr. Weed Sci. 17:20-26 (1969).
29. Young JA and RA Evans. Weed Sci. 23:358-364 (1975).
30. Evans RA and Young JA. Weed Sci. 23:354-357 (1975).
31. Young JA, RA Evans, and BL Kay. Agron. J. 63:551-555 (1971).
32. Chepil WA. Sci. Agric. 26:307-346 (1946).
33. Lee GA. Weed Sci. 34:2-6 (1986).

6

Weed Dynamics on Rangeland

Robert H. Callihan and John O. Evans

Abstract

Historical evidence indicates that western ranges once had much more perennial grass as dominant species in their plant communities. The degradation of these ranges can be traced from the 19th century with the wide-scale grazing of domestic livestock. Alien plants intentionally or accidentally introduced to western rangelands exhibit extreme competitive abilities and dominate vast areas weakened by unfavorable management strategies or stresses to native plant population from environmental events such as wild fires. Much western rangeland has such low residual grass populations at present that revegetation is essential.

Introduction

Few noxious weed species are native to western rangelands. Although there are important exceptions, most noxious weeds are alien to the lands they infest. Weed invasion has resulted in radical conversion of most western rangelands and has altered the manageability of those lands to the point where reclamation cannot return the vegetation to its native condition.

Rangeland west of the Rocky Mountains was, in its native state, uniquely sensitive to abusive grazing. Unlike grasslands of Europe, the eastern U.S., and the Great Plains, the primary grasses were bunchgrasses well adapted where the annual precipitation ranges from 150 to 600 mm per year and falls during autumn, winter, and spring. These native bunchgrasses, such as bluebunch wheatgrass (*Agropyron spicatum*) and Idaho fescue (*Festuca idahoensis*), expend energy on perenniality rather than reproduction by seeds. They do not have high seedling vigor and do not recover rapidly after grazing.

Discussion

The irreversible alteration of western rangeland occurred largely during the mid-to-late 19th century. The process occurred in two stages. The first stage began during and after the U.S. Civil War, when the livestock industry used grassland resources heavily, resulting in severe overgrazing. In the early 20th century, ranchers began to rely on grazing deferral to provide the time necessary for recovery sufficient to produce the next grazeable crop. The second stage of range deterioration began when downy brome (*Bromus tectorum*), medusahead (*Taeniatherum caput-medusae*), sixweeks fescue (*Festuca myuros*), and other winter annual grasses were inadvertently introduced in the late 19th century and the rangeland species composition changed dramatically. The annuals were opportunists with high seedling vigor and were thus able to germinate, establish, and complete life cycles before summer drought. The perennials were no longer able to regenerate from seed due to seedling competition from new annual grasses. Established native perennial grass plants were overgrazed, and their vigor and populations declined since they were unable to compete successfully with the winter annuals. This rangeland degeneration was further aggravated by tendencies for managers to shift grazing practices to better utilize the new annual species. Extensive early spring grazing was practiced, which further hampered the survival of perennial forage grasses to the point of nearly complete disappearance of native communities from many rangelands. Many ranges were thus converted to predominantly annual grasses and this left these ranges open to the next wave of invaders, the dicots.

Noxious dicotyledonous species have apparently been introduced continuously. Many have winter annual characteristics, as do the annual grasses, and are consequently well adapted to the western rangeland climate. Seedlings of noxious annual and short-lived perennial species such as yellow starthistle (*Centaurea solstitialis*), diffuse knapweed (*Centaurea diffusa*), spotted knapweed (*Centaurea maculosa*), dyer's woad (*Isatis tinctoria*), and others have vigorous taproot systems that penetrate the soil more deeply and rapidly than do the fibrous roots of grasses. These dicot weeds are thus far more aggressive than annual or perennial grasses and quickly dominate rangelands. Livestock avoid these weeds and consume mostly grasses, thus favoring rapid shifts toward complete dominance by the noxious dicot species. Since these plants extract more moisture from deep within the profile, they remain green longer into the summer than do the grasses. Thus escaping competition, they are able to be more prolific seed producers.

New weed species have been introduced at an average of six or more per year into the Intermountain states over the last 100 years. Species new to Idaho have been reported at the rate of seven to nine per year in recent years. Some of these are well adapted to rangelands, and each species adds a new element of complexity to land management.

Spread of noxious weeds occurs at varying rates on rangelands. Major factors that determine rate of spread include seed productivity, natural dispersal mechanisms, site availability, range condition, plant phenology, and transport of seeds by man. Most noxious weed infestations for which documentation is available appear to be expanding at average rates of from 3 to 60% per year, depending mostly on niche availability.

Yellow starthistle has increased in newly infested areas at the rate of 60% per year. However, in northern Idaho where the main body of infestation exists on about 120,000 ha, the average rate of expansion has been only 3% per year recently. This is because most of the susceptible land around the main infestation is already occupied. Since the susceptible area is surrounded by forests and cropland, there are fewer opportunities for infestation of adjacent lands. Yellow starthistle appears to be adapted to any habitat in which downy brome will grow, and normally produces 750 to 1,000 kg/ha of dry matter where populations are 5-8 million plants/ha. Although this species is not difficult to suppress on accessible sites, low-productivity rangeland may not yield sufficient returns to pay for suppression treatment if annual grasses are the only species to benefit.

The rate of spread of noxious weed species does not appear to be directly correlated with natural morphological dispersal mechanisms such as presence or type of pappus, but it is obviously dependent upon seeds in their various forms. Seed longevity does not seem to be a substantial factor in dispersal rate. The apparent expansion rate of common crupina (Crupina vulgaris) infestations is not greatly different from that of yellow starthistle, even though crupina seeds remain viable only two years compared to the decade for which yellow starthistle seeds may survive.

Persistence of a weed species, once seeds are dispersed, is highly dependent upon seed longevity, however. Common crupina is not difficult to eradicate by simply preventing seed production and plant growth for two consecutive growing seasons. On the other hand, since a small proportion of a crop of yellow starthistle seeds will continue to germinate over a period of many years, survival of the species on an infested site is virtually assured unless a much more costly, persistent program of repeated actions is maintained over a period of many years. For these reasons, a rugged terrain infested with a thousand acres of yellow starthistle is not likely to inspire an eradication program, whereas if the invading species were common crupina, eradication would be a viable option.

Where the invading noxious species has become irreversibly established, integrated weed management (IWM) offers the only sensible solution to the problem. IWM includes removal of the weed species, normally by a selective herbicide treatment, followed by or concomitant with a range renovation and/or rehabilitation action to enhance the competitive abilities of desirable forage species. Well-managed introduced perennial grasses are able to retard (but not prevent) the reinvasion or resurgence of alien species. They provide that suppressive resistance against not only the weed species of greatest concern but all others as well, thus reducing the necessity and frequency of suppressive actions such as herbicide application. Alien species will continue to invade, but normally at dramatically lower rates of invasion.

The net result of an IWM scheme is that the overall productivity and profitability often exceeds that when the range was in a degraded state and dominated by common annual grass weeds. The arrival of noxious weeds has forced the choice between initiation of a higher level of management technology and abandonment of grazing as a land use.

In some cases, however, management technology has not been able to cope with the noxious weeds. Many species of noxious weeds have proven too difficult to manage economically when both weed suppression and range revegetation are necessary components of an IWM program. Species such as leafy spurge (Euphorbia esula), yellow hawkweed (Hieraceum pratense), rush skeletonweed (Chondrilla juncea), and field bindweed (Convolvulus arvensis) are difficult to suppress in the habitats they dominate. Steep canyonlands and other difficult terrain often preclude efficient management operations.

In view of the imminence of noxious weed introductions, prudence dictates that rehabilitation practices be performed prior to their arrival. When rehabilitation has only to contend with recovery of residual perennial grasses and suppression of annual weedy grasses, the cost of management is less and the economic returns to management are greater. The rate of noxious weed infestation is slower when such weeds do arrive, the cost of noxious weed suppression is far less, and the interruption of rangeland utilization is reduced.

Many grass species are known to be well adapted for reseeding western rangelands. Wheatgrasses, fescues, and wildryes are among the most commonly used. Normally, selection of a forage grass for reseeding is based primarily on productivity and palatability with secondary consideration given to seedling vigor, season of growth, ecological and physiological amplitude, and stand life. When rough terrain precludes intensive management and noxious weeds must be considered, the selection criteria change. Competitive ability and survivability are primary factors. For example, hard fescue (*Festuca ovina duriuscula*), regardless of its low productivity, may be best suited for steep, rocky terrain precisely because it is a long-lived species not highly palatable and less likely to be overgrazed. Rhizomatous species such as intermediate wheatgrass (*Agropyron intermedium*) are good choices where noxious weeds are a consideration because they do not depend on seeds to fill in where plants are lost from the stand. Rhizomatous grasses thus compensate for their disadvantage in seedling competition against vigorous annual weed seedlings.

Much western rangeland has such low residual grass populations that revegetation is warranted. Revegetation in the face of not only common annual weeds (especially grass weeds) but also noxious weeds is both costly and risky. Risk is increased where terrain is too rugged for operation of ground-operated equipment.

Several time-critical operations are necessary in revegetation of rangeland dominated by downy brome and noxious dicot weeds. The operations necessary are destruction of noxious weed growth, suppression of the annual grass, planting the intended forage, and deferral of grazing. Occasionally, removal of litter, rodent suppression, fertilization, and seedbed preparation are involved. Correlation of all of these practices with both climate and weather are necessary to maximize soil moisture, to achieve the best suppression of weeds, and to ensure good forage seedling establishment.

Many noxious dicot weeds are susceptible to selective hormone-type herbicides such as picloram, clopyralid, dicamba, and 2,4-D. The annual weedy grasses are susceptible to atrazine and herbicides with related modes of action. Annual grasses and many annual and perennial dicots are also susceptible to glyphosate. All of these herbicides are useful in range renovation and can be critically important in stand establishment. Substantial soil activity from residual herbicides is exceptionally valuable for maintenance of weed suppression during stand establishment. Picloram has been especially useful in this regard.

A program in which a herbicide is to be used may begin with suppression of noxious weeds with a soil-residual selective herbicide in the fall of the year preceding fall planting. This herbicide should reduce the noxious weed's seed production the following year and should persist at levels sufficient to maintain control of new weed seedlings the second autumn, yet be degraded below levels toxic to the new forage seeding. The following year, after autumn rains have caused winter annual grasses and dicots to emerge, and prior to seeding, a glyphosate treatment will destroy grasses and suppress most other weeds to allow the forages to emerge and establish with little competition for moisture, light, and nutrients. Litter removal is helpful if litter is so dense that forage seeds are unable to establish sufficient soil contact to ensure germination and if rodent cover and forage removal are appropriate. Burning in mid- to late fall can be safely done prior to fall weed germination. Glyphosate may be applied a week or more after seeding and can be done any time during the winter where conditions permit. Grazing must be deferred on most arid to semiarid ranges for a period of one to two years or more, until the forage grasses have become vigorous and have produced at least one substantial seed crop.

Seeding by drilling methods is always preferable to aerial seeding, but where terrain cannot be negotiated by ground equipment, aerial seeding may be the only option. The cost of the operations described is obviously high, but may be necessary in order to decrease the high risk of stand failure that is attendant with aerial seeding. Typically, a range that has been revegetated by aerial seeding requires an additional year or two without grazing to ensure maximum establishment.

The procedures described are as important for renovation of range where biological control by parasites is to be used as where chemical control is planned. Where such biological control is intended, the temptation is often to omit renovation or rehabilitation and to rely only on the weed parasites for noxious weed suppression. The problem with this approach is that even if the parasites suppress the target weed, the vegetation simply reverts back to the annual grasses and other common weeds. In this case, the range is susceptible to the next invading noxious species, and its productivity continues at a low level as before. Only when the range is renovated can biological control with parasites contribute toward greater economic returns. In any case, even where biological control with parasites is planned, range improvement under conditions of serious noxious weed infestation is best done in a rapid, precipitous manner, after which the parasites can maintain suppression of the target weed in a productive pasture or range.

Elimination of noxious weeds is never without substantial cost; that is why they are given the legal designation as noxious weeds. Prospective land buyers and current managers of land should be aware that land values and returns to investment are drastically reduced unless maintenance costs are drastically increased when noxious weeds take over.

Noxious weed dynamics are subtle and slow, but relentless. When noxious weeds move to new lands, the legal landowners may pay the taxes, but the weeds possess the land if the landowner does not wage an active offensive battle.

Bibliography

Blaisdell JP, RB Murray, and ED McArthur. USDA, Forest Service Gen. Tech. Rep. INT-134, 41 pp, 1982.

National Res. Council./National Acad. Sciences. *Developing Strategies for Rangeland Management.* Westview Press. 2018 pp. Boulder, CO.

Roché BF and CT Roché. *Proc. Pacific Northwest Range Management Short Course.* 85 pp. 1989.

Sharp LA and KD Sanders. Univ. of Idaho Misc. Publ. No. 6., 1978.

Vallentine JF. *Range Development and Improvement, 2nd Ed.,* 545 pp. BYU Press. Provo, UT, 1980.

7

Global Change and Vegetation Dynamics

Herman S. Mayeux, Hyrum B. Johnson, and H. Wayne Polley

Abstract

The current range weed and brush problem is largely the result of dramatic increases in the density and abundance of mostly native species. Essentially all broadleafed weeds and woody species utilize the C3 photosynthetic pathway, while the warm-season perennial grasses they replaced over much of the West are C4 plants. Recent research indicates that increasing atmospheric CO_2 levels confer a physiological advantage upon C3 species that is manifested as increased growth and competitive ability. If today's higher CO_2 levels favor C3 over C4 plants, the phenomenon represents a more satisfactory explanation for the replacement of open C4 grasslands and savannahs by C3 shrublands than overgrazing, suppression of fire, and climate change. The hypothesis that increased atmospheric CO_2 causes vegetation change on rangelands implies that C3 weeds and shrubs have become the inherently dominant functional group, in contrast to the classical climax concept, and may partially explain why woody plants increasingly dominate rangelands despite concerted efforts to control them.

Introduction

Global change, a phrase that appears with increasing frequency in the scientific literature, usually refers to the impacts of increasing concentrations of radiation-active gases in Earth's atmosphere, especially

CO_2. The CO_2 concentration of the atmosphere has increased from as low as 265 ppm only 125 years ago (1-3) to about 350 ppm at present (4). The increase to date has been substantial, almost 30%, and atmospheric levels are expected to double by the end of the next century (5). Possible effects of this phenomenon on vegetation fall into two broad categories: direct effects on plants caused by "CO_2 fertilization" and indirect effects associated with possible climate change, the "greenhouse effect." Emphasis here is placed upon the direct effects of increasing CO_2 levels on vegetation.

A dramatic example of vegetation dynamics is the increase in abundance and density of perennial weeds and woody shrubs on open grasslands which also occurred over the last 125 years or so. This phenomenon seems best expressed in the southern Great Plains and Southwest (Figure 7.1) but has also occurred elsewhere in North America. Shrubs were natural components of the floras of the American West in presettlement times but were often relegated to widely separated sites, such as drainageways or rocky outcrops, or occurred as widely separated individuals in otherwise essentially pure grasslands. A host of woody species moved out of these limited sites and became established across the broad range of soils and topographical situations available. These include widely distributed shrubs such as mesquites (*Prosopis* spp.), creosotebush (*Larrea divaricata*), and junipers (*Juniperus* spp.), many shrubby Composites like the sagebrushes (*Artemisia*), rabbitbrushes (*Chrysothamnus*), and snakeweeds (*Gutierrezia*), and a host of less geographically important plants. In some areas, shrubs like big sagebrush (*Artemisia tridentata*) occurred at high densities prior to settlement (6); these stands have since thickened and the extent of sagebrush-dominated rangeland has increased (7,8).

Causes of Vegetation Change

Several explanations for the change in structure of wildland plant communities have been proposed. They include the destruction of the vigor and competitive ability of grasses by droughts (9) or by prolonged and intensive grazing following the introduction of large numbers of livestock in the nineteenth century (10), distribution of hard-coated seed of woody species to new sites after ingestion by roving livestock (11,12), climate change (13), elimination by man of the periodic wildfires which removed woody seedlings without destroying grasses (14), or combinations of the effects of grazing, fire, and changing weather patterns (15-17). These factors undoubtedly helped shape contemporary native plant communities in at least some situations, but this has been

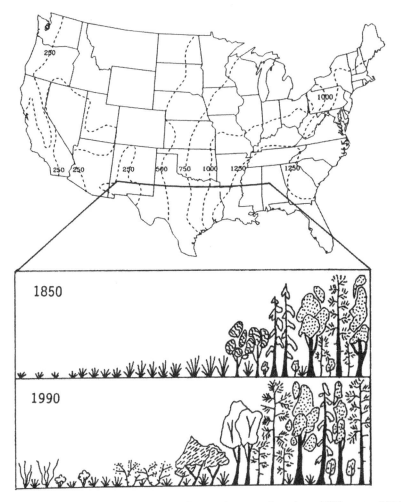

FIGURE 7.1 Changes in the structure of natural vegetation since 1850 over a 2000 km latitudinal (33 N) transect from the desert grasslands to the Southeastern mixed forests on the right. Isohyets indicate average annual precipitation (mm).

difficult to demonstrate. The evidence is reviewed by Branson (18) and described as based more on opinion and observation than on quantitative measurements.

Conflicting opinions are offered regarding the role of grazing. Early observers like Wooten (19,20), who witnessed both the destruction of grass cover soon after desert grasslands were stocked and the beginnings of shrub encroachment, felt that the two were directly related. Hastings and Turner (13) note examples of a lack of temporal and spatial associations between intensive grazing and the onset of vegetation change and conclude that "livestock have made important contributions to the changing vegetation of the desert region, but have not been the primary agent of change." York and Dick-Peddie (21) concluded that lightly grazed desert grasslands supported less woody cover than heavily grazed areas in southern New Mexico, but others (22,23) documented increases in mesquite density and cover on areas protected from grazing for many decades in the same area. Mesquite increased rapidly on grasslands that were closely grazed, moderately grazed, or protected from grazing for extended periods in Arizona (24-27). Similar results have been reported with other woody species (28-30), even when grazing of the unprotected areas was controlled in a three-pasture rest-rotation system (31).

Several authors have proposed that fire is one of the most important ecological factors in maintaining open grasslands (32-34); the first, apparently, was Griffiths (35). However, according to Cook (36), woody plants replaced herbaceous vegetation in southern Texas in the face of frequent burning. Weniger (37) attributes the shrinking of Texas' prairies to intentional burning as well as grazing by domestic animals. Hastings and Turner (13) concluded that fire failed to pass two necessary tests; it had not been shown to be sufficiently effective in killing woody plants and fires did not occur frequently enough to be a factor. However, they emphasized that each area must be considered individually because of differences in the flora, topography, climate, and habits of native peoples.

Historical shifts in the abundance of shrubs have been attributed to changes in weather patterns. The most compelling example documents increasing temperatures and decreasing winter precipitation in Arizona and New Mexico (13). Causal effects of climate depend upon the existence of a change of sufficient duration and magnitude and in the direction necessary to favor shrubs at the expense of grasses. Fundamental processes or mechanisms must be identified which translate small differences in weather patterns into such dramatic changes in the structure of vegetation as the replacement of long-dominant grasses by woody species. Hastings and Turner (13), for instance, assume that increasing aridity associated with higher mean temperatures and lower precipitation favors shrubs over grasses and that such trends over a period of decades are sufficient to cause dramatic change.

We cannot adamantly deny that these traditional explanations contributed to vegetation change on rangelands, but we do contend that no compelling evidence exists which substantiates that any single one or combination of them is responsible. A major difficulty with traditional explanations is the ubiquity of increases in the abundance of woody plants across the extreme heterogeneity of the entire American West, from tallgrass prairie to desert grassland. Furthermore, vegetation change of the kind observed in North America is actually a global phenomenon. Johnston (32) pointed out that similar shifts have been reported on other continents. Muello (in 32) described such changes in Argentina, and the shift from open grasslands to shrublands continues today in Mexico (38) and South America (39-41), Australia (42), Africa (43,44), and India (45). In the lowveld of Rhodesia, shrubs increased at the expense of perennial grasses more in the absence of domestic livestock than when grazed lightly or moderately (46).

The CO_2/Vegetation Change Hypothesis

Future changes in community structure were predicted as consequences of continued increases in atmospheric CO_2 in the next century (47,48), as were shifts in the boundaries of vegetation types in the Southwest (49). However, the lack of satisfactory explanations for and the global nature of vegetation dynamics on rangelands encouraged us to consider whether effects of global increases in atmospheric CO_2 levels are already evident as shifts from open grasslands to shrubs. Increasing CO_2 may have already begun to influence productivity of forests (50-53). After a thorough analysis of annual growth rings of two species of pine trees, LaMarche et al. (54) concluded that "greatly increased tree growth rates observed since the mid 19th century exceed those expected from climatic trends but are consistent in magnitude with global trends in CO_2, especially in recent decades." Historic increases in wheat (55) and soybean (56) yields may be partly due to CO_2 increases already experienced.

Increasing the CO_2 concentration of the atmosphere in which plants grow increases carbon assimilation rates and has favorable effects on other physiological processes of all functional groups of plants (57). For instance, a highly positive relationship between elevated CO_2 levels and water use efficiency is commonly reported (58). The mechanism is the reduction of transpiration through partial stomatal closure while the CO_2 supply to the site of fixation is maintained by the increased CO_2 gradient (59). The magnitude of this response is species specific but appears to be best expressed in plants growing in arid environments (60,61). Loblolly pine (*Pinus taeda*) and sweetgum (*Liquidambar styraciflua*) seedlings exhibited lower water stress and higher water use efficiency at elevated CO_2 (62). Exposure of white clover (*Trifolium repens*) swards to elevated CO_2 not only doubled photosynthetic rates but increased canopy water use efficiency by 63% (63). The beneficial effect of increasing CO_2 on water use efficiency provided the mechanism for the prediction by Idso and Quinn (49) that changes in vegetation boundaries would occur in the arid Southwest. Oechel and Strain (60) predicted that individual chaparral plants will grow larger as CO_2 continues to increase because their growth is water-limited. Johnston (32) attributed the loss of open prairies to increases in the stature as well as the density of woody species, writing that woody species on open grasslands in south Texas were "stunted" in pre-industrial times. Scifres (64) considers the "brush problem" to be a reflection of increases in stature as well as density of woody species.

While increasing CO_2 appears to improve water use efficiency of all plants, experimental evidence suggests that effects on other physiological processes are more strongly expressed in plants that possess the C3 photosynthetic pathway than in those with C4 photosynthesis. Carbon assimilation rates vary widely with species and environmental conditions, but are thought to be higher in C4 than C3 species at current CO_2 levels (65). Black et al. (66) recognized that photosynthetic rate is only one of a number of traits that affect ecological success, but proposed that the higher photosynthetic rates of C4 plants associated with their unique carbon conservation system conferred competitive superiority, especially in habitats with high light intensity and high temperatures. Conversely, Pearcy and Ehleringer (67) argued that C3 and C4 plants of comparable physiognomy growing in similar environments have similar photosynthetic rates.

In either event, physiological sensitivity of plants to changing CO_2 is greatest at concentrations approaching the photosynthetic compensation point, as suggested by curves relating photosynthetic rates to CO_2 concentrations (68). The greatest response of cucumbers to increasing CO_2 levels ranging from 100 to 8000 ppm occurred at levels well below current ambient (Dauncht in 69). Photosynthetic response curves published in these and other references demonstrate that C4 plants have an advantage in this regard at low CO_2 levels, such as the ambient concentration of pre-industrial (presettlement) times. But as CO_2 levels increase, photosynthetic rates of C3 plants increase more rapidly than those of C4 plants and the advantage of the C4 pathway is lost as photosynthetic rates converge at higher CO_2 levels (i.e., 67). Our comparisons of net carbon assimilation rates of little bluestem (*Schizachyrium scoparium*), a C4 native perennial bunchgrass widely distributed across the Great Plains, and the C3 woody invader honey mesquite (*Prosopis glandulosa*) at increasing CO_2 levels indicate that photosynthetic rates of the grass are about 20% higher at CO_2 levels characteristic of 150 years ago (Figure 7.2). However, net photosynthetic rate of the shrub equals that of the grass at today's atmospheric CO_2 level, 350 ppm, and will exceed that of the grass as ambient CO_2 level continues to increase. Thus, the C3 mesquite has realized a greater relative advantage than the C4 grass as CO_2 increased over the last 150

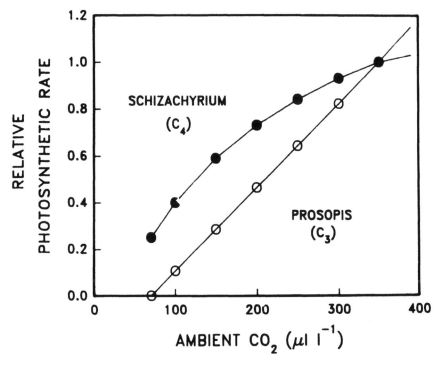

FIGURE 7.2 Relative net photosynthetic rates of little bluestem (*Schizachyrium*) and honey mesquite (*Prosopis*) at subambient atmospheric CO_2 concentrations. Assimilation rate of both species was about 30 µmoles CO_2 m^{-2} s^{-1} at 350 µl l^{-1} (350 ppm) CO_2.

years, if increasing photosynthetic capacity improves performance at the whole-plant or higher levels.

Quantum yield, the efficiency of light energy conversion in the photosynthetic process, increases greatly with atmospheric CO_2 level in C3 plants but not in C4 plants (70) because the internal CO_2-concentrating mechanism of C4 plants allows CO_2 assimilation to be less dependent on the CO_2 concentration outside the leaf (57). Increasing CO_2 may also reduce the negative effect of photorespiration in C3 plants, especially over the range of CO_2 increases of the last century (71). Small increases in the leaf's internal CO_2 concentration increase carboxylation by the enzyme RUBISCO (ribulose 1,5-biphosphate carboxylase/oxygenase) at the expense of oxygenation in C3 plants. In C4 plants, the energy-incorporating carboxylation mediated by RUBISCO follows carboxylation by phosphenolpyruvate carboxylase and occurs in the highly CO_2-enriched environment of the bundle sheath cells; almost no photorespiration occurs.

Increasing CO_2 suppresses "dark respiration" or the normal respiration which occurs continuously and is essential for maintenance and growth. Increasing atmospheric CO_2 from current ambient to about twice that level decreased respiration by 45% in wheat (*Triticum aestivum*) (72), 16% in tomato (*Lycopersicon esculentum*), 24% in soybean (*Glycine max*) (73), and 64% in roots of American beech (*Fagus grandifolia*) seedlings (74). All of these are C3 species, and decreases in respiration represent comparisons between current and much higher CO_2 levels. However, the limited amount of information available suggests that this effect is better expressed in C3 than C4 plants, and high sensitivity of wheat respiration to 140 ppm CO_2 has been reported (75).

Differences in physiological responses of C3 and C4 plants to additional CO_2 are reflected in growth rates, biomass and leaf area accumulation, and other whole-plant parameters associated with fitness and competitive ability, at least at CO_2 levels representative of the future. Patterson and Flint (76) compared responses of C3 and C4 crop and weed species and found that biomass of the C3 species increased more than that of C4 species in response to elevated CO_2. The C4 herb *Amaranthus retroflexus* produced less biomass at elevated CO_2 than at ambient CO_2, while C3 competitors attained maximum biomass under elevated CO_2 at mesic soil water levels (77). Elevated CO_2 not only increases growth rate and ultimate size, but also root/shoot ratios, numbers of stems and branches, and leaf thickness (78-80). Timing and output of reproductive processes are generally accelerated (81).

The response of C3 herbs to increasing CO_2 at subambient levels representative of the past can be dramatic. In a recent experiment in our lab, oats (*Avena sativa*) and wild mustard (*Brassica kaber*) were grown in a continuous CO_2 gradient from 150 ppm to current ambient, about 350 ppm. Over the range of increasing CO_2 from 250 to 350 ppm, representative of the change in the last 150 years, net carbon assimilation increased by over 50%. Leaf area and oven-dry weight of topgrowth increased by the same extent or more, indicating that historical changes in CO_2 may have had profound effects on the growth of C3 herbs.

C3 woody plants respond to elevated CO_2 to the same or a greater extent as C3 herbs (62,69,82,83). Conifers exhibit a pronounced growth increase (84,85), suggesting that CO_2 enhancement may have played a role in recent increases in the extent of the *Pinion-Juniper* type and the abundance of a number of species of *Juniper* throughout North America (27,86-91).

Physiological and whole-plant responses to increasing CO_2 must be reflected in the competitive ability of plants if they provide a mechanism for the replacement of one functional group by another. In studies of competition between four species of annuals, response of C4 species to elevated CO_2 was depressed by the presence of C3 species whereas the reverse was not observed (47). The greater sensitivity and competitive superiority of C3 than C4 herbs to enriched CO_2 was also demonstrated by others (76,82,92,93). A C3 aster (*Aster pilosus*) was a much stronger competitor than a C4 grass, broomsedge (*Andropogon virginicus*), at elevated CO_2 levels in the presence or absence of drought stress (94).

Limited experimental evidence is available addressing the question of whether differences in responses of C3 and C4 plants to increasing CO_2 can be observed at the community level in nature. No significant effect on productivity was associated with CO_2 enrichment in a C3 arctic sedge community (61). However, doubling atmospheric CO_2 increased the density and biomass and delayed senescence of a C3 sedge (*Scirpus olneyi*) but had no such effect on associated C4 species, saltmeadow cordgrass (*Spartina patens*) and a saltgrass (*Distichlis spicata*) (95).

The hypothesis that CO_2 effects on rangelands are expressed as changes in community structure is based on these differences in physiological responses and consequent growth and competitive ability of the two functional groups, woody shrubs and the warm-season perennial grasses. Essentially all broadleafed weeds, shrubs, and trees utilize the C3 photosynthetic pathway, while warm-season grasses are C4 plants.

Kramer (96) postulated that the response of vegetation to elevated CO_2 would be greatest where nutrients, light, water, or temperature were not limited. Recent research concerning interactions between CO_2 enrichment and availability of water, nutrients, and light (77,92,93,97) indicates that the extent to which limiting resources controls plant response to increasing CO_2 varies widely, even within functional groupings like the photosynthetic pathway. However, Idso (98) considers these and other reports and concludes that CO_2 enrichment generally ameliorates negative effects of stress. Amelioration of stress could have a profound effect on the structure of vegetation on arid and semiarid rangelands.

Implications of the Hypothesis

Favorable effects of increasing CO_2 that apply to all functional groups of plants, especially improved water use efficiency and amelioration of stress, suggest that overall productivity of rangelands will increase. However, increased productivity will probably continue to be reflected in increased biomass of less desirable C3 weeds and woody vegetation, as opposed to C4 warm-season perennial grasses, where the two functional groups occur together. It also seems reasonable to expect improved productivity of C3 cool-season grasses, and to consider whether increasing atmospheric CO_2 is influencing ecosystem structure and function in environments where they predominate.

The hypotheses that historical increases in atmospheric CO_2 conferred competitive superiority upon C3 weeds and shrubs and rendered them inherently better adapted to today's rangelands than C4 grasses implies that, by definition, current "climax" is characterized by shrub dominance. This seems true not only for the present but increasingly so in the future as CO_2 levels continue to climb. Efforts to define range condition and trend relative to an historical species composition in which shrubs were poorly represented seem unrealistic in light of both the hypothesized CO_2 effect and the well-documented success of woody plants over recent decades.

The possibility of a mechanism by which C3 woody vegetation tends to dominate rangeland in the absence of other factors has implications for the way we view vegetation management in a practical sense. Efforts to reduce the density of weeds and brush on rangelands with herbicides, fire, and mechanical practices have been described as expediting secondary succession and reversing the trend caused by mismanagement. If the CO_2 hypothesis is correct, succession proceeds in the opposite direction of that intended and shifts from open grasslands to shrublands cannot be fully attributed to mismanagement. These considerations suggest that a lack of understanding of the fundamental causes of vegetation dynamics on rangelands has probably been the primary constraint to the development of totally successful approaches to dealing with the range weed and brush problem.

References

1. Delmas RJ, JM Ascancio, and M Legrand. Nature 284:155157 (1980).
2. Neftel A, E Moore, H Oeschger, and B Stauffer. Nature 315:45-47 (1985).
3. Raynaud D and JM Barnola. Nature 315:309-311 (1985).
4. Keeling CD, RB Bascastow, and TP Whorf. In: *Carbon Dioxide Review: 1982* (WC Clark, Ed.), pp. 377-385. Oxford Univ. Press, 1982.
5. Trabalka JR, JA Edmonds, JM Reilly, RH Gardner, and LD Voorhees. In: *Atmospheric Carbon Dioxide and the Global Carbon Cycle"* (JR Trabalka, Ed.), pp. 247-286. US Dept. Energy DOE/ER-0239 (1985).
6. Vale TR. J. Range Manage. 28:32-36 (1975).
7. Blackburn WH and PT Tueller. Ecology 51:841-848 (1970).
8. Young JA; RA Evans, and PT Tueller. In: *Holocene Environmental Change in the Great Basin"* (R Elston, Ed.), pp. 187-215. Nevada Archaeological Survey Res. Paper No. 6, Reno, NV, 1976.
9. Buffington LC and CH Herbel. Ecol. Monogr. 35:139-164 (1965).
10. Ellison L. Bot. Rev. 26:1-78 (1960).

72

11. Bogusch ER. Texas J. Sci. 4:85-91 (1952).
12. Brown JR and S Archer. Vegetatio 73:73-80 (1987).
13. Hastings JR and RM Turner. *The Changing Mile*. Univ. Arizona Press, Tucson, 1965.
14. Humphrey RR and LA Mehrhoff. Ecology 39:720-726 (1958).
15. Branscomb BL. J. Range Manage. 11:129-133 (1958).
16. Rogers GF. *Then and Now; A Photographic History of Vegetation Change in the Central Great Basin Desert*. Univ. Utah Press, Salt Lake City, 1982.
17. Vale TR. Amer. Midland Nat. 105:61-69 (1981).
18. Branson FA. Range Monogr. No. 2, Soc. Range Manage., Denver CO, 1985.
19. Wooton EO. N. Mex. Agric. Exp. Sta. Bull. 66 (1908).
20. Wooton EO. USDA Bull. 211 (1915).
21. York JC and WA Dick-Peddie. In: *Arid Lands in Perspective* (WG McGinnies and BJ Goldman, Eds.), pp. 157-166. Univ. Arizona Press, Tucson, 1969.
22. Hennessy JT, RP Gibbens, JM Tromble, and M Cardenas. J. Range Manage. 36:370-374 (1983).
23. Wright RA. J. Arid Environ. 5:277-284 (1982).
24. Brown AL. J. Range Manage. 3:172-177 (1950).
25. Glendening GE. Ecology 33:319-328 (1952).
26. Parker KW and SC Martin. USDA Circ. No. 908 (1952).
27. Smith DA and EM Schmutz. J. Range Manage. 28:453-458 (1975).
28. Anderson JE and KE Holte. J. Range Manage. 34:25-29 (1981).
29. Rice B and H Westoby. J. Range Manage. 31:28-34 (1978).
30. Williams K, RJ Hobbs, and SP Hamburg. Oecologia 72:461-465 (1987).
31. Hughes LE. Rangelands 2:17-18 (1980).
32. Johnston MC. Ecology 44:456-466 (1963).
33. Vogl RJ. In: *Fire and Ecosystems* (TT Kozlowski and CE Ahlgren, Eds.), pp. 139-194. Academic Press, New York, NY, 1974.
34. Wright HA and AW Bailey. *Fire Ecology: United States and Southern Canada*. John Wiley and Sons, New York, NY, 1982.
35. Griffiths D. USDA Bur. Plant Indus. Bull. No. 177 (1910).
36. Cook OF. USDA Bur. Plant Indust. Circ. No. 14 (1908).
37. Weniger D. *The Explorer's Texas - The Lands and Waters*. Eakin Pub., Austin, TX, 1984.
38. Fierro LC. Rangelands 11:166-171 (1989).
39. Feldman I. Trabajos y Resumenes, III Congreso Asociacion Latino Americana de Malezas "Asam 4:46-55 (1976).
40. Schofield CJ and EH Buchler. Trends Ecol. Evol. 1:78-80 (1986).
41. Veblen TT and V Markgraf. Quat. Res. 30:331-338 (1988).
42. Walker J and AN Gillison. In: *Ecology of Tropical Savannas* (BJ Huntley and BH Walker, Eds.), pp.5-24. Springer-Verlag, New York, NY, 1982.
43. Taylor HC, SA MacDonald, and IAW MacDonald. S. Afr. J. Bot. 51:21-29 (1985).
44. van Vegten JA. Vegetatio 56:3-7 (1983).
45. Singh JS and MC Joshi. In: *Management of Semiarid Ecosystems* (BH Walker, Ed.), pp. 243-273. Elsevier, Amsterdam, 1979.
46. Kelley RD and BH Walker. J. Ecol. 64:553-676 (1976).

47. Bazzaz FA and K Garbut. Ecology 69:937-946 (1988).
48. Overdieck D and F Reining. Acta Oecol./Oecol. Plant. 7:357-366 (1986).
49. Idso SB and JA Quinn. Climat. Publ. Paper No. 17, Arizona State Univ., Tempe, 1983.
50. Busing RT. Bull. Torrey Bot. Club 116:282-288 (1989).
51. Hari P and H Arovaara. Scand. J. For. Res. 3:67-74 (1988).
52. Kienast F and RJ Luxmoore. Oecologia 76:487-495 (1988).
53. Parker ML. In: *Proc. Intl. Symp. Ecol. Aspects of Tree-Ring Analysis* (GC Jacoby, Jr. and JW Hornbeck, Eds.), pp. 511-521. US Dept. Energy Rep. CONF-8608144, 1987.
54. LaMarche VC, DA Gray, HC Fritts, and MR Rose. Science 225:1019-1021 (1984).
55. Gifford RM. Aust. J. Plant Physiol. 6:367-378 (1979).
56. Allen LH Jr., KJ Boote, JW Jones, PH Jones, BA Valle, HH Rogers, and RC Dahlman. Global Biogeochem. Cycles 1:1-14 (1987).
57. Pearcy RW and O Bjorkman. In: *CO_2 and Plants: The Response of Plants to Rising Levels of Atmospheric Carbon Dioxide* (ER Lemon, Ed.), pp. 65-105. Amer. Assoc. Advance. Sci. Select. Symp. No. 84, Westview Press, Boulder, CO, 1983.
58. Kimball BA and SB Idso. Agric. Water Manage. 7:55-72 (1983).
59. Farquhar GD and TD Sharkey. Ann. Rev. Plant Physiol. 33:317-345 (1982).
60. Oechel WC and BR Strain. In: *Direct Effects of Increasing Carbon Dioxide on Vegetation* (BR Strain and JD Cure, Eds.), pp. 118-154. US Dept. Energy, DOE/OE-0238, 1985.
61. Tissue D and WC Oechel. Ecology 68:401-410 (1987).
62. Tolley LC and BR Strain. Canad. J. Bot. 62:2135-2139 (1984).
63. Nijs I, I Impaens, and T Behaeghe. J. Exp. Bot. 40:353-359 (1989).
64. Scifres CJ. *Brush Management: Principles and Practices for Texas and the Southwest.* Texas A&M Univ. Press, College Station, 1980.
65. Osmond CB, O Bjorkman, and DJ Anderson. *Physiological Processes in Plant Ecology.* Springer-Verlag, New York, NY, 1980.
66. Black CC, TM Chen, and RH Brown. Weed Sci. 17:338-344 (1969).
67. Pearcy RW and J Ehleringer. Plant, Cell and Environ 7:1-13 (1984).
68. Hesketh JD. Crop Sci. 3:493-496 (1963).
69. Kimball BA. USDA, Agric. Res. Serv., Water Conserv. Lab. Rep. 14, Phoenix, AR, 1983.
70. Ehleringer J and O Bjorkman. Plant Physiol. 59:86-90 (1977).
71. Sharkey TM. Physiol. Plant. 73:147-152 (1988).
72. Gifford RM, H Lambers, and JIL Morison. Physiol. Plant. 63:351-356 (1985).
73. Bunce JA. Ann. Bot. (in press) (1990).
74. Reid CD and BR Strain. Supp. Bull. Ecol. Soc. Amer. 70:239 (1981).
75. Gifford RM. Aust. J. Plant Physiol. 4:99-110 (1977).
76. Patterson DT and EP Flint. Weed Sci. 28:71-75 (1980).
77. Bazzaz FA and RW Carlson. Oecologia 62:196-198 (1984).
78. Dahlman RC, RR Strain, and HH Rogers. J. Environ. Qual. 14:1-8 (1985).
79. Lemon ER (Ed.). AAAS Selected Symp. 84. Westview Press, Boulder, CO, 1983.

80. Strain BR and JD Cure (Eds). US Dept. Energy, Carbon Dioxide Res. Div. DOE/ER-0238 (1985).
81. Strain BR. Biogeochem. 1:219-232 (1985).
82. Rogers HH, JF Thomas, and GE Bingham. Science 220:428429 (1983).
83. Tinus RW. Tree Planters Notes 23:12-15 (1972).
84. Hellmers H, N Sionit, BR Strain, and CH Jaeger. Bull. Ecol. Soc. Amer. 66:191 (1985).
85. Higgenbotham KO, JM Mayo, SL Hirondelle, and DK Krystofiak. Canad. J. Forest Res. 15:417-421 (1985).
86. Holthuijzen AMA, TL Sharik, and JD Fraser. Canad J. Bot. 65:1092-1095 (1987).
87. Johnsen TN. Ecol. Monogr. 32:187-207 (1962).
88. Livingston RB. Ecology 53:1141-1147 (1972).
89. Snook EC. In: *Proc. Eastern Red Cedar in Oklahoma Conference* (RR Wittmer and DM Engle, Eds.), pp. 45-52. Coop. Ext. Serv. Bull E-849. Oklahoma State Univ., Stillwater, OK, 1985.
90. Steinauer EM and TB Bragg. Amer. Midland Nat. 106:358365 (1987).
91. Wilson J and T Schmidt. Rangelands 12:156-158 (1990).
92. Patterson DT, EP Flint, and JL Beyers. Weed Sci. 32:101-105 (1984).
93. Zangerl AR and FA Bazzaz. Oecologia 62:412-417 (1984).
94. Marks S and BR Strain. New Phytol. 111:181-186 (1989).
95. Curtis PS, BG Drake, PW Leadley, W Arp, and D Whigham. Oecologia 78:20-26 (1989).
96. Kramer PJ. BioScience 31:29-33 (1981).
97. Patterson DT and EP Flint. Weed Sci. 30:389-394 (1982).
98. Idso SB. *Carbon Dioxide and Global Change: Earth in Transition.* Institute for Biospheric Research Press, Tempe, AZ, 1989.

8

Noxious Weed Management Strategies

Celestine A. Lacey

Abstract

Noxious weeds are a serious problem on millions of ha of range and grazeable woodland in western North America. Management of extensive weed populations requires the implementation of three basic management techniques. These include: (1) weed inventories; (2) selection and application of weed control techniques; and (3) follow-up management. Inventories provide the basis for all management decisions and must be considered the initial step in any weed control effort. The second step is selection of control techniques based on environmental factors, growth characteristics of the target plant, management objectives, and size of the infestation. The third step is implementing follow-up management such as specialized grazing systems and reseeding disturbed sites to slow the rate of reinvasion of noxious weeds.

Introduction

Noxious weeds infest millions of ha of rangeland throughout western North America. For example, yellow starthistle (*Centaurea solstitialis*) infests 3.2 million ha in California and Idaho, spotted knapweed (*Centaurea maculosa*) 2.8 million ha in Idaho and Montana (1), and 1 million ha are infested with leafy spurge (*Euphorbia esula*) (2).

The size of the weed infestations, longevity of weed seed viability, occurrence of weeds in riparian and other environmentally sensitive sites, and cost of controlling weeds on extensively-managed lands combine to reduce the feasibility of weed control efforts on public and private lands. Therefore, a strategy must be developed and followed to optimize effectiveness of treatments and reduce the impact of noxious weeds.

This report outlines a strategy for controlling noxious weeds on rangeland. Specifically, inventory methods, control techniques, and follow-up activities are reviewed.

Inventory and Assessment

Inventories are the initial phase of all weed management programs. They provide information on acreage, density, and location of weed infestations. Inventory data can also be interpreted to rank sites according to their susceptibility to invasion. For example, inventories in Montana indicate that spotted knapweed initially colonizes disturbed sites such as roadsides (Figure 8.1), trails, and railroad rights-of-way (3,4). The susceptibility of specific sites to various weed species allows public and private land managers to concentrate monitoring and "field scouting" efforts in areas with a high potential for invasion.

FIGURE 8.1 Spotted knapweed initially colonizes disturbed sites such as roadsides, trails, and railroad rights-of-way.

Weed infestations can be inventoried by field surveys, aerial photography (color and color-IR), and/or geographic information systems. The optimal method varies with the target specie and the level of mapping detail required to achieve the desired goal. Advantages and disadvantages of each technique are discussed below.

Field surveys are an on-the-ground approach for mapping noxious weed infestations. Orthophoto quads or topographic maps at 1:24,000 scale or aerial photographs at 1:24,000 or 1:1320 scale are suitable for locating weed infestations. Symbols, numbers, or color codes should be assigned to each weed specie included in the survey. Symbols can also be assigned to indicate the size of a weed infestation such as an " X " for less than 0.1 ha, a "triangle" for 0.1 to 0.5 ha, and a "square" for 0.5 to 2 ha (Figure 8.2). Infestations greater than 2 ha should be delineated on the

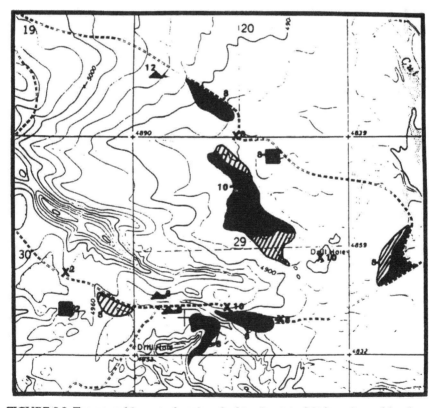

FIGURE 8.2 Topographic map showing the location weed infestations. Numbers adjacent to symbols represent different species such as "10" for spotted knapweed and "8" for diffuse knapweed. Symbols correspond to the number of hectares infested: X = <0.1 ha, triangle = 0.1 to 0.5 ha, and square = 0.5 to 2 ha.

map and acreage determined by a dot grid or planimeter. Infestations that follow roads or trails can be calculated as linear meters (or kilometers) multiplied by the average width of the infestation.

When conducting a field survey, each section should be traversed several times by foot, horseback, or all-terrain vehicle (ATV) to insure location of all infestations. The use of an ATV will reduce field survey time by about one-half compared to walking or horseback. Light aircraft have also been used successfully for locating weed infestations in inaccessible areas.

Waterways are often difficult to access and can support relatively large infestations of noxious weeds. Rafts and/or canoes have been found to be the most expedient (and enjoyable) method for conducting inventories on navigable streams. In addition, float trips can also be an effective tool for educating the general public about the impact of noxious weeds in riparian areas.

The optimum time to conduct a field survey is when the target weed is in full bloom. However, this timing is difficult to achieve since the bloom period for some species is relatively short and often more than one weed specie is included in an inventory. In the Northwest, field inventories on most of the problem weed species can be effectively conducted from May through September.

The biology of the target plant must be considered in an inventory. Survey efforts should be most intensive on sites that are highly susceptible to invasion. Particular attention should be paid to trails, roads, railroad rights-of-way, and other disturbed areas where most noxious weeds are introduced.

The quality of the field survey will depend directly on the personnel conducting the work. Personnel assigned to the inventory must be physically able to walk 15 to 20 km per day in rough terrain. The person must also have a basic understanding of: plant biology; weed identification in all stages of growth; map reading skills; and compass reading. A team of two people is often more efficient than a single person, especially if large areas must be covered on foot or horseback.

Field inventories provide the most accurate and detailed information. However, they are more labor intensive than other survey techniques.

Aerial photography can be a useful tool in locating and mapping weed infestations (5). Recommendations on photograph scale, accuracy, and cost trade-offs between different types of photographs are available for leafy spurge (6), spotted knapweed (7), and several other weed species.

Photo scale, film type (color vs color-IR), and season (full flower or fall) must be determined prior to initiating large-area surveys. A photograph scale of 1:24,000 to 1:1320 appears to be most effective for weed inventories. Scales up to 1:660 can be used if more detail is required in the inventory. Photographic scales less than 1:25,000 do not provide enough detail to meet most inventory objectives.

Reconnaissance, photographic analysis, and ground checking are recommended when mapping vegetation from aerial photographs (8). These steps must be followed to ensure the accuracy of the weed inventory. Reconnaissance involves correlating field observations with appearance of weed infestations on the photograph. Photographic analysis or interpretation involves the use of magnifying stereoscopes to locate and identify weed infestations. Field checking involves confirmation of aerial photographs on the ground. If adequate reconnaissance has been conducted, ground checking may be confined to infestations that could not be verified in the office (8).

The use of color-IR photography can expedite inventory and assessment. Collection of ground data is critical with either color or color-IR photographs to insure that the signature from the target plant does not mimic another specie (5).

Geographic Information Systems (GIS) are being used more frequently to monitor resources on public and private lands. It has been used effectively to map areas occupied by noxious weeds (9). Infestations as small as 30.4 m^2 can be digitized onto maps. Once the infestations are mapped, high potential sites for that specie can be identified (9).

GIS also provides land managers the opportunity to study the location of weed infestations in relation to soil, groundwater, and nontarget vegetation and develop weed management options that are environmentally safe. A pilot GIS project that involves mapping leafy spurge infestations, sensitive groundwater areas, and soil types is presently being conducted in Teton County, MT. By showing present and future impacts from noxious weeds, GIS may become an important tool to gain legislative and public support for weed management programs (10).

Methods of Weed Control

Selection and application of weed control techniques is the second phase of a range-weed management strategy. Chemicals (herbicides), biological agents (insects, pathogens, or grazing animals), and/or cultural control methods (mechanical or hand removal, mowing, or burning) may be used to control, eradicate, or contain a weed. Integrated weed management (IWM) is a multidisciplinary, ecological approach to managing weed populations. IWM involves the use of several control techniques in a planned, coordinated program, to limit the impact of weeds on range production. An IWM approach is critical when managing extensive weed infestations that occupy a variety of ecological niches such as riparian areas, foothill, and mountain slopes. There is no example of an extensive noxious weed infestation being eradicated by a single method.

Control information must be evaluated in terms of the landowner's objective and the extent and nature of the weed infestation. The control method or methods selected will depend upon the following seven factors:

1. *Control Objectives*: The objective or goal for a weed management program, i.e., eradication, stand reduction, or containment, must be determined. The goal should be realistic and achievable within a defined period of time. For example, eradication of 40 ha of leafy spurge is not a realistic goal given our present technology.
2. *Effectiveness of the Control Technique*: The effectiveness of the control method on the target specie needs to be considered. For example, will the target weed be reduced, controlled, or eliminated? What is the impact on nontarget species? Will retreatment be necessary? For example, many herbicides are selective, others are nonselective. Burning favors grasses and forbs over shrubs. Hand pulling is not a realistic control option on rhizomatous weeds.
3. *Environmental Factors*: Surface and groundwater features, adjacent nontarget vegetation, sensitive crops, soil texture and structure, physiography, and other environmental factors will impact the selection of a control tool. For example, mechanical tillage is not feasible on many rangeland sites. Sandy soils and high water tables will limit the use of some herbicides.
4. *Land Use*: The present and projected use for the site needs to be evaluated. For example, burning or spraying big sagebrush in a sage grouse area is not recommended. In addition, livestock grazing may not be a viable control option in recreational areas.
5. *Time Frame*: The length of time required to meet control objectives for a given area needs to be understood. If relatively fast results are necessary, such as elimination of a new infestation, then herbicides will provide the most feasible solution. In contrast, control with biological agents is usually much slower.
6. *Economics*: The amount of revenue available to achieve control objectives and cost effectiveness of the treatment must be determined. Generally, control programs are more feasible on the more productive or "best" land.
7. *Size of the Weed Infestation*: The size of a weed infestation will determine the selection of control methods. IWM must be utilized to manage extensive weed infestations. In contrast, a single control

technique, such as herbicides or hand pulling, may be used to control or eradicate newly invading weed species.

The effectiveness of various herbicides, cultural techniques, and biological control agents on noxious weed species has been evaluated. Information regarding the response of the target plant to a specific control technique is needed prior to initiating a program.

Chemical Control: Recommendations regarding the proper herbicide, rate, and time of application for a specific weed can be obtained from published reports. Herbicide characteristics such as environmental fate (longevity, mobility), potential toxicity to humans and animals, and selectivity to nontarget species also need to be considered.

Biological Control: The availability and effectiveness of insects and pathogens in addition to their site adaptation must be known. The compatibility of biological agents with herbicides and grazing animals may restrict or limit an IWM approach. Most insects and pathogens will be well-adapted to some sites and poorly adapted to others.

The possibility of managing livestock to utilize a target specie is attractive to many land owners. A good example is the use of sheep to manage leafy spurge infestations (11-13). Grazing must be timed to have the greatest impact on the target plant. The economic feasibility of grazing certain kinds of livestock may be limited by predators (i.e., coyotes on sheep). Potential modification of diet or grazing behavior to increase utilization of a weed specie is not fully understood. The possibility of introducing nontraditional kinds of livestock to graze weeds (i.e., camels on tarbush) should not be overlooked (14).

Cultural Control: Prescribed burning is often used to control big sagebrush (*Artemisia tridentata*) and juniper (*Juniperus* spp.). Burning must be conducted at the proper time to minimize damage to nontarget plants. Burning has little effect on most annual and herbaceous species and has actually been reported to cause an increase in diffuse knapweed (*Centaura diffusa*) (15).

Some herbaceous species can be controlled by hand pulling. Generally, pulling is effective in areas of initial colonization when native species are present to fill the niches. It is not recommended on high-density infestations.

Cultivation either by hand or mechanical methods is effective on some species. Impact on nontarget plants, soil, and water resources must be considered. Reseeding of desirable plants is usually necessary.

Mowing can reduce seed production of some weed species. The timing and frequency of mowing necessary to reduce seed production will vary with each specie. For example, Popova (15) reported an increase in density of diffuse knapweed when mowed. However, Watson and

Renney (16) measured a reduction in seed production when plants were mowed in the flowering stage. Davis (17) noted a decrease in seed production of spotted knapweed when plants were mowed at the late bud to early bloom growth stage. Response of a plant to mowing may vary with environmental conditions, especially the amount of rainfall, that occur after the mowing event.

Follow-Up Management

Follow-up management is the third phase of a weed management strategy. Follow-up management determines the longevity of control obtained with chemical, biological, or cultural weed control methods. Because most noxious weeds have persistent and tenacious growth characteristics and seeds that remain viable for years, long-term control programs must be implemented. These include retreatment with herbicides or continued cultural control practices to maintain low levels of weed populations.

Range improvements such as grazing systems, cross fencing, and water developments will facilitate a more competitive plant community and help retard the reinvasion of most weed species. Sites that are devoid of desirable species may have to be reseeded as part of the total management program (18).

Some weeds, such as the knapweeds and Canada thistle (*Circium arvense*), have a negative response to shading (19). Losensky (19) recommends that shrub and tree growth be maintained or established along roadsides and other high-risk areas within forested sites to help reduce weed reinvasion.

Minimizing disturbance to native vegetation will reduce the risk of establishment and spread of most weed species. There has been a direct correlation measured between spotted knapweed infestations and the degree of soil disturbance (16). Roads, trails, overgrazing by livestock, and construction activities have been cited as causing initial weed infestations. Strategies to prevent disturbance or to mitigate the impact by reseeding and monitoring of disturbed sites should be developed prior to and following a control program.

Summary

In summary, noxious weeds are a serious problem. The key to a viable management program is to: (1) inventory the location and acreage infested with noxious weeds; (2) apply control techniques that most effectively facilitate land management goals; and (3) implement follow-up management practices that reduce reinvasion of the weed specie. Perennial weeds require a perennial solution.

References

1. Lacey CA. In: *Proc. Knapweed Symposium*, pp. 1-6. Bozeman, MT, 1989.
2. Lacey CA, PK Fay, RG Lym, CG Messersmith, B Maxwell, and HP Alley. Montana State Coop. Ext. Ser. Cir. 309, 1985.
3. Lacey CA. In: *Montana Weed Control Assoc. Ann. Conf. Proc.* Great Falls, MT, 1987.
4. Tyser RW and CH Key. Northwest Sci. 62:151 (1988).
5. Richard Myhre, personal communication.
6. Myhre RJ. In: *Proc. Leafy Spurge Symposium*, p. 52. Sundance, WY, 1983.
7. Daniel Long, unpublished report to Silver Bow County Weed Board, Butte, MT, 1985.
8. Spurr SH. In: *Photogrammetry and Photo-Interpretation, 2nd Ed.* The Ronald Press Company, 1960.
9. Steven Dewey, personal communication.
10. Dewey S. Proc. Western Soc. Weed Sci. 43:119 (1990).
11. Johnston A and RW Peake. J. Range Manage. 13:192 (1960).
12. Landgraf BK, PK Fay, and KM Havstad. Weed Sci. 32:348 (1984).
13. Lacey CA, RW Kott, and PK Fay. Rangelands 6:202 (1984).
14. Dean M. Anderson, personal communication.
15. Popova AY. Bot. Zl., SSSR 45:1207 (1960).
16. Watson AK and AJ Renney. Can. J. Plant Sci. 54:687 (1974).
17. Edward Davis, personal communication.
18. Berube DE and JH Myers. J. Range Manage. 35:459 (1982).
19. Losensky BJ. Unpublished report to USDA-FS Lolo Forest, Missoula, MT (1987).

9

Biological Control of Rangeland Weeds

P.C. Quimby, Jr., W.L. Bruckart, C.J. DeLoach,
Lloyd Knutson, and M.H. Ralphs

Abstract

Weedy forbs and brush cost America's range managers at least $1.7 billion/year. Biological controls, or "the planned use of living organisms to reduce the vigor, reproductive capacity, density, or effect of weeds," should be considered and included in pragmatic integrated weed management systems for rangelands. Various approaches to biocontrol under that definition are discussed. These include foreign exploration and introduction of exotic insects, mites, and plant pathogens as biocontrol agents; augmentation of native biocontrol agents, especially plant pathogens; grazing systems; and positive and aversion conditioning for various classes of livestock to use against troublesome weeds or brush or to avoid palatable poisonous weeds. USDA's Agricultural Research Service has at least nine laboratories, worldwide, devoted to research on various aspects of biocontrol of exotic and native rangeland weeds. The usual goal of biocontrol is to improve ecological systems by using biotic agents to restore target plant species to lesser competitive intensities or to negate their effects so that they do not overwhelm plant communities or cause damage to livestock. The usual results of biocontrol are: improved agricultural production, improved ecosystem functions and status in terms of species richness and diversity of plant and animal communities, and improved protection of rare species. Regardless of whether target weeds are introduced or native, researchers must make balanced evaluations of risks, benefits, and the potential for success in developing biological control programs.

Introduction

Weedy forbs and brush cause more losses on America's approximately 252 million hectares of rangeland than all other pests combined. An estimated 136 million hectares are infested with weeds and brush and the annual cost is at least $1.7 billion (1). This estimate may be low by at least an order of magnitude when we consider the recent example of just one species' total economic impact. Thompson et al.(2) reported that 400,000 ha of leafy spurge in North Dakota cost that state $105 million annually. Losses to poisonous plants were estimated in 1989 to exceed $340 million/year in the 17 western states (3). Thus, control of weeds and brush is vital to overall vegetation management.

Biological controls should be considered and included in pragmatic integrated weed management systems, which are in turn essential to integrated rangeland vegetation management systems. The purpose of this paper is to review the background of biological control of weeds of rangelands, review the steps required in an introductory (classical) biological control program, discuss the use of plant pathogens, discuss grazing and aversion/positive conditioning of livestock as approaches, and provide some general information about ongoing projects.

Definition of Biological Control of Weeds

Peter Harris (4) of Agriculture Canada defined biocontrol of weeds as "the use of undomesticated organisms that feed on the pest for the purpose of reducing its density, vigor, or reproduction."

For the purpose of this paper, we will use a modified version of a broader definition (5): "the planned use of living organisms to reduce the vigor, reproductive capacity, density, or effect of weeds." Under this broader definition of biological control, we can employ various approaches: introduction of exotic biocontrol agents (classical); augmentation of native biocontrol agents (such as use of mycoherbicides); grazing systems and positive conditioning that enable livestock of various classes to consume troublesome weeds or brush; grazing desirable forage in ways that help keep weeds in check; aversion conditioning of livestock to avoid palatable poisonous weeds; and the use of superior, fast-growing forages that can successfully compete with troublesome weeds (the latter is discussed in other papers during the symposium).

TABLE 9.1 Biological control of weeds: ARS labs and their major rangeland weed targets.

Argentina-Hurlingham	Snakeweed (*Gutierrezia* spp.), creosote-bush (*Larrea* spp.), tarbush (*Flourensia* spp.)
California-Albany	Starthistles (*Centaurea* spp.), toadflax (*Linaria* spp.), gorse (*Ulex europaeus*); potential new targets = dyer's woad (*Isatis tinctoria*), Scotch thistle (*Onopordum acanthium*), poison hemlock (*Conium maculatum*)
Italy-Rome	Knapweeds and starthistles (*Centaurea* spp.), leafy spurge (*Euphorbia esula*), thistles (*Carduus, Cirsium*), saltcedar (*Tamarix chinensis*)
Maryland-Fort Detrick	Knapweeds, thistles, starthistles, leafy spurge (plant pathogens)
Missouri-Columbia	Cropland weeds and leafy spurge (soil microbes) (in cooperation with Sidney, MT)
Montana-Bozeman	Knapweeds, leafy spurge, thistles, St. Johnswort (*Hypericum perforatum*), larkspur (*Delphinium* spp.); potential new targets = dyer's woad (*Isatis tinctoria*), whitetop (*Cardaria draba*), field bindweed (*Convolvulus arvenis*)
Montana-Sidney	Leafy spurge
Texas-Temple	Snakeweed, saltcedar, creosotebush, field bindweed, cocklebur (*Xanthium* spp.), bitterweed (*Hymenoxys* spp.)
Utah-Logan	Larkspur, and various other weeds through aversion and preference conditioning of different classes of livestock

Also new satellite labs in Russia and China: many targets

Adapted from Quimby et al. (5).

Research is currently in progress at nine ARS laboratories to control at least 16 genera and more than 30 species of rangeland weeds (Table 9.1). Any or all of the above approaches to biological control can be used alone or in combination as components in integrated range management systems that might also employ cultural practices such as burning, fertilizing, chaining, and treating with herbicides. Although many scientists from other countries, State Agricultural Experiment Stations, the Extension Service, and other federal and state agencies are involved in various aspects of research and implementation in these areas, this paper will focus primarily on contributions and activities of USDA's Agricultural Research Service.

Biological Control Approaches

Introduction of Exotic Biocontrol Agents (Classical Approach)

This approach to biological control has recently been reviewed (6-8). Traditionally, the "classical" approach to biological control has involved the introduction of *exotic* natural enemies for control of *exotic* weeds and brush, but more recently has been applied to control native weeds. In 1985, Johnson (9) examined, from the point of view of plant ecology, the effects of species removals and introductions on ecosystem function. He found that the paleoecological record clearly shows that species composition of communities always has been, and still is, changing. However, at the ecosystem level, these changes in species composition (such as past natural or accidental catastrophic reductions in species abundance or in biocontrol successes) have not noticeably affected gross ecosystem structure and functional processes such as primary production and energy flow. Also, Johnson (9) could not distinguish any basic ecological differences between introduced weeds and native species that had weedy tendencies except that the introduced species lack natural herbivores. Thus, Johnson's thoughtful analysis has provided interesting scientific counterpoint to earlier concern (10) that biological control of native weeds might cause unwanted serious consequences in ecosystems. Moreover, no weed nor non-target plant species has been pushed near an endangered status by the introduction of a successful biocontrol organism (6,11,12).

Through environmental and economic necessity, the Agricultural Research Service has broadened the scope of its program in the introductory approach to biological control of weeds to include native target species. Most of the native species of weeds and brush targeted for biological control are in the Southwest, while exotic species are in the Central Plains and the Northwest (Table 9.2).

Thus, we can apply George B. Vogt's (personal communication) goal for biocontrol to both exotic and native weeds, viz., "to establish

TABLE 9.2 Type and distribution of major weeds of western U.S. rangelands.[a]

Type weed	Number of Species		
	Northern	Central	Southern
Introduced			
Herbaceous	11	5	0
Woody	0	1	2
	11	6	2
Native			
Herbaceous	9	13	3
Woody	6	12	16
	15	25	19

[a]   summarized from Platte (13)

ecological systems so that the presence of biotic agents restores the target plant species to reduced competitive intensities where they do not overwhelm plant communities." Applying principles of introductory biological control for native weed targets is more difficult than for introduced targets because foreign insects introduced to control indigenous plants are less likely to be host-specific to the native target weed since the biotic agents occur on different host plants in their land of origin (14). On the other hand, "new associations" of biotic agents and target plants can sometimes result in greater efficacy of control because of a lack of co-evolutionary resistance on the part of the target (15). We should also cite Goeden and Kok (16), who have provided examples of successful biocontrol of weeds projects that do not represent "new associations." Regardless of whether the target weeds are introduced or native, researchers must consider and deal with conflicts of interest, which have been discussed in detail by Turner (17).

One of the major conflicts of interest over biocontrol of weeds programs has been concern about potential damage to endangered and threatened plant species closely related to target species (17). While no intentional introduction of a weed biocontrol agent into a given country has placed any known endangered plant species at risk, practitioners of the art and science of biocontrol have to be aware of the potential for that to happen (18). For example, the natural dispersal of *Cactoblastus cactorum* into Florida from intentional establishments in the Caribbean has created fears that this moth species may attack endangered *Opuntia* spp. (19). Every effort must be expended to introduce biotic agents that will provide the greatest benefit with the least risk. Harris (8) has suggested that introduced stenophagous insects pose little risk to rare species because of their tendency to use the most abundant host.

Pemberton (20) has cited Cates (21) who reported that about half of 22 "monophagous" insect species preferred the "rarest" of all the plant species on the same sites. Cates' definitions need to be considered in his work with "polyphagous," "oligophagous," and "monophagous" species. For example, by his definition, "oligophagous" herbivores "are restricted to feeding on two or more genera in a family or closely related families." Biocontrol workers would generally consider an oligophagous insect phytophage as one feeding on more than one plant species within a genus. Moreover, Cates' "rarest" plant species in his tests were not really "rare" in the sense of being endangered. Also, Cates did not indicate what effect the insects had on the density of the host plants; were certain plant species at low densities because of insect attacks? Further, Cates' data showed a wide range of variability in responses which leave the interpretations open to some question. The bottom line is that much more research is needed to settle this vital question of rare host plant/insect interactions.

Current procedural steps in the introductory approach are designed to take advantage of the knowledge we possess to produce maximum protection of the environment. These steps are as follows for USDA-ARS scientists: (1) select target and submit proposal for approval of the ARS National Program Staff Biological Control Matrix Team; (2) after receiving ARS approval, submit proposal to conduct research on the candidate target weed for review and approval of USDA, APHIS-PPQ's interagency advisory committee, i.e., the Technical Advisory Group for the Introduction of Biological Control Agents for Weeds (TAGIBCAW) (22); (3) examine herbaria/museum collections for information on potential biocontrol agents, study biogeography of weed/biotic agents in land of origin and in their new home; (4) explore foreign lands for host-specific natural enemies; (5) characterize biology/taxonomy/ecology/host range of candidate biological control agents [note: some of this research may be conducted under quarantine which requires petitions as regulated by USDA, APHIS-PPQ advised by TAGIBCAW and a similar group in Canada]; (6) release experimentally over range of habitats/time (note: releases are regulated more strictly than introduction into quarantine in that they require petitions to TAGIBCAW *and* environmental assessments as well as applications for field release permits); and (7) release for implementation and evaluate progress with time, monitor efficacy, and determine economic effect. Steps 1-5 are research functions and are typically performed by federal and/or state research agencies. Steps 6-7 are research/implementation functions and the lines of responsibility may overlap between federal/state research agencies and federal/state action agencies. Private companies are sometimes involved in Step 7. Cooperation and communication are essential, especially in the early stages of a release program, to obtain maximum benefit from often scarce numbers of biocontrol agents.

Our knowledge of the principle and procedures involved in the various steps of the research protocol has increased greatly during the expansion of biological control of weeds research in North America and worldwide, especially since 1945, beginning with the review of Huffaker (23) in 1957. Since then the methodologies of target weed selection (14), foreign exploration (24-26), selection of the most promising control agents (27,28), and host-range testing (29,30) have been greatly refined. Newly developing technologies such as enzyme electrophoresis and DNA analysis add valuable tools to determine phylogenetic relationships, to match control agent and weed biotypes, and to determine the biochemical determinants of specific attraction.

We need to determine host range relative to the degree of relatedness between a target weed and nontarget species. This becomes more difficult (and more important) if nontarget natives (potential hosts) are closely related to the target weed.

Vogt (31) has reported that susceptibility of plants to biotic agents is a continuum, i.e., it involves a spectrum of degree in responses. Behavioral changes often precede morphological changes in evolution; this adds to the complexity of selecting and testing "strains" and "isolates" of natural enemies. We are conducting research to help us understand the molecular basis for these plant/biotic agent interactions, so that we will be better able to assess benefits and risks of introductions. Coupled with this, then, is an ongoing research need to develop more complete information on taxonomic and biogeographic relationships worldwide among target and nontarget hosts and among "strains" or "isolates" of biotic agents. Also, this information will help us to determine objectively the relative nativeness[1] of target and nontarget hosts as a factor in benefit/risk analysis.

Harris (7) has provided recent information on the costs of conducting biological control programs against weeds. Finding and testing a biotic agent may cost about $400,000 for each agent. Establishing biological control of one weed may take 20 scientist years (SY's) and cost about $4 million. Benefit:cost ratios for successful projects typically exceed 50:1 (32,33). As DeLoach et al. (34) point out, total cost of a program remains constant regardless of the area treated, but cost/unit area decreases as

---

[1]Johnson (9) used the terms "new natives" and "old natives" to describe the relativity of nativeness in plant origins.

larger areas are controlled, and decreases still more with each added year of control. This makes the method particularly attractive for rangelands with low economic return.

In several cases, control of a given weed by one insect species or pathogen has been dramatic. However, introductory biological control is not a "silver bullet" that will solve all weed problems. When we consider the arid western rangelands since 1940, introductory biological control has been attempted for 23 species of weeds; only six (26%) of these weed species so far have been completely or substantially controlled in larger areas (12). These include St. Johnswort (*Hypericum perforatum*), puncture vine (*Tribulus terrestris*), tansy ragwort (*Senecio jacobaea*), musk thistle (*Carduus nutans*), skeletonweed (*Chondrilla juncea*), and to a lesser extent, toadflax (*Linaria vulgaris* and *L. dalmatica*). In fact, research is still continuing on all the above except for puncture vine and tansy ragwort. We need a comprehensive analysis of these successes and failures. Some are still in progress as incomplete, unfinished projects. We should determine *why* success has not been achieved on some of the targets and *how* success can be realized on all or most of the targets.

The usual effect of controlling a dominant introduced weed with an introduced control agent is that the weed is replaced by a mixed community of plants that include the weed at a low density (8). For example, the control of St. Johnswort in California allowed ranges to return to their bunch grass climax and increased plant diversity by at least 35% (35). The weed probably will, however, continue to spread geographically into its range of climatic and edaphic adaptability.

The objective of some land managers of complete absence of the weed is not necessary if biological control is in effect. The presence of a few plants does not signal a future serious infestation. However, in the case of some weeds highly toxic to livestock, biocontrol agents may not be able to maintain weed populations below the economic loss threshold and spot treatment with other types of control (such as herbicides) may be necessary.

The development of an introductory biological control program for a given weed may take several years, and may not even be possible for some weeds. In the meantime, land managers must use other practices to manage weeds. However, land managers should be willing to have a part (often only a small area) of their land used as establishment sites for biocontrol agents and they should manage their options to protect and favor them. The land managers may have to extend the protected areas if and when the agents become effective and spread. More research will be required on integrated approaches to provide guidance to those managers on cultural practices and grazing systems that will be compatible with biological control agents.

Research at Overseas Biological Control Laboratories

Overseas research on biological control of rangeland weeds is currently underway at ARS laboratories in Rome, Italy, Hurlingham, Argentina, and Beijing, Peoples Republic of China (PRC). This research is coordinated by the USDA-ARS-NPS Matrix Team for Biological Control and administered by ARS International Activities. These laboratories have a group of permanent scientists and technicians who conduct explorations for new natural enemies (insects, mites, and pathogens), conduct research, especially on the life cycles and host ranges of candidate control agents, and ship approved living control agents to U.S. quarantine facilities for further testing and eventual release in the field. They assist and cooperate with federal, state, and university biocontrol of weeds specialists. The U.S. cooperating scientists often visit for periods of a few days to a few months to establish projects, make explorations, and review progress.

Research at Rome

Since being established in Rome in 1959, with substations in Thessaloniki (1981) and Montpellier (1989), the USDA-ARS Biological Control of Weeds Laboratory-Europe (BCWL-E) has emphasized U.S. western rangeland weeds of Euroasian origin. The laboratory explores for new natural enemies (insects, mites, and pathogens); conducts research, especially life cycle studies and host-specificity testing; and ships extensive living material to stateside colleagues and works with them to establish new natural enemies (36).

The laboratory serves as a focal point for U.S. biocontrol of weeds interests in Europe, Africa, the Middle East, and the USSR. Laboratory programs are coordinated with IIBC, Delémont, Switzerland; CSIRO, Montpellier, France; PL-480 projects in the Plant Protection Institute, Yugoslavia; and other biocontrol of weeds units. Explorations occur primarily in southern and eastern Europe, and since 1989 in the USSR and PRC. Biological studies are carried out in the field and laboratory and various host specificity tests are conducted in quarantine. Several insect species as weed biocontrol agents have been approved recently for release or study in U.S. quarantine, and shipment of living material to ARS, APHIS-PPQ, and state collaborators has been emphasized.

Leafy spurge is the highest priority target weed. Other prime targets are knapweeds, yellow starthistle, musk thistle, and recently, tamarisk (*Tamarix* spp.). Gorse (*Ulex europaeus*), dyer's woad (*Isatis tinctoria*), whitetop (*Cardaria draba*); common toadflax (*Linaria vulgaris*); weedy species of grasses, Spanish broom (*Spartium junceum*), cutleaf geranium

(*Geranium dissectum*), and others are being considered as new targets. The natural enemies being researched are covered in other papers in this proceedings.

Future emphasis will include (depending on funding and staffing) more experimental studies on competition between natural enemies and on the interrelationships of the members of natural enemy guilds on target weeds; quantification of impact on target weeds; site specificity based on natural enemy habitat characterization, both biologically and physically at micro and macro habitats; and genetic characterization of biotypes by electrophoretic and other methods.

Buenos Aires

The ARS Biological Control laboratory was established near Buenos Aires in 1960. Early research resulted in the successful control of the aquatic weeds alligatorweed (*Alternanthera philoxeroides*), waterhyacinth (*Eichhornia crassipes*), and waterlettuce (*Pistia stratiotes*). Since 1974, research at Hurlingham (a suburb of Buenos Aires) has concentrated on control of native rangeland weeds of the Southwestern U.S. Several of these most important weeds are from genera with disjunct distributions that include different species native in each area. The natural enemies found and introduced are necessarily "new associations" (15) as discussed above.

The laboratory serves as the focal point for research in all of southern South America, and excellent contacts exist with weed scientists, entomologists, and taxonomists in local universities and federal and provincial research stations in these countries. Explorations are made in countries from Brazil and Bolivia southward and detailed biological and host range studies are conducted in the field in Argentina and in the laboratory at Hurlingham.

Primary target weeds at present are snakeweeds and broomweeds (*Gutierrezia* spp.), creosotebush (*Larrea* spp.), and tarbush (*Flourensia* spp.). Much exploration was previously done on mesquite (*Prosopis* spp.) and *Baccharis* spp. Several other weeds could be considered for future research, such as bitterweed (*Hymenoxys* spp.), sneezeweed (*Helenium* spp.), dogfennel (*Eupatorium* spp.), loco (*Astragalus* spp.), whitebrush (*Aloysia* spp.), alfombrillo (*Drymaria* spp.), and others, plus many important crop weeds.

Plant Pathogens for Biological Control

In nature, most plants are attacked by several to many plant pathogens when climatic conditions are correct. Many of these appear to be promising biocontrol agents in greenhouse studies, but some are unsuccessful in field tests because their microclimatic conditions for infections may not be met there. In the 1970s, Australian scientists had great success in controlling the European skeletonweed (*Chondrilla juncea*) with an introduced European rust, *Puccinia chondrillina*. This stimulated much research and *P. chondrillina* was released in California and the Northwest in 1978; it has been the most effective of three biocontrol agents in California (37). In addition, two crop weeds have been controlled with native pathogens, northern jointvetch (*Aeschenomene virginica*) in Arkansas rice fields and stranglervine (*Morrenia odorata*) in Florida citrus orchards. Recent research has produced breakthroughs that can overcome many of the early difficulties encountered in using plant pathogens in the field.

Evaluation of foreign plant pathogens for introduction into the U.S. is conducted in a containment greenhouse located in Frederick, MD (38). This laboratory is screening more than 150 acquisitions of stem, leaf, and soil-borne (including wilt) pathogens collected by Rick Bennett (ARS), David Sands from Montana State University, and Genevieve DeFago at the Eidgenossische Technische Hochschule in Zurich, Switzerland. These acquisitions are from several countries in Eastern Europe and the Soviet Union. Philosophy and procedures for the evaluation of these candidate organisms have been described (39).

Major targets for biological control using introduced plant pathogens are: leafy spurge (*Euphorbia esula-virgata* complex), musk thistle (*Carduus thoermeri*), the knapweeds (*Centaurea diffusa* and *C. maculosa*, and the starthistles (*C. calcitrapa* and *C. solstitialis*).

Leafy spurge. Presently, most work has been on the evaluation of an *Alternaria* sp., a leaf-spotting fungus first isolated in North Dakota by Krupinski (40). Since then, another *Alternaria* sp. collected in Nebraska has been included in the evaluations. Development of disease in preliminary greenhouse studies required at least a 24 hr dew period, which is a situation rarely found in the field. However, a formulation of the pathogen in a water-in-oil invert emulsion (41), which was modified and applied through standard air pressure sprayers, allowed infection in the absence of a dew period. This formulation was used in field trials in Nebraska and North Dakota in 1990.

Musk thistle. Of seven *Carduus* species introduced into North America from Eurasia, musk thistle is the most widespread. Dunn (42) considered it economically important in 10 % of the counties in the U.S. *Puccinia carduorum* has been found causing a rust disease throughout the distribution of musk thistle in Eurasia since 1978. A strain from Ankara,

Turkey, is among the most virulent and was selected for extensive study. Although host range tests revealed the pathogen to be most aggressive on musk thistle, minor infections occurred in the greenhouse on one of three additional species of *Carduus*, on artichoke (*Cynara scolymus*), and on 8 of 16 species of *Cirsium* (43). Seedlings and young plants of these species were the most susceptible and plants six to seven weeks old (from planting) were very difficult to infect. Also, reinoculation of symptomatic plants usually did not result in new infections; in a few cases, disease severity was greatly reduced.

A proposal to APHIS to conduct a limited field evaluation of *P. carduorum* was initiated in 1984, amended in 1985, and approved in October 1987 following completion of an Environmental Assessment. The first inoculations were made at a site near Blacksburg, VA, in cooperation with scientists at Virginia Polytechnic Institute late in the same month. Musk thistle in the plots developed high levels of disease and diseased plants were found at least 100 m outside of the plot area. None of the five *Cirsium* species developed symptoms of infection, and only one small pustule was found on one of 35 artichoke plants late in the second year. A proposal has been made to APHIS for permission to use this pathogen without restriction for biological control of musk thistle in the U.S.

Starthistles and knapweeds. These plants are susceptible to a rust disease caused by *Puccinia jaceae* introduced from Europe. Like the musk thistle rust, it produces some nontarget effects, i.e., a limited infection on safflower (*Carthamus tinctorius*) (44). In quarantine tests, *P. jaceae* caused much less infection on safflower than did *P. carthami* (which causes an inconsequential safflower rust in North America), and resistance to safflower rust may confer resistance to *P. jaceae* (44). In Canada, severity of disease caused by the diffuse knapweed strain of *P. jaceae* declined as the age of safflower plants increased at time of inoculation (45,46).

A strain of *P. jaceae* from yellow starthistle is under consideration by APHIS for limited field evaluation in California. Proposals are being developed for release of three other strains of *P. jaceae* from purple starthistle, diffuse knapweed, and spotted knapweed.

Another approach to control knapweeds, being investigated by David Sands at Montana State University, is the augmentation of *Sclerotinia sclerotiorum*, a fungus with a broad host range (47). Among several mutants tested, one does not produce overwintering sclerotia and so it would not persist and pose a threat to subsequent desirable forbs in a treated location. If this approach can be developed, it would provide a biological substitute for the commonly used 2,4-D herbicide.

Livestock as Biological Control Agents for Weeds

Differential grazing habits, preferences, and selective abilities of livestock species may allow them to exert selective grazing pressure against some weeds that would result in effective control while providing an economic return. Brock (48) listed four conditions which must be met for livestock to be effective biocontrol agents:

1. The target plant must be accepted by livestock as forage.
2. The target plant must have a differential susceptibility to grazing at some time of the year to aid in the control strategy.
3. Other forage plants must be present to replace the target species.
4. Livestock must be controlled closely.

Different animal species are suited to utilize different types of plants. Hanley (49) proposed a framework for forage selection strategies among animals based on morphological characteristics: body size, type of digestive system, ruminoreticular volume/body weight ratio, and mouth size. Cattle, as the largest domestic species, have large rumens and long retention times that enable them to digest low-quality, high-cellulose diets. Their large mouths render them less selective, thus they are suited to ingest large amounts of low-quality fibrous forage.

Goats, as the smallest domestic species, have small rumens, a short retention time, and a high relative nutrient requirement. Their small mouth and nimble lips allow them to be very selective. They are suited to selecting shrubs, digesting the cell contents and rapidly passing undigestible lignin. Sheep are intermediate, having a small body size, yet a large ruminoreticular volume/body size ratio. Their small mouths and nimble lips enable them to be selective, yet they have the rumen capacity to digest low-quality roughage.

The morphological characteristics that enable a goat to utilize shrubs and their observed preference for shrubs have led to their use in brush control systems. In Texas where goats are a viable economic enterprise, they have been used to control resprouts of shin oak (*Quercus mohariana*), liveoak (*Q. virginiana*), mesquite (*Prosopis qlandulosa*) and juniper (*Juniperus ashei* and *J. pinchoti*) following chaining to knock down mature trees (50,51). Goats have been successful in defoliating regrowth of brush species in burned over and seeded areas of chaparral in Arizona (52) and in fuel breaks in California (53). Heavy goat browsing of gambel oak (*Q. gambelii*) removed current year's growth in Utah (54) and in Colorado (55). Goats have been effective in converting brush-covered abandoned farmland to productive pastures in New England (56), and in controlling gorse (*Ulex europaeus*), the major scrubweed in productive pastures in New Zealand and Australia (57,58). Sheep may also be effective in reducing brush and understory plant growth in coastal Douglas fir (*Pseudotsuga menziesii*) plantations; the reduced competition resulted in more rapid growth of commercial trees (59).

Sheep have the propensity to graze forbs selectively. One of the earliest recommendations for control of leafy spurge (*Euphorbia esula*) in the northern Great Plains was sheep grazing (60,61). Sheep selectively grazed small, young leafy spurge plants in an infested crested wheatgrass (*Agropyron cristatum*) seeding and continued to graze the mature plants into the summer as crested wheatgrass matured and became unpalatable (62). In controlled grazing behavior studies, sheep did not readily graze leafy spurge for the first 1-3 weeks, but increased consumption of spurge to 40-50% of their diets by midsummer (63,64). Physical trampling damage by cattle in an intensive short-duration grazing system also controlled leafy spurge (65).

Sheep also appear successful in controlling tansy ragwort (*Senecio jacobaea*). Sheep defoliated tansy ragwort plants and prevented this biannual plant from going to seed (66). Grazed plants did not continue to grow the next year. West and Farah (67) reported that sheep grazed dyer's woad (*Isatis tinctoria*), but did not utilize it heavily enough late in the growing season to prevent seed production.

Cattle are less selective but can utilize areas not suited for sheep [such as bitterweed (*Hymenoxys odorata*), sneezeweed (*Helenium hoopesii*), cheatgrass (*Bromus tectorum*) and medusa head rye (*Taeniatherum caput-medusae*) ranges]. The potential also exists to train livestock to preferentially graze particular weed species. Animal species and individuals within a species exhibit a wide range of preference among plants. Preferences can be modified by experience, thus providing the opportunity to adjust selection patterns to graze particular plants to meet management objectives (68). Preferential grazing habits developed early in life influence subsequent diet selection. Provenza and Balph (69) describe three mechanisms to help young animals learn to select specific foods.

1. *Food imprinting.* During the period around weaning, when animals are in the transition between maternal care and independence, they apparently learn persistent food preferences. Squibb et al. (70) reported that 4- to 8-wk old lambs exposed to the shrub, mountain mahogany (*Cercocarpus montanus*), consumed more of the shrub when tested at five months and one year later than lambs exposed at younger or

older ages. Lambs 4-8 wk of age are functional ruminants and are reducing their dependence on mothers' milk.

2. *Social learning.* Young animals learn what to eat from observing and participating with social models (mothers and "respected" adults). Lambs exposed to novel foods (rolled barley or serviceberry (*Amelanchier alnifolia*) with their mother or another adult consumed more of the novel food when tested 4 weeks later (71). Lambs ate more of the respective novel food when exposed with their mothers, an intermediate amount when exposed with another adult, and least when exposed alone.

3. *Individual learning.* Garcia and Koelling (72) coined the term cue-consequence specificity, meaning the taste of food can be related to its post-ingestive consequences (either positive or negative). Taste neurons converge with neurons from the gut in the emetic center of the brain. Feedback from the gut may be relayed along pathways recently activated by taste cues and associations are formed. If the consequence is positive (nutrient loading), the food acquires a palatable value. If consequences are negative (nausea or illness), the food becomes unpalatable. The taste of the food takes on hedonic value and varies in acceptability.

Principles of aversive conditioning have been developed for livestock (73,74). Cattle have been trained to avoid eating larkspur, a palatable poisonous plant (75,76). Sheep have been trained to avoid eating common foods (barley, corn, wheat, and shrubs in a nursery) (71,73,77).

Positive conditioning is also possible. Mere exposure to a food increases an animal's acceptance of it (78). Pairing of beneficial consequence with taste of a food will increase the relative preference for that food. Most positive conditioning research has been performed on rats (79). Pairing a taste with caloric input, recovery from nutritional deficiencies, and recovery from post-ingestive distress will increase preference for that taste. There is no reason to assume preferences cannot be likewise shaped in livestock. The potential exists through diet training early in life to condition groups of livestock to graze specific weeds preferentially and thus be more effective biological agents in the fight against noxious weeds.

One major advantage of using various classes of livestock to graze noxious weeds preferentially is that a resource is being utilized and some income or gain can be derived from the practice, although training animals may have a cost, also. If the only goal is chemical control, then expenditures to achieve that may or may not be recouped with increased forage. A land manager would have to evaluate very carefully the economics and environmental consequences of various options for weed control.

Summary

The "take-home" message that we would like to impart is that biological control can be and should be an important part of integrated weed management systems on North America's rangelands. Achieving successful management of troublesome species of rangeland weeds will require a continuous research effort coupled with effective, coordinated communication and cooperation among public research and action agencies and private land managers. The goal of weed management should be to economically complement other rangeland management practices that foster stable, complex ecosystems which have high energy flows, a wide diversity of plants and animals, and sound mineral and water cycles.

References

1. Anonymous. In: *Report of the ARS (USDA) Research Planning Conference on Weed and Brush Control and Related Integrated Research on Rangelands*, p. i, College Station, TX, 1982.
2. Thompson F, FL Leistritz, and J Leitch. Agric. Econ. Report No. 257, North Dakota State Univ., Fargo (1990).
3. Nielsen DB, NR Rimbey, and LF James. In: *The Ecology and Economic Impact of Poisonous Plants on Livestock Production* (LF James, MH Ralphs, and DB Nielsen, Eds.), pp. 5-15. Westview Press, Boulder, CO, 1988.
4. Harris P. BioScience 38:542-548 (1988).
5. Quimby PC, AL Christy, RD Child, and RS Soper. In: *Proc. National Grazinglands Weed Management Conference* (E Lehnert, Ed.), pp. 81-86, 1990.
6. Schroeder D. In: *Recent Advances in Weed Research* (WW Fletcher, Ed.), pp. 41-78. CAB, Slough, 1983.
7. Harris P. In: *Rangeweeds Revisited* (BF Roché Jr and CT Roché, Eds.), pp. 61-68. Proc. Range Manage. Shortcourse, Wash. State Univ., Pullman, 1989.
8. Harris P. In: *Critical Issues in Biological Control* (M Mackauer, LE Ehler, and J Roland, Eds.), pp. 289-300. Intercept Ltd, Andover, Hants, UK, 1990.
9. Johnson HB. In *Proc. VI Int. Symp. Biol. Contr. Weeds* (ES Delfosse, Ed.), pp. 27-56. Canadian Government Publishing Centre, Ottawa, 1985.
10. Andres LA. In: *Proc. V Int. Symp. Biol. Contr. Weeds* (ES Delfosse, Ed.), pp. 11-20. CSIRO, Melbourne, 1981.

11. Goeden RD. In: *Introduced Parasites and Predators of Arthropod Pests and Weeds: A World Review* (CP Clausen, Ed.), pp. 357-414. USDA-ARS Agric. Handbook 480.

12. Julien MH. *Biological Control of Weeds: A World Catalogue of Agents and Their Target Weeds, Second Ed.* CAB Intl. Inst. Biol. Contr., Wallingford, UK, 1987.

13. Platte KB. J. Range Management 12:194-200 (1959).

14. DeLoach CJ. In: *Proc. V Int. Symp. Biol. Contr Weeds* (ES Delfosse, Ed.), pp. 175-190. CSIRO, Melbourne, 1981.

15. Hokkanen H and D Pimental. Canad. Entomol. 116:1109-1121 (1984).

16. Goeden RD and LT Kok. Canad. Entomol. 118:51-58 (1986).

17. Turner CE. In: *Proc. VI Int. Symp. Biol. Contr. Weeds* (ES Delfosse, Ed.), pp. 203-225. Canadian Government Publishing Centre, Ottawa, 1985.

18. Turner CE, RW Pemberton, and SS Rosenthal. Environ. Entomol. 16:111-115 (1986).

19. Kass H. Plant Conservation 5:2 (1990).

20. Pemberton RW. In: *Proc. VI Int. Symp. Biol. Contr. Weeds* (ES Delfosse, Ed.), pp. 365-390. Canadian Government Publishing Centre, Ottawa, 1985.

21. Cates RG. Oecologia 48:319-326 (1981).

22. Coulson JR and RS Soper RS. In: *Plant Protection and Quarantine, Vol. III Special Topics* (Kahn RP, Ed.), pp. 357-414. CRC Press, Inc., Boca Raton, FL, 1989.

23. Huffaker CB. Hilgardia 27:101-157 (1957).

24. Room PM. In: *Proc V Intl. Symp. Biol. Control Weeds* (ES Delfosse, Ed.), pp. 113-124. Brisbane, Australia, 1980.

25. Sands DPA and KLS Harley. In: *Proc. V Intl. Symp. Biol. Control Weeds* (ES DelFosse, Ed.), pp. 81-89. Brisbane, Australia, 1981.

26. Wapshere AJ. In: *Proc. V Intl. Symp. Biol. Control Weeds* (ES DelFosse, Ed.), pp. 75-79. Brisbane, Australia, 1981.

27. Harris P. Canad. Entomol. 105:1495-1503 (1973).

28. Goeden RD. Prot. Ecol 5:287-301 (1983).

29. Zwölfer H and P Harris. Ann. Rev. Entomol. 16:159-178 (1971).

30. Wapshere AJ. Ann. Appl. Biol. 77:201-211 (1974).

31. Vogt GB. In: *Insects of Panama and Mesoamerica: Selected Studies* (DQ Arias and A Aiello, Eds.), (In Press). Oxford Univ. Press, Oxford, 1991.

32. Harris P. Weed Science 27:242-250 (1979).

33. Andres LA. Aquatic Botany 3:111-123 (1977).

34. DeLoach CJ, PE Boldt, HA Cordo, HB Johnson, and JP Cuda. In: *Management and Utilization of Arid Land Plants* (DR Patton, CE Gonzalez, AL Medina, LA Segura T, and RH Hamre, Eds.), pp. 49-67. Saltillo, Mexico, 1986.

35. Huffaker CB and CE Kennett. J. Range Manage. 12:69-82 (1959).

36. Fornasari L and L Knutson. In: *Pesticides and Alternatives* (JE Casida, Ed.), pp. 515-530. Elsevier Science Publishers, Amsterdam, 1990.

37. Supkoff DM, DB Joley, and JJ Marois. J. Appl. Ecol. 25:1089-1095 (1988).

38. Melching JS, KR Bromfield, and CH Kingsolver. Plant Disease 67:717-722 (1983).

39. Bruckart WL and WM Dowler. Weed Science 34 (Suppl. 1):11-14 (1986).

40. Krupinski JM and RJ Lorenz. Weed Science 31:86-88 (1983).
41. Quimby PC, FE Fulgham, CD Boyette, and WJ Connick. In: *Pesticide Formulations and Application Systems: Eighth Volume* (DA Hovde and GB Beestman, Eds.), pp. 264-270. Amer. Soc. for Testing & Materials, Philadelphia, PA, 1989.
42. Dunn PH. Weed Science 24:518-524 (1976).
43. Politis DJ, AK Watson, WL Bruckart. Phytopathology 74:687-691 (1984).
44. Bruckart WL. Plant Disease 73:155-160 (1989).
45. Watson AK and I Alkoury. In: *Proc. V Int. Symp. Biol. Contr. Weeds* (ES Delfosse, Ed.), pp. 301-305. CSIRO, Melborne, 1981.
46. Mortensen K. In: *Proc. VI Int. Symp. Biol. Contr. Weeds* (ES Delfosse, Ed.), pp. 447-452. Canadian Government Publishing Centre, Ottawa, 1985.
47. Ford EJ. In: *Proc of the 1989 Knapweed Symp.* (PK Fay and JR Lacey, Eds.), pp. 182-189. Montana State Univ., Bozeman, 1989.
48. Brock JH. Rangelands 10:32-34 (1988).
49. Hanley TA. J. Range Manage. 35:146-151 (1982).
50. Merrill LB and CA Taylor. Rangemans J. 3:74-76 (1976).
51. Wiedemann HT, CH Meadors, and CE Fisher. Texas Agric. Exp. Sta. Consolidated Progress Report 3665 (1980).
52. Knipe OD. Rangelands 5:252-255 (1983).
53. Green LR, CL Hughes, and WL Graves. In: *Proc. First International Rangeland Congress* (DN Hyder, Ed.), pp. 451-455. Soc. Range Manage., Denver, CO, 1979.
54. Riggs RA and PJ Urness. J. Range Manage. 42:354-360 (1989).
55. Davis GG, LE Bartel, and CW Cook. J. Range Manage. 28:216-218 (1975).
56. Wood GM. Agron. J. 79:319-321 (1987).
57. Radcliffe JE. New Zealand J. Exp. Agric. 13:181-190 (1985).
58. Harradine AR and AL Jones. Aust. J. Exp. Agric. 25:550-556 (1985).
59. Sharrow SH, WC Lieninger, and B Rhodes. J. Range Manage. 42:2-4 (1989).
60. Wood HE. Manitoba Dept. Agric. Immig. Pub. 194 (1944).
61. Wood HE. Manitoba Dept. Agric. Immig. Pub. 200 (1945).
62. Johnston A and RW Peake. J. Range Manage. 12:192-195 (1960).
63. Landgraf BK, PK Fay, and KM Havstad. Weed Science 32:348-352 (1984).
64. Bartz SJ, EL Ayers, and KM Havstad. In: *Proc. Leafy Spurge Symp.*, p. 24. Sundance, WY, 1983.
65. Parman M. Rangelands 8:183 (1986).
66. Sharrow SH and WD Mosher. J. Range Manage. 35:480-482 (1982).
67. West NE and KO Farah. J. Range Manage. 42:5-10 (1989).
68. Provenza FD and DF Balph. Appl. Anim. Behav. Sci. 18:211-232 (1987).
69. Provenza FD and DF Balph. J. Anim Sci. 66:2356-2368 (1988).
70. Squibb RC, FD Provenza, and DF Balph. J. Anim Sci. 68:987-997 (1990).
71. Thorhallsdottir AG, FD Provenza, and DF Balph. Appl. Anim. Behav. Sci. 18:327-340 (1987).
72. Garcia J and RA Koelling. Psychon. Sci. 4:123-124 (1966).
73. Lane MA, MH Ralphs, JD Olsen, FD Provenza, and JA Pfister. J. Range Manage. 43:127-131 (1990).
74. Ralphs MH and JD Olsen. J. Anim. Sci. 68:(In Press) (1990).

75. Thorhallsdottir AG, FD Provenza, and DF Balph. Appl. Anim. Behav. Sci. 25:25-33 (1990).
76. Burritt EA and FD Provenza. J. Anim. Sci. 68:1003-1007 (1990).
77. Ralphs MH. J. Range Manage. (In Press) (1991).
78. Zajonic RB. J. Personality and Social Psychology, Monograph Supp. 9:1-27 (1968).
79. Provenza FD, JA Pfister, and CD Cheney. J. Range Manage. (In Press) (1991).

10

Principles of Chemical Control

Rodney W. Bovey

Abstract

Identification of the problem weed species is necessary to apply proper management and control measures. Several herbicides are available for controlling most weeds and brush. Herbicides are applied to foliage (sprays) or roots (sprays or pellets), depending on the species to be controlled and the herbicide to be used. Sprays can be applied to weeds and brush with hand-carried equipment, power equipment, or aircraft. Selection of application equipment depends on the density of weeds and brush, the types of species, and size of the area to be covered. In many cases, rough terrain and tall vegetation limit the use of ground or hand-carried equipment, especially on large areas. Where stands are scattered, individual plants can be treated with herbicides by foliar sprays, cut surface and injection treatment, and soil treatment. Special wiping devices also are used to treat individual plants in small or scattered stands of weeds and brush. All equipment must be calibrated to deliver the proper amount of herbicide. Too little herbicide will result in poor results and too much will be costly and may injure desirable vegetation. The user must know the most favorable time for treatment and the factors affecting results.

Introduction

Herbicides are an important means of weed and brush control on rangelands. Compared to mechanical practices, herbicides usually are less expensive, less damaging to the environment, and often more effective. Herbicide sprays, however, are subject to drift that may damage susceptible crops or valuable vegetation on nearby areas if improperly applied. A variety of herbicides and herbicide combinations are available commercially. It is essential that the user understand the properties and the effects of herbicides to use them safely and effectively. Herbicides commonly used for weed and brush control include 2,4-D, dicamba, bromacil, picloram, hexazinone, triclopyr, amitrole, and tebuthiuron (1, 3). Combinations of herbicides such as 2,4-D + triclopyr, triclopyr + picloram, and 2,4-D + picloram or dicamba also have been used (2).

Characteristics of Herbicides

Benzoics. Dicamba, a selective translocated herbicide, controls many broadleaf weeds in pastures and cropland and some woody plants. Similar to the phenoxy herbicides in activity and use, it is absorbed through roots as well as through foliage. Dicamba can be applied either by ground or aerial sprays or as granules, depending on the weeds to be controlled and their proximity to susceptible crops. Dicamba can be applied in mixtures with 2,4-D to broaden the spectrum of weeds controlled.

Dicamba has a low order of toxicity to wildlife, fish, livestock, and humans. It rapidly degrades and does not accumulate in the environment. Dicamba has a low corrosion hazard to spray equipment. It is formulated as the dimethylamine or sodium salt and is sprayed in a water carrier. In granular formulation, it is an acid.

Bipyridyliums. Diquat and paraquat are desiccant and defoliant herbicides used for general contact activity against weeds and brush. In some situations, they are used as selective herbicides. Woody species usually will resprout from foliar sprays of diquat and paraquat.

Paraquat is water soluble and is inactivated by soil contact. Paraquat is registered for suppression of existing sod and emerged undesirable weeds and grasses to permit pasture and range seeding. Paraquat can be used to control winter weeds in dormant warm-season grasses. Protective clothing and respirators should be used when applying paraquat because it is highly toxic when ingested.

Phenoxy herbicides. Phenoxy herbicides such as 2,4-D, 2,4-DB, dichlorprop, and MCPA have been used for more than 40 years and are effective for controlling many weed and brush species. They are used to produce changes and shifts in plant cover that are beneficial for livestock use and wildlife habitat. The phenoxys are not toxic to livestock or man at dosages labeled for weed control, and they disappear rapidly from soil, vegetation, and water. They do not accumulate in the food chain.

Ester formulations of phenoxy herbicides are moderately toxic to certain fish. They are only slightly toxic to lower aquatic organisms, birds, and wild mammals under laboratory conditions.

Susceptible vegetation, like cotton (*Gossypium* spp.), can be damaged from spray drift or from volatilization (especially high volatile esters). Following label instructions and making applications during favorable weather should prevent drift and volatilization problems.

The phenoxy herbicides selectively control broadleaf weeds in grasslands or grass crops (3). Rates of 0.28 to 2.2 kg/ha effectively control many broadleaf plants. High rates of phenoxys can injure grasses.

The phenoxy compounds are relatively inexpensive and easy to apply. They usually are marketed as liquid concentrates of salts or esters. The ester formulations often are more effective as foliar sprays on trees and brush than are the salts.

Esters are classified as high volatile or low volatile, depending on how readily they vaporize. Methyl, ethyl, isopropyl, and butyl esters of 2,4-D are examples of 2,4-D formulations that are highly volatile when sprayed on hot days. Such phenoxy compounds are identified on the herbicide container or package label. Examples of low volatile phenoxy esters include the propylene glycol butyl ether, butoxyethyanol, and isooctyl esters. Low volatile esters should be used in areas where sensitive crops or vegetation are grown. The tendency for a herbicide to vaporize (volatilize) is important because it may be carried as fumes from the target area and may damage nearby valuable vegetation, especially under windy conditions on hot days.

Leaves readily absorb the phenoxy compounds which are translocated throughout the plant along with the products of photosynthesis. Oil soluble formulations, usually esters, applied in kerosene or diesel oil will penetrate the bark of most woody plants and can be used as basal or foliar sprays on individual plants. The phenoxy herbicides, however, are more commonly applied broadcast to large areas containing dense stands of weeds.

Sulfonylureas. Chlorsulfuron, metsulfuron, and sulfometuron have activity on a broad spectrum of weeds. Metsulfuron has label use for rangelands. Common broomweed (*Gutierrezia dracunculoides*), bitter sneezeweed (*Helenium amarum*), woolly croton (*Croton capitatus*), and perennial broomweed (broom snakeweed; *Gutierrezia sarothrae*) are included on the label. The compounds are effective at extremely low rates and can be mixed with 2,4-D and other herbicides for western ragweed (*Ambrosia psilostachya*) control.

Triazines. Atrazine applied at low rates (0.9 to 1.1 kg/ha) in the fall is a soil active herbicide that controls downy brome and other weeds. Perennial grasses can be seeded the following fall and will establish under much reduced weed competition. Atrazine also can be used to renovate existing stands of certain perennial range grasses in the western region of the United States (1).

Ureas and uracils. These compounds include bromacil, diuron, fenuron, fenuron-TCA, hexazinone, monuron, monuron-TCA, siduron, and tebuthiuron.

Ureas and uracil-type herbicides can be selective at low rates and nonselective at high rates. They usually are formulated as wettable powders for water sprays or as granules or pellets for dry application.

Bromacil (a uracil) will control a variety of woody species. If rates above 5.6 kg/ha are used, it also will kill many desirable grasses and forbs on grazing lands.

Fenuron is no longer produced commercially, but a fenuron-TCA combination is available for control of certain woody plants and weeds on noncrop areas. Monuron and monuron-TCA combinations are available commercially, but they have limited use for brush control because high rates are required for effectiveness. The monuron-TCA and fenuron-TCA combinations generally are used for nonselective, temporary sterilization in noncrop areas (2).

Siduron is not used for woody plant control but is mentioned because it selectively controls germinating annual weed grasses in newly seeded plantings of perennial pasture grasses and turf.

Hexazinone is a relatively new compound that shows promise for woody plant control and for use in noncrop areas. Hexazinone is recommended for forestry site preparation and for pine release where loblolly (*Pinus taeda*), longleaf (*P. palustris*), shortleaf (*P. echinata*), slash (*P. alliottii*), and Virginia (*P. virginiana*) pines are grown. The pelletized product can be spread in a grid pattern for hardwood brush control. It also shows promise for total vegetation control, including control of perennial grasses. It is highly water soluble, and a new liquid formulation has federal registration on rangeland for individual plant treatment to control woody plants.

Tebuthiuron excels in controlling a variety of undesirable woody plants. This herbicide, formulated as pellets containing 20 or 40% active ingredient, is commercially available for both rangeland use and weed control in noncrop areas.

The ureas and uracil herbicides are absorbed primarily through the roots of plants. They can be applied in spring or fall when weeds and brush are actively growing and when rainfall is adequate to leach them into the soil.

Fall, winter, and early spring applications of hexazinone and tebuthiuron can be timed to reduce injury to forage plants and to eliminate drift hazards. These compounds may kill trees at a considerable distance from the point of application, depending on the size of the root system and whether it extends into the treated area. The compounds are nonvolatile and do not corrode equipment.

Most urea and uracil herbicides can be injurious to some forage species when applied broadcast, especially as sprays. Applying herbicides as pellets or spheres to confine the herbicide to a few spots in the treated area reduces exposure to desirable forage plants. Most of the ureas and uracils persist in the soil for several months at rates used for brush control. They are low in toxicity to warm-blooded animals.

Other organic herbicides. Amitrole is effective against poison ivy and poison oak. If amitrole is accidentally sprayed on desirable plants, it is less likely to cause severe injury than other herbicides.

Amitrole is available as a powder containing 50% ai or as a liquid formulation. Another formulation, Amitrole-T, contains equal amounts of ammonium thiocyanate in addition to amitrole in liquid form. Amitrole is effective through the roots and tops of plants but cannot be used where there is any possibility of residues on food or feed crops.

Clopyralid, an auxin-type selective herbicide related to picloram, is being investigated for control of many annual and perennial weeds and woody plants in crops and rangeland (2). It is highly effective against members of Polygonaceae, Compositae, and Leguminosae families but shows poor activity against grasses, crucifers, and many woody plants. It is highly phytotoxic to honey mesquite (*Prosopis glandulosa*).

Fosamine is a liquid herbicide for control of woody vegetation. Response to late summer or early fall foliage spray is not observed until the following spring when budbreak is prevented in some species in certain areas.

Glyphosate, a nonselective herbicide, is effective against both grasses and broadleaf plants. It is readily translocated from leaf and stem tissue to roots, killing a high percentage of many weeds. Glyphosate is inactivated by contact with the soil and should not injure newly seeded forages planted in treated soil. It is sprayed in a water carrier and is not corrosive to equipment. Glyphosate is registered for use for noncrop and pre-till weed control and as a directed spray for orchards, plantations, Christmas trees, and many other crops. Broadcast sprays over woody species will damage desirable forage plants, so applications should be made to individual plants on noncrop areas or areas to be renovated. Glyphosate has a low order of mammalian toxicity.

Imazapyr has demonstrated activity with residual control of a wide variety of annual and perennial weeds, deciduous trees, vines and brambles in noncropland situations. Additional research is being conducted to support a granular product and the use of the chemical in site preparation and pine release applications in forestry. Imazapyr is available in an oil soluble formulation for cut stump, basal bark and frilling use.

Picloram, a selective, translocated herbicide, effectively controls many weed and brush species on grasslands. It can be applied to the soil or foliage and is effective as an injection/cut surface treatment on many undesirable trees. Picloram can be applied in liquid sprays and as pellets to brush in the spring and fall. Foliar sprays of picloram are absorbed by both foliage and roots. Most perennial grasses are resistant. Its high activity at moderate rates against many woody plants makes it desirable for brush control.

Care must be taken to prevent drift of picloram to desirable plants. Picloram is relatively persistent in soil, especially in cool climates. Since it is water soluble, care must be taken to prevent its movement into water used for irrigation. It should not be applied where it can be leached or moved to sensitive crop areas by rainfall. Picloram has low mammalian toxicity. It is only slightly corrosive to spray equipment.

Triclopyr is a relatively new selective postemergence herbicide for use on rights-of-way and industrial and forestry sites. Triclopyr, used experimentally on weeds and brush, recently has been approved by EPA for control of honey mesquite. It also can be used to kill trees and brush by injection/cut surface treatments. Triclopyr is readily translocated in plants and is reportedly more effective than 2,4,5-T on some species. It is moderately toxic to warm-blooded animals. It degrades rapidly in soil. Ester formulations are available (2).

Oils. Diesel oil and kerosine are commonly used to control honey mesquite and huisache (*Acacia farnesiana*). From 0.25 to 4 L of oil are used per tree depending on tree size. The oil is applied around the base of the tree during dry weather when the soil is pulled away from the trunk. Application at this time enables oil to penetrate to the lower buds on the stem.

Oils alone are not very effective herbicides when applied to the foliage of woody plants. Diesel fuel is commonly used as a diluent and carrier for oil soluble herbicides for individual plant treatments. It is also used as a carrier in aircraft spraying.

Application Methods

Formulated herbicides are of little value if application methods and equipment are not available to treat weeds in a safe and practical manner. Equipment for applying chemicals must disperse small quantities of herbicide over large areas. Spray applications can be classified according to volume. An ultra-low-volume signifies a total spray volume no greater than 4.7 L/ha. An ultra-low-volume treatment may be nondiluted herbicide. Low-volume sprays are from 9.4 to 280 L/ha of spray solution per acre. High-volume sprayers apply 280 or more L/ha.

Low-volume sprays. Several herbicides control weeds and brush when applied at low volume as broadcast foliar sprays. Low-volume spraying is usually quicker, easier, and less expensive than high-volume spraying. Low-volume sprays are often applied by aircraft.

Ground equipment is also practical for applying low-volume sprays to low-growing weeds and brush on uncleared land or to regrowth on land that has been mechanically cleared. Low-volume sprays applied in swaths up to 15 m wide with either a spray boom with several nozzles or a large, boomless nozzle have produced satisfactory coverage of plant foliage. Spray booms on ground sprayers should be mounted to clear the tallest weeds and brush by 1 m. An operating pressure of 207 to 276 kPa (30 to 40 psi) generally is effective. Boomless nozzles should be mounted to clear all weeds and brush by 1 m. Spray pressures from 207 to 414 kPa are used. They are most effective under low wind conditions.

Aerial spraying is most efficient for treating large areas or tall and dense stands of brush on rough terrain. Both fixed wing aircraft and helicopters are used. Aerial spraying can give excellent coverage in most brush. Sprays applied in tall dense brush often fail to kill understory plants. A second aerial spraying may be necessary a year to two after the first time.

High-volume sprays. Best control of some woody or resistant species may require complete coverage of the plant with a high-volume spray. High-volume sprays are used to apply herbicides to weeds and brush along roads, rights-of-way, fence rows, and other similar areas or to individual plants on pastures and rangeland. All foliage and twigs are thoroughly wetted. High-volume sprays usually are applied to foliage with power sprayers that maintain pressures up to 690 kPa. This pressure is enough to force the spray through the foliage to the tops of tall trees. Pressure higher than 690 kPa forms fine spray droplets that may drift and may damage nearby susceptible crops.

Because of the large volume of spray necessary for this method, it is impractical for treating large areas where the water supply is limited or areas are inaccessible to truck- or tractor-mounted sprayers.

Individual Plant Treatments

Treatment of individual plants is especially useful for controlling undesirable hard-to-kill woody species that occur in scattered stands. Individual plant treatments include foliar sprays, basal sprays, cut surface and injections, and soil treatment (1).

Foliar sprays. Foliar sprays of 2,4-D, dicamba, and picloram usually are applied in spring and summer to individual plants or groups of plants when they are actively growing and after the leaves have reached full size. Low- or high-volume sprays can be used. Ester formulations usually are more effective on woody species than amine formulations and are less likely to be washed off should rainfall occur soon after application. If complete coverage is necessary to kill a given species, high-volume sprays that wet all foliage, twigs, and terminal stems are necessary.

Basal sprays. Basal sprays are used to treat bark at the base of individual plants to the point of runoff. Mixtures of ester formulations of 2,4-D or triclopyr with diesel oil or kerosene carriers or oil alone commonly are used. Basal sprays can be used to kill brush and trees with main stems up to about 13 cm in diameter any time of the year. Effectiveness of the treatment may be reduced if the bark or soil is wet. The stem should be sprayed from the soil line to 45 cm above ground. Four liters of spray will treat 50 trees 5 cm in diameter. Compressed air sprayers, knapsack sprayers, or power sprayers can be used.

Low-volume and streamline basal applications of herbicides for control of brush were introduced to Texas rangeland in late 1987 by the Extension Service. Since then, many result demonstrations have been established utilizing these two techniques. Most of the demonstrations have had mesquite as the target species although other brush species have been treated.

Low-volume basal uses a solution containing 25% herbicide and 75% diesel fuel oil. The solution is applied to the lower 30-46 cm of the stem with fan or hollow-cone nozzle to wet the stem but not to the point of runoff. Thus, less total volume is used per plant than in the conventional basal treatment that applies sufficient solution to allow runoff with puddling at the base of the plant. With low-volume basal, the herbicide solution must be applied completely around the stem. Because the mixture is applied only to wet the stem, the herbicide must penetrate the bark of the plant.

Streamline basal application utilizes a mixture of 25% herbicide, 10% penetrant and 65% diesel fuel oil or 25% herbicide and 75% diesel fuel oil. The penetrant that has been tested extensively is d,1 limonene. It is sold under the product names of Cide-Kick, Cide-Kick II, and Quick Step II. It is a naturally occurring chemical derived from by-products of the citrus industry and the pine industry. The mixture is sprayed in a band (7-10 cm wide) with a straight stream nozzle to 1 or 2 sides of the stem near ground level or at the line dividing young (smooth) and mature (corky) bark. The material must go completely around the stem. Best results have been obtained on stems less than 10 cm in diameter and with smooth bark.

Streamline basal and low-volume basal application of herbicides are very useful methods for maintenance brush control. They may be used for treating a thin stand of brush, for spot treatment and for controlling brush near desirable plants.

Cut surface injection treatment. Trees larger than 13 cm in diameter often have bark too thick for basal sprays to penetrate. Herbicides can be applied to sapwood of trees either through frills or notches cut into the bark or by mechanical injection.

Frills are cuts made in the sapwood. They encircle the tree trunk and act as cups to hold the herbicide. The frill is made with ax cuts that penetrate the sapwood at least 0.6 cm. Spacing depends on the species, herbicide, and time of year. The frill can be filled with either the same solution as used for basal sprays or undiluted herbicide. A notch also can be cut into one or more places at the base of the tree and sprayed with the appropriate herbicide solution.

The same spray equipment used for basal sprays can be used for frills and notches. The solution also can be applied with a properly labeled plastic squeeze bottle or with a small can that has a pouring spout or lip.

Injection treatments are effective for killing woody plants, but the labor requirements make them expensive. Also, to be effective, the correct volume of herbicide solution must be placed in properly spaced cuts. In winter, resistant species may require higher herbicide concentrations or closer spaced cuts with herbicide than in summer. Commercial injection tools are available.

If trees are felled, the freshly cut surface of the stump should be treated with herbicide to prevent sprouting of certain species. It is more efficient to prevent sprouting than to kill the sprouts. Herbicides mixed with oils, an undiluted herbicide, a solution of 0.25 to 0.5 kg AMS per L of water, or AMS crystals are applied directly to the fresh cut. If oil solutions are used, the cut surface and bark are thoroughly drenched from the cut to the ground line.

Soil Treatment

Certain soil-active herbicide pellets or sprays either can be applied to individual plants, can be placed in grids or bands, or can be applied broadcast to kill plants. Rainfall soon after application may be required for satisfactory results. Dry herbicide materials commonly are formulated as extruded pellets 0.13 to 0.38 cm diameter, spherical granules from 0.13 to 1.3 cm diameter, or tablets of varying size up to 2.5 cm long. The use of soil active herbicides has become more common since the development of new herbicides effective for brush control (1).

The smaller herbicide granules can be broadcast or applied in bands in brush-infested areas. The larger spherical pellets and tablets are effective when applied in grids spaced on 1.8 to 2.7 m centers. The various species respond to grid applications according to differences in their root systems. Gambel oak (*Quercus gambelii*), which has a deep root system, does not respond well to grid pattern applications. Because of their shallow, widespread lateral roots, shrub live oak (*Quercus germinata*) and Utah juniper (*Juniper osteosperma*) respond well to a grid pattern in which the herbicide pellets are spaced on 0.9, 1.8, or 2.7 m centers (1). Damage to forage plants is minimized using grid or band placement. Snakeweed control is best with broadcast application of granules. Pellets can be broadcast readily by aircraft. Ground equipment can apply herbicide pellets in bands spaced at 1.8 to 2.7 m apart.

Wipers and Special Techniques

Herbicides can be applied directly to weeds and brush using specially designed tractor-mounted rope-wick or carpeted applicators. Herbicide solutions are conducted through a porous medium and wiped onto the unwanted plants. Present designs allow treatment of weeds and brush not exceeding 1.8 to 2.1 m tall. The ropes or carpets are saturated with herbicide either undiluted or diluted in water and wiped onto unwanted plants. Handcarried wipers for small jobs also have been constructed. These wiping devices are available commercially and enable better placement of the herbicide on weeds with less exposure to valuable plants. Mayeux and Crane (4) suggested that the brush roller, constructed of household carpet stretched over a 25-cm-diameter by

2-m-long polyvinyl chloride (PVC) cylinder can be used effectively to treat weeds and brush on rangeland. The carpeted roller is tractor mounted, is saturated with herbicide as needed, and is rotated continuously against the direction of travel. Treatment of small honey mesquite has been highly successful.

Factors Affecting Results

When herbicides are applied as foliar sprays, they usually are effective only during certain periods of the year, commonly in the spring after the leaves of woody plants have fully expanded and plants are growing actively. Also, weeds may be susceptible for only a short time during their life cycle (seedling stage); they may be virtually unaffected later. Foliar damage from insects, hail, or drought also may render the treatment unsuccessful due to poor absorption and translocation of the chemical. Consequently, foliar sprays of herbicides should be applied when plants have developed under favorable soil moisture and temperature and environmental conditions. Each weed or brush species, however, may respond differently. Consult the proper reference or agricultural authority before wasting money on an improper treatment.

Spraying the proper rate of herbicide onto the foliage at the right time is the first step in successful application. Proper application equipment is essential. One of the most important factors is application at low wind velocity - below 16 km/h for ground equipment and below 10 km/h for aerial equipment. Applying herbicide when winds are above these velocities may result in loss of chemical from the target area and damage to adjacent areas from spray drift. If some wind is encountered, be certain the wind direction is away from susceptible crops or valuable vegetation.

Good coverage of foliage with herbicides is essential for satisfactory results. Small spray droplets of 100 μm in diameter or less give better coverage and results than do large spray droplets of 300 to 400 μm. Small droplets are subject to drift; therefore, a balance between small and large droplets is desired. Spray droplet size can be increased by modifying spray solution viscosity, nozzle type, and spray pressure.

Once on the stems and leaves, the herbicide must be absorbed in large enough amounts to be translocated throughout the plant. Barriers to absorption include the waxy outer covering of leaf surfaces and the bark on woody plants. Absorption of plant surfaces sometimes can be improved by oil-in-water carriers or by the use of surfactants or wetting or penetrating agents in water carriers. Plants differ in their ability to absorb herbicides. Absorption usually is more rapid in young plants.

Translocation of the herbicide throughout the plant or to the site of action (stem and root tissue) is necessary for control. Most herbicides move best within plants during favorable environmental conditions when the plant is growing actively. Too high rates of some foliar applied herbicides such as the phenoxys may reduce translocation and overall control. These herbicides can disrupt the tissues responsible for transport in the plant.

Rainfall before and soon after treatment is desirable to leach pelleted herbicides into the root zone of the weeds and brush as well as stimulate growth. Herbicides are absorbed by the roots and move up through the plant to kill it. Tebuthiuron applied to the soil as pellets can be applied in the Southwest at any time of year for control of certain woody species. Tebuthiuron is not readily broken down by sunlight and will remain intact on the soil surface until adequate rainfall leaches it into the soil. Picloram and dicamba are broken down by sunlight; avoid application during hot, dry months.

Esters of triclopyr or similar herbicides in diesel oil carrier commonly are used as basal sprays. These treatments kill trees up to 13 cm in diameter. Treatments can be applied any time of the year. If the bark or soil is wet, effectiveness is reduced. For trees larger than 13 cm in diameter, bark often is too thick to penetrate with sprays, therefore, the herbicide can be applied to the sapwood through frills or notches cut into the bark. In winter, difficult to control species may require higher herbicide concentrations or closer spacings of cuts around the tree.

References

1. Bovey RW, AF Wiese, RA Evans, HP Morton, and HP Alley. *Control of weeds and woody plants on rangelands.* USDA and Univ. Minn. AB-BU-2344 (1984).

2. Humburg NE (Ed.). *Herbicide Handbook, 6th Ed.* Weed Science Society of America, Champaign, IL, 1989.

3. Klingman DL, RW Bovey, EL Knake, AH Lange, JA Meade, WA Skroch, RE Stewart, and DL Wyse. *Systemic herbicides for weed control - phenoxy herbicides, dicamba, picloram, amitrole and glyphosate.* USDA and Univ. Minn. AD-BU-2881 (1983).

4. Mayeux HS Jr and Crane RA. Rangelands 5:53-56 (1983).

11

Timing of Herbicide Application for Effective Weed Control: A Plant's Ability to Respond

Ronald E. Sosebee and Bill E. Dahl

Abstract

Herbicidal control of weedy species has often been erratic and unsatisfactory. Some of the reasons underlying the lack of success include our inattentiveness to appropriate environmental conditions, phenological stage, or physiological condition of the plant at the time of herbicide application. Plants can respond only if the herbicides are applied at a time during which growing points can be affected. Annuals and non-sprouting perennials are relatively easy to control if herbicides are applied while the plant is vegetative and "actively growing", but before it shifts toward reproduction. Sprouting perennials, regardless of whether they are herbaceous or woody, are controlled only when the herbicide is translocated to and kills the perennating buds that otherwise would be activated with shoot mortality. We have been extremely successful in achieving total plant mortality of sprouting perennial weeds (herbaceous, suffrutescent, and woody) when we time the herbicide application to total nonstructural carbohydrate (TNC) trends and environmental conditions (e.g., soil temperature) that favor translocation of photosynthate and, consequently, herbicides to the perennating organ and buds. If a plant resprouts following herbicide, fire, or mechanical treatment, subsequent control of sprouts is more difficult and herbicides must be applied when the phenological stage and physiological condition have shifted from long shoots (vegetative) to short shoots (reproductive).

Weed Control

Weed control has been practiced intensively in parts of the western United States for the past 50 years and in some local areas more than 50 years. Yet, even today, we are not as successful in controlling weedy species as we would like to be. Two reasons for our erratic success have been (1) our inattentiveness to the plant's ability to respond to the herbicides (i.e., physiological status of the plant), and (2) to understand the environmental conditions conducive to herbicide effectiveness. A third possible reason for our erratic success has been our lack of knowledge of the specific mechanism of action of the herbicide on the target species. We often know more about the mechanism of action of herbicides than we know about the physiological status of the plant necessary to respond to herbicides or the environmental conditions necessary to induce a herbicide response.

The paucity of physiological, phenological, and environmental information related to weed control in the literature prior to 1980 is surprising. Even today, it is often difficult to identify from the literature the most susceptible stage of growth or the most appropriate environmental conditions at which to apply herbicides. It has been pointed out that there is a need for more studies of phenological development of weedy species to increase the efficiency of control methods used (1). In addition to knowing the phenological stage, one should know the physiological condition that corresponds to the various stages of phenological development (2). Assuming that sufficient herbicide is absorbed (3) by the target species and that the target species is susceptible to the herbicide used, then effective control depends upon the ability of the plant to respond to the herbicide.

Timing of Application

This paper primarily addresses timing of application of aerial, foliar-applied herbicides. Timing of soil-applied herbicides and individual plant treatment (IPT) is important, but from a different perspective. Generally, timing of soil-applied herbicides is related to sufficient precipitation to allow the herbicide to infiltrate the soil and to be incorporated into the soil solution. If the herbicide target is a dicot and the herbicide is also injurious to grass, time of application should consider the pattern of precipitation during the dormant or quiescent phase of the grasses to prevent grass damage from the herbicide (4). Usually, timing of herbicide application in IPT closely corresponds to aerial, foliar-applications, but it is not as critical as for aerial, foliar-application.

Annuals and Nonsprouting Perennials

Growth habits of targeted plants significantly influence herbicide effectiveness (2). Generally, summer annuals, winter annuals/biennials, and nonsprouting perennials are relatively easy to control if sprayed while they are in the vegetative (or rosette) stage prior to bolting. However, environmental conditions at the time of spraying are important. Often we attribute degree of control of sprayed annuals to available soil water. However, if targeted annuals are vegetative (vs. reproductive), soil water is usually not a limiting factor in herbicide effectiveness. Bitterweed (*Hymenoxys odorata*) was effectively controlled when sprayed in the late winter/early spring by Bunting and Wright (5) when air temperatures were approximately 20 C; however, when sprayed at about the same Julian date with air temperatures approximately 10 C, control of bitterweed was ineffective, even with plentiful soil water (6). Presumably, leaf temperatures in the latter study were too cool for either adequate absorption or translocation of growth-regulating herbicides.

Nonsprouting perennial plants that cannot regenerate following shoot mortality are usually easier to control than sprouters (plants that can regenerate from perennating buds following shoot mortality). Knowledge of the mechanism of herbicidal action and site of herbicide activity are particularly important for effective control of nonsprouting perennials. About one-half of the herbicides used today inhibit some aspect of the photosynthetic process (7,8). Other herbicides interrupt amino acid synthesis (9) or disrupt membranes (10). For effective control with herbicides, nonsprouting perennials should be actively growing and photosynthesizing.

Sprouting Perennials

One must assume a different philosophy toward those plants that resprout from perennating buds following shoot mortality. It is important that the plant be physiologically active to allow incorporation of the herbicide into the photosynthate stream (11-13) and for transport to the perennating organ where control must ultimately be effected. However, in sprouting species it is not necessarily important whether photosynthesis is inhibited, membranes disrupted, or amino acid synthesis interrupted. In fact, if the plant is "actively growing" when sprayed, the herbicide will probably be translocated to the region of new growth (usually the shoot). The new growth will be affected and perhaps killed, but the plant will probably resprout from the unaffected perennating buds. Therefore, herbicides should be applied when the

118

plant can transport them to the site of perennating buds. It is also important that environmental conditions, especially soil water, soil temperature, and air temperature, be optimum for herbicide translocation to the perennating buds, which often are located below ground.

Mesquite Control

Mesquite (*Prosopis glandulosa*) has been a nemesis for rangeland managers in the Southwest for many years (14). Attempts to control it chemically date back at least to the 1920s (15). Yet, results have always been erratic. Recommendations for controlling mesquite by aerial applications have been to spray the plants 50-80 days after the first leaves appear in the spring (16). Additional studies (17-21) led to the conclusion that mesquite is most susceptible to growth-regulating herbicides on the average 42-63 and 72-84 days post bud-break when the total nonstructural carbohydrates (TNC) are being replenished in the roots and basal bud zone (Figure 11.1) (22). The interim period of 63-72 days post bud-break usually corresponds to pod elongation (not maturation, but elongation from 5-20 cm) which apparently becomes the major sink for photosynthates.

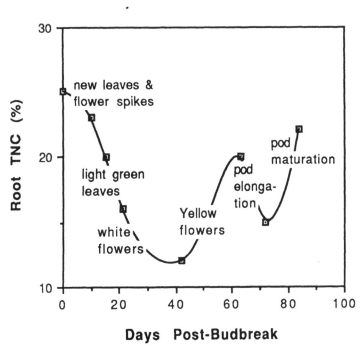

FIGURE 11.1 Average nonstructural carbohydrate (TNC) trends in mesquite.

TNC trends alone do not provide sufficient information on which to base the decision to spray mesquite. Soil temperature is also very important (23). Soil temperature influences vitamin concentrations (which are necessary in carbohydrate synthesis and translocation) (24) and photosynthesis (25). If soil temperature is not above 24 C at 30- to 45-cm soil depth (Figure 11.2), regardless of TNC trend, herbicidal control of mesquite is nil and resprouts are abundant. However, if soil temperature exceeds 24 C during the two periods when TNC concentrations are accumulating in the roots and root crowns (root reserves being replenished), mesquite can be effectively controlled with growth-regulating herbicides.

Broom Snakeweed Control

Broom snakeweed (*Gutierrezia sarothrae*), a suffrutescent shrub, is a major weed in the southwestern United States, and is a local problem throughout the western United States, northern Mexico, and southern Canada (26). Snakeweed becomes established primarily during wet falls and winters when the dominant warm season grasses are dormant. In

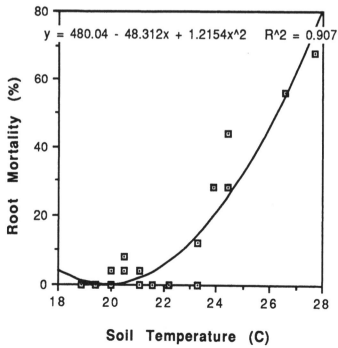

$$y = 480.04 - 48.312x + 1.2154x^2 \qquad R^2 = 0.907$$

FIGURE 11.2 Influence of soil temperature, 30-cm depth, on mesquite mortality (23).

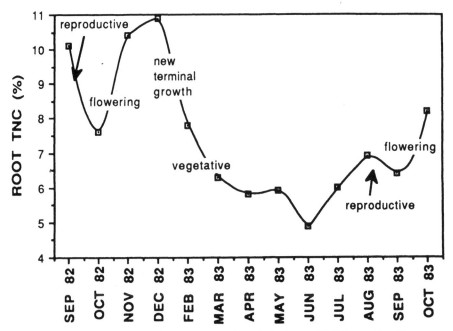

FIGURE 11.3 Average total nonstructural carbohydrate (TNC) trends of broom snakeweed (36).

a few years it dominates many plant communities even to the exclusion of grasses. Traditionally, snakeweed was sprayed in the spring when it was actively growing; the results were erratic and usually unsatisfactory. Description of TNC trends (Figure 11.3) revealed that post flowering (in the fall in the southwestern United States) was the most appropriate time to control snakeweed with aerially-applied, growth-regulating herbicides. During the late winter and spring, TNC trends rapidly decline when the plants produce new stems and leaves; i.e., the photosynthate (hence, herbicide) apparently goes into structural compounds rather than to perennating organs. Broom snakeweed can sometimes be effectively controlled by spraying in the spring or summer, but the risk of failure is greater because the TNC trends are more strongly influenced by local environmental conditions and phenological stage (Figure 11.4).

Silverleaf Nightshade Control

Silverleaf nightshade (*Solanum elaeagnifolium*) is a perennial herbaceous plant that is also difficult to control with herbicides. It has an extensive root system from which new plants arise when the shoots are destroyed. Since it is primarily a weed of cultivated crops, the shoots are often destroyed by plowing or by herbicide, but rarely have we killed the

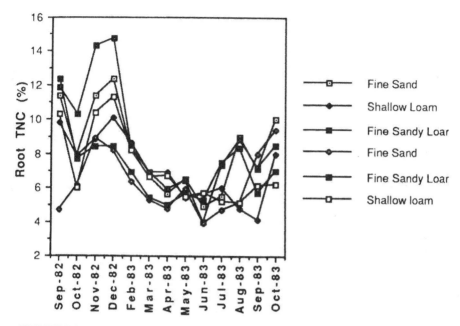

FIGURE 11.4 Influence of range site on total nonstructural carbohydrate (TNC) trends in broom snakeweed (36).

plant. Yet, when Stubblefield (27) correlated TNC trends in silverleaf nightshade to the degree of herbicidal control, he found a very high mortality (>80% root-kill) when the herbicide was applied during the green to yellow berry stage when root TNC concentrations were increasing (Figure 11.5). Only a top-kill was obtained when the plants were sprayed during other parts of the growing season.

Artemisia Control

Artemisia caudata is a short-lived perennial that invades sandy rangeland in the Southwest following sand shinnery oak (*Quercus havardii*) control. It can become so dense that it essentially eliminates all grass production. Like silverleaf nightshade, it is a herbaceous perennial that dies back to the soil surface during the late fall and early winter. However, it produces rosette leaves that remain green until new shoot growth is initiated in the spring (28). It is easily controlled with herbicides if sprayed during the rosette stage before bolting occurs (C. S. Brumley and R. E. Sosebee, unpublished data).

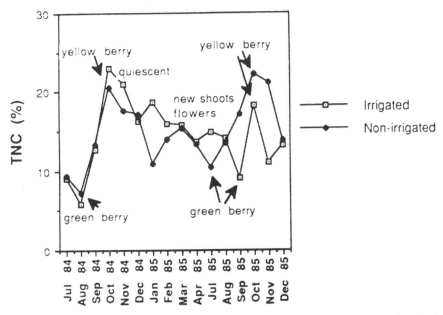

FIGURE 11.5 Average total nonstructural carbohydrate (TNC) trends in silverleaf nightshade growing under irrigation and no irrigation (27).

Pricklypear Control

Potter et al. (29) have obtained satisfactory control of pricklypear cactus (*Opuntia* sp.) by observing TNC trends as they were influenced by phenological development. The greatest control from aerially-applied herbicides was achieved when the plants were sprayed after the flowering process was complete and fruits were produced. This phenological stage corresponded to distribution of TNC throughout the plant (and consequently, herbicide distribution throughout the plant).

Summary--Control of Perennial Plants

It is important that perennial plants with the ability to regenerate via basal sprouts or from stolons and rhizomes be sprayed when the plant's physiological conditions allow translocation of the herbicide to the perennating buds. Pitelka (30) compared differences in energy allocation of annual and perennial lupines (*Lupinus* sp.). He found that most of the energy produced by *L. nanus* (annual) was allocated to reproductive tissues and seeds (Table 11.1). He also found that *L. variicolor* (perennial

TABLE 11.1 Energy allocation of three *Lupinus* species (annual, herb, and shrub). Adapted from Pitelka (30).

| Plant Part | Energy Allocation (%) | | |
	L. nanus (annual)	*L. variicolor* (herb)	*L. arboreus* (shrub)
Reproductive tissues	61	18	20
Seeds	29	5	6
Stems	--	--	50
Roots	3-4	40	--

herb) allocated a major portion of the photosynthate to the roots, the "reserve" organ where the perennating buds occur. Likewise, the majority of the photosynthates in *L. arboreus* (shrub) was allocated to the stems where the perennating organs occur.

Therefore, if one waits until an annual becomes reproductive (bolts and initiates the flowering process), aerial spraying will adversely affect seed production, but will have minimal effect on the entire plant. Likewise, if sprouting perennials are sprayed with a herbicide that is not translocated to the perennating organs (e.g., one that only inhibits photosynthesis) or at a time when either the plant is not responsive physiologically or the environmental conditions are not conducive for a plant to respond, then application will result only in a top-kill and not a total plant-kill.

In herbaceous perennials, new shoots (sprouts) are derived from inactive buds at nodes on stolons and rhizomes or from adventitious buds in root tissue. Sprouting in woody plants arises from dormant or inactive buds in basal crowns (transition zone between the root and shoot) or from adventitious buds in root tissue.

Sprouting presents another problem relative to control of woody plants. In woody plant control with growth-regulating herbicides, it is important to ascertain the shoot order of the plant's actively growing shoots. Most authors that describe variations in woody plant shoots recognize differences between long and short shoots. Long shoots have leaves separated by distinct internodes and bear lateral buds, some of which develop into more long shoots and others into short shoots. Internode elongation is much reduced in short shoots; it is essentially lacking in some species, such as mesquite, and their leaves cover the shoot tip. Short shoot lateral buds are either lacking or they develop into flowers (31,32). Wilson (31) classified branches in a way that shows why long shoots are more prevalent in young trees than in older trees. The main stem (a long shoot) is a first order shoot, laterals to the main stem are second order, etc. As the shoot order increases, vigor generally decreases and higher order shoots produce a smaller proportion of long shoots. By the fourth or fifth order, the shoots bear no lateral long shoots--only short shoots. Short shoots bear no laterals so branches higher than fifth or sixth order usually do not exist in the tree crown.

The reason proportions of long vs. short shoots are important in herbicidal plant control is because of major differences in location of carbohydrate sinks (sites of photosynthate use or storage). Kozlowski (32) reported that shoot elongation of mature, bearing apple (*Malus pumila*) or pear (*Pyrus communis*) trees with predominantly short shoots usually requires 6-8 weeks. However, shoots of young trees, with higher proportions of long shoots, continue elongation for 10-12 weeks. Bedell and Heady (33) studied mature plants, hedged plants, and root crown sprouts of chamise (*Adenostoma fasciculatum*) in the California chaparral. Twig elongation ceased by the first day of summer on mature plants, by midsummer on hedged plants, and not until late autumn with root crown sprouts. Zimmerman and Brown (34) showed that in immature and actively elongating shoots carbohydrates are translocated primarily upward from the older leaves to the shoot tip and younger leaves. Only when shoot elongation was complete was photosynthate translocation primarily downward. Thus, root-kill from translocated herbicides is rarely possible on trees with a high proportion of long shoots where active shoot extension is occurring (18).

Root crown and stem shoots resulting from browsing, burning, cutting, chaining, herbicides, and stand thinning are invariably long shoots that exhibit continuous growth, i.e., the apical meristem of the twig grows constantly, forming new leaves throughout the growing season (35). Foliar herbicide application to plants with a high proportion of such shoots results in little, if any, root mortality.

Conclusions

Weed control is very often less satisfactory than we as land and resource managers prefer. Reasons for our variable degrees of success, and indeed failure to synchronize the timing of herbicide application to the physiological stage of the plant, are related to the plant's ability to respond to the herbicides and to the environmental conditions that are conducive to the plant's response. It is also important that we understand the mechanism of herbicide action and site of herbicide activity to plan for appropriate timing of application.

Summer annuals, winter annuals/biennials, and nonsprouting perennials are relatively easy to control, but they should be sprayed while they are vegetative and "actively growing". Once they become reproductive, they are more difficult to control because the herbicides are translocated to the reproductive tissues. Consequently, reproduction may be affected, but the plant is not killed, or control is very difficult.

Sprouting perennials (herbaceous or woody) are more difficult to control because not only must the shoot be controlled but also the perennating buds from which new shoots arise following destruction of apical dominance. Therefore, it is important that herbicide application be timed according to when translocation to the perennating buds and storage organs will occur. A convenient and very reliable predictor is nonstructural carbohydrate (TNC) trends in the plants. Herbicides applied when TNC concentrations increase in perennating organs and storage tissue correspond to a high degree of control. Herbicides applied when TNC concentrations in perennating organs and storage tissue are decreasing usually result in unsatisfactory control.

Environmental conditions must also be conducive for control when TNC are accumulating in perennating organs (trends are in the appropriate direction). Soil water must be sufficient and soil temperature must be appropriate for ongoing physiological processes. However, higher soil water during the growing season often means colder soils and colder soils adversely affect photosynthesis and/or translocation which adversely affects herbicide translocation to desired sites of physiological activity in plants.

Timing of herbicide application coincides with the appropriate phenological stage, physiological condition, and environmental conditions.

References

1. McCarty MK. Weed Sci. 34:218 (1986).
2. Sosebee RE. In: *Brush Management Symp., Proc.* (KC McDaniel, Ed.), pp. 27-43. Soc. Range Manage., Albuquerque, NM, 1983.
3. Bovey RW and RE Meyer. In: *Brush Management Symp., Proc.* (KC McDaniel, Ed.), pp. 45-52. Soc. Range Manage., Albuquerque, NM, 1983.
4. Sosebee RE, WE Boyd, and CS Brumley. J. Range Manage. 32:179 (1979).
5. Bunting SC and HA Wright. J. Range Manage. 27:381 (1974).
6. Ueckert DN, CJ Scifres, SG Whisenant, and JL Mutz. J. Range Manage. 33:465 (1980).
7. Bartels PG. In: *Weed Physiology: Herbicide Physiology, Vol. 2* (SO Duke, Ed.), pp. 63-90. CRC Press, Inc., Boca Raton, FL, 1985.

126

8.   Black CC Jr. In: *Weed Physiology: Herbicide Physiology, Vol. 2* (SO Duke, Ed.), pp. 1-36. CRC Press, Inc., Boca Raton, FL, 1985.
9.   Duke SO. In: *Weed Physiology: Herbicide Physiology, Vol. 2"* (SO Duke, Ed.), pp. 91-112. CRC Press, Inc., Boca Raton, FL, 1985.
10. Balke NE. In: *Weed Physiology: Herbicide Physiology, Vol.2* (SO Duke, Ed.), pp. 113-139. CRC Press, Inc., Boca Raton, FL, 1985.
11. Crafts AS and CE Crisp. *Phloem Transport in Plants.* WH Freeman & Co., San Francisco, CA, 1971.
12. Peel AJ. In: *Transport in Plants I; Phloem Transport* (MH Zimmerman and JA Milburn, Eds.), Encyclopedia of Plant Physiology, New Series, Vol. 1, pp. 171-195. Springer-Verlag, New York, NY, 1975.
13. Ziegler H. In: *Transport in Plants I; Phloem Transport* (MH Zimmerman and JA Milburn, Eds.), Encyclopedia of Plant Physiology, New Series, Vol. 1, pp. 59-100. Springer-Verlag, New York, NY, 1975.
14. Fisher CE. In *Mesquite* (BB Simpson, Ed.), US/IBP Synthesis Series 4, pp. 177-188. Dowden, Hutchinson and Ross, Inc., Stroudsburg, PA, 1977.
15. Scifres CJ, RW Bovey, CE Fisher, and JR Baur. In *Mesquite: Growth and Development, Management, Economics, Control, Uses,* Research Monogr. No. 1, pp. 24-32. Texas Agr. Exp. Sta., College Station, 1973.
16. Fisher CE, CH Meadors, R Behrens, ED Robinson, PT Marion, and HL Morton. Texas Agr. Exp. Sta. Bull. No. 935, College Station, 1959.
17. Wilson RT, DR Krieg, and BE Dahl. J. Range Manage. 27:202 (1974).
18. Wilson RT, BE Dahl, and DR Krieg. J. Range Manage. 28:286 (1975).
19. Fick WH and RE Sosebee. J. Range Manage. 34:205 (1981).
20. Goen JP and BE Dahl. J. Range Manage. 35:533 (1982).
21. Seipp WW. MS Thesis, Texas Tech Univ., Lubbock, 1982.
22. Sosebee RE and BE Dahl. Management Note 2, Dept. Range and Wildlife Management, Texas Tech Univ., Lubbock, 1983.
23. Dahl BE, RB Wadley, MR George, and JL Talbot. J. Range Manage. 24:210 (1971).
24. Brumley CS and RE Sosebee. Plant and Soil 50:13 (1978).
25. Wan C and RE Sosebee. J. Range Manage. 43:171 (1990).
26. Lane MA. Syst. Bot. 10:7 (1985).
27. Stubblefield RE. MS Thesis, Texas Tech Univ., Lubbock, 1986.
28. Mounsif M. MS Thesis, Texas Tech Univ., Lubbock, 1986.
29. Potter RL, JL Peterson, and DN Ueckert. Weed Sci. 34:361 (1986).
30. Pitelka LF. Ecology 58:1055 (1977).
31. Wilson BF. *The Growing Tree.* The Univ. of Mass. Press, Amherst, 1970.
32. Kozlowski TT. *Growth and Development in Trees, Vol. I.* Academic Press, New York, 1971.
33. Bedell TE and HF Heady. J. Range Manage. 12:116 (1959).
34. Zimmerman MH and CL Brown. *Trees-Structure and Function.* Springer-Verlag, New York, 1971.
35. Dahl BE and DN Hyder. In: *Rangeland Plant Physiology* (RE Sosebee, Ed.), pp. 257-290. Soc. for Range Manage., Range Sci. Series No. 4, 1977.
36. Courtney RW. MS Thesis, Texas Tech University, Lubbock, 1984.

12

Allelopathy

Kenneth L. Stevens

Abstract

Five allelopathic plants found in range environments are discussed. Knapweeds produce both sesquiterpene lactones and polyacetylenes, some of which have been shown to be phytotoxic. Broom snakeweed has been reported to be phytotoxic; however, its agressive behaviour may be attributed to competition for water rather than liberation of allelochemicals into the environment. The creosote bush produces copious quantities of nordihydroguaiaretic acid (NDGA) which has been demonstrated to have an adverse effect on many plants. Likewise, Canada thistle has been reported to be allelopathic; however, the chemicals involved have not been isolated or identified. The interaction of small everlasting with leafy spurge has been extensively investigated. Small everlasting produces hydroquinone which is toxic to leafy spurge. Hydroquinone's differential toxicity toward leafy spurge and small everlasting has been ascribed to the "ability" of small everlasting to transform hydroquinone to arbutin more efficiently than leafy spurge. The future of allelopathy for managing range plants is discussed.

Introduction

Allelopathy may be defined as the direct or indirect effect of one plant species on another through the production of chemical compounds that escape into the environment. In practice it is often difficult to distinguish allelopathy from competition which produces the same symptoms in a plant community, i.e., one plant adversely affecting another plant. However, competition for light, water, and nutrients can be eliminated in controlled experiments and if any residual effect is observed between plants, it can logically be ascribed to allelopathy, especially if the offending compounds(s) are identified and reapplied in other controlled experiments to duplicate the field observations. The concept of allelopathy is not new. In fact, Theophrastus (1) in his "Enquiry into Plants and Minor Works on Odors and Weather Signs", written before 285 B.C., observed that chick-pea does not reinvigorate the ground as other legumes but "exhausts" the soil. He goes on to say "it destroys weeds and above all and soonest caltrop" (*Tribulus terrestris*). Three centuries later, Pliny (Plinius Secundus, 1 A.D.) in book 17 of his "Natural History" reported that chick-pea, barley, fenugreek, and bitter vetch all "scorch up" cornland (2). He also described the deleterious effect of walnut trees on plants within their vicinity (book 17). These and other references down through the ages attest to the observational skill of the ancients and are known today to be the result of allelopathy. In 1832 DeCandolle (3) recognized soil sickness as a problem which might be associated with chemical exudates from crops. He also was the first to suggest that crop rotation might alleviate the problem even though crop rotation and allowing land to "fallow" had been practiced for centuries (4).

The word allelopathy was first used by Molisch (5) in 1937 to describe the chemical interactions among all plants, including microbes and higher plants and included both mutual benefit as well as harmful effects. This has led to some confusion since the word literally means "mutual suffering." For this reason many authors have used the term allelopathy to refer to those chemical interactions which are detrimental to the recipient plant. This is probably an unrealistic definition since many plant growth regulators are injurious at high concentrations but become stimulants at hormonal levels (*vide infra*, Figure 12.1).

The study of allelopathy has become a major discipline in recent years as evidenced by the number of publications and a plethora of books on the subject. For the most part, plant-plant interactions that are being studied are related to those involving crops and associated weeds and to a smaller extent the role of allelopathy in forestry. An example of allelopathy with a practical application centers around walnut trees. Beginning with the report by Pliny the Elder on the poisonous effect of the *Juglans* genus, many workers have confirmed his earlier account. In 1925, Massy (6) suggested the pigment, juglone (1), to be the phytotoxic agent responsible for the observed effects of the walnut trees. This was later confirmed by a number of workers in a variety of experiments. The utilization of this information has taken a rather interesting turn. In attempts to reforest plantations and wooded areas denuded of hardwoods

Juglone (1)

with the rapidly growing and economically favorable black walnut (*Juglans nigra*), it was noticed that growth was rather slow and indeed appeared to be much more rapid in the presence of leguminous plants or other plants known to have a symbiotic relationship with nitrogen-fixing bacteria (7-9). Consequently, plantations and other areas needing reforestation with black walnut were co-planted with the European black alder (*Alnus glutinosa*) which acted as a nurse species by contributing nitrogen to the soil via the symbiotic nitrogen-fixing soil microorganisms belonging to the genus *Frankia*. After the walnut trees became well established and were quite vigorous, the juglone released into the environment by the walnut trees began to take effect. The nurse trees began to decline and ultimately died between 8 and 13 years after planting. The overall effect was to allow the black walnut to grow to its full potential since adequate space was now available.

The number of confirmed allelopathic relationships found within the context of rangeland is rather small and becomes even smaller when the constraint is noxious weeds. For the purpose of this discussion we will look at five plants which are to some degree associated with rangeland and have been shown to have an allelopathic effect on the surrounding plant community.

Knapweeds

The knapweeds, [spotted knapweed (*Centaurea maculosa*), Russian knapweed (*Acroptilon repens*), and diffuse knapweed (*Centaurea diffusa*)] are introduced weeds which have expanded to occupy large areas of range- and cropland and are considered one of the more serious weed pests in the United States. Several reports have been published attesting to the allelopathic characteristics of these weeds (10-13). Fletcher and Renney (10) reported that Russian knapweed was more inhibitory than either spotted or diffuse knapweed. Our own work (11-13) has centered on an investigation of Russian knapweed. The aerial parts of the plant, including the flowers have for the most part an array of sesquiterpene lactones, while the roots have a high concentration of polyacetylenes. At the present time, we have isolated and identified fourteen sesquiterpene lactones from Russian knapweed.

All of the compounds have guaianolide type structures, i.e., they have a fused 5- and 7-membered ring system with the lactone ring attached to the 6 and 8 positions. Phytotoxicity tests were run on several of the compounds and show that none of the sesquiterpene lactones inhibit germination nor do they retard the growth of the hypocotyl in lettuce seedlings after 48 hours. However, the growth of the root of the lettuce seedlings is markedly altered.

Figure 12.1 shows the mean root length in mm of lettuce seedlings at various concentrations of centaurepensin (2). Approximately a 50% reduction in mean root length is observed at 80 ppm of centaurepensin. Other isolated sesquiterpene lactones show similar results. At the present time, we have no data to support or suggest that these compounds are exuded into the environment to have an allelopathic effect on the surrounding plant community. As a side issue, Russian knapweed has been implicated in the equine neurological disease called "chewing disease" (equine nigropallidal encephalomalacia, ENE) which causes necrosis of brain neural cells in the substantia nigra. Upon ingestion of Russian knapweed, horses develop Parkinsonian-like symptoms. The interesting fact is that Parkinson's disease affects the same brain area as ENE. Recently, Riopelle et al. (14) tested several of the sesquiterpene lactones and found potent activity against chick embryo neural cells, especially repin (3), which had a TD_{50} of 80nM. This high degree of activity against neural cells makes it a prime candidate for the causative agent in ENE disease. A much more potent and active series of compounds are found in the roots of Russian knapweed, viz. the polyacetylenes (12). These very reactive and unstable compounds have

Centaurepensin (2)   Repin (3)

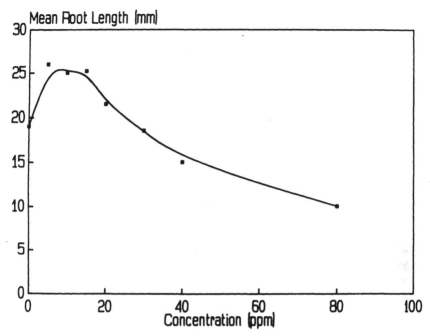

FIGURE 12.1 Effect of centaurepensin on root length of lettuce.

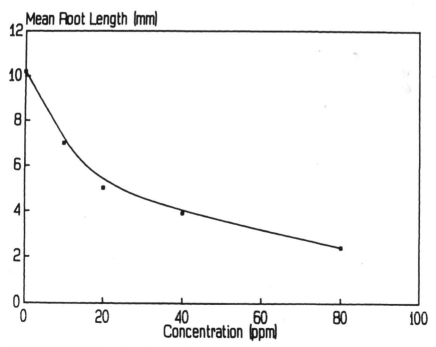

FIGURE 12.2 Effect of polyacetylene $\underline{4}$ on root growth of lettuce.

$$CH_3-(C{\equiv}C)_2-\underset{S}{\text{[thiophene]}}-C{\equiv}C-\underset{\underset{OH}{|}}{CH}-CH_2Cl$$

4

been tested for phytotoxicity, one of which shows activity in our bioassay system, e.g., the chlorohydrin 4 shows substantial activity in the inhibition of root length elongation for a number of different species. A typical curve is shown in Figure 12.2 which plots the root length (mm) of lettuce seedlings against concentration of the polyacetylene. A 50% reduction in root length is observed at 12 ppm of compound 4. The escape of the substance into the environment was also examined by extracting the soil around the roots of live Russian knapweed plants. A concentration ranging between 4 and 5 ppm of 4 was found. The concentration did not vary significantly over the growing season.

Since the assay method involved the isolation of 4, the 4-5 ppm represents a minimum amount due to the instability of the compound. This amount of material in the soil around the roots of the Russian knapweed would be sufficient to have an appreciable effect on the surrounding plant community.

Muir et al. (15) examined the allelopathic potential of spotted knapweed and concluded that in both field trials and pot cultures it failed to show any phytotoxic effects. Their conclusion was that allelopathy is not a factor in the spread of spotted knapweed. To our knowledge no one has thoroughly investigated the allelopathic potential of diffuse knapweed.

Broom Snakeweed

Broom snakeweed (*Gutierrezia sarothrae*) is an aggressive weed common on Western rangelands from Canada to Mexico. In 1959 Platt (16) estimated perennial snakeweeds infested more than 57 million hectares of rangeland in the United States. Dense stands of snakeweed can reduce grassland forage production by 70% or more (17,18), causing an economic loss of major proportions. In addition, broom snakeweed is toxic to cattle, sheep, goats, and several other species of animals (19). The most common problem associated with ingestion of broom snakeweed is abortion in cattle which occurs to the extent of 60% when the cattle feed on the weed growing on sandy soil.

In 1981 McDaniel (20) suggested allelopathy as a possible explanation for the small number of grass and snakeweed seedlings under established snakeweed stands. However, it is well known (21) that juvenile broom snakeweed plants compete vigorously for water and deplete water stores from the upper 15-45 cm. This would certainly have a profound effect on any grass attempting to compete. The possibility of allelopathy having a secondary effect has not been completely ruled out.

Creosote Bush

Although the creosote bush (*Larrea tridentada*) is not classified as a noxious weed, it can be a dominant shrub of the North American deserts occurring in almost pure stands of up to 6000 plants/hectare (22). This characteristic and a report by Gergel in 1980 (23) that if the creosote bush was sprinkled on a train bed no weeds will grow, led to an investigation (24) of the allelopathic potential of the creosote bush.

Nordihydroguaiaretic acid (NDGA, 5) a tetrahydroxy lignan, makes up 5-10% of the dry weight of the leaves of the creosote bush (25) and is known to possess numerous biological activities (26). The phytotoxic characteristic of NDGA was investigated and found to have substantial activity against several weed and crop plants. For example, Figure 12.3 shows the effect of NDGA on the hypocotyl growth of lettuce seedlings. A 50% reduction in root length was observed at 10 ppm of NDGA. Since NDGA occurs in high concentrations in the creosote bush and has a relatively high activity, it is probably the major contributor to the allelopathic properties of this dominant bush.

Nordihydroguaiaretic Acid (5)

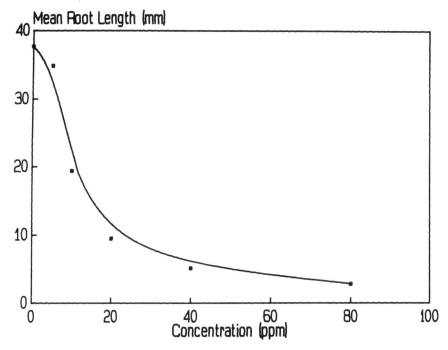

Figure 12.3 Effect of NDGA on root growth of lettuce.

Thistles

The number of reports on the allelopathic nature of thistles are rather limited (27-30), all dealing with the Canada thistle (*Cirsium arvense*). Stachon and Zimdahl (28) performed several experiments in 1980 in an attempt to confirm the allelopathic nature of Canada thistle and to verify the bioassay procedures. They found in field trials that annual plants in general did not grow with Canada thistle while perennial grasses and rushes tended to grow with the thistle. Canada thistle litter, ground roots, and ground foliage added to soil under greenhouse conditions inhibited the growth of pigweed and green foxtail. The addition of nutrients, water and adjustment of osmotic potential did not reverse the effect. From these experiments it was concluded that Canada thistle does have an allelopathic influence on the surrounding plant community; however, neither the chemical agents nor the mechanism involved were elucidated.

OH

OH

Hydroquinone (6)

OH

OGlucose

Arbutin (7)

HO

HO—⬡—CH=CH—COOH

Caffeic acid (8)

Leafy Spurge

The allelopathic interaction between leafy spurge (*Euphorbia esula*) and small everlasting (*Antennaria microphylla*) was first reported in 1972, by Selleck (31), who suggested that the diminutive plant produces and exudes antibiotics which are detrimental to the growth of leafy spurge. In 1985, Manners and Galitz (32) isolated and identified three phytotoxic constituents from small everlasting, viz., hydroquinone (6), arbutin (7), and caffeic acid (8), and demonstrated their toxicity towards leafy spurge. Arbutin, the monoglucoside of hydroquinone, was somewhat less active than hydroquinone itself; however, it is easily hydrolyzed to hydroquinone in the moist soil surrounding the plants.

In order to examine the allelopathic potential of small everlasting in controlling leafy spurge, an investigation was initiated to elaborate the biochemical mechanism(s) of the allelopathic interaction. To better accomplish this and to fully understand the biochemical transformations taking place, both leafy spurge and small everlasting were grown in callus and suspension cultures. The allelopathic effect was also observed in suspension and callus cultures, i.e., suspension cells of small everlasting produced hydroquinone which was toxic to leafy spurge cells. Differences in the toxicity of hydroquinone towards small everlasting and leafy spurge has been shown (33) to be the result of small everlasting's ability to biotransform hydroquinone to the less toxic arbutin more efficiently than leafy spurge. Hogan and Manners (34) have recently isolated and identified the specific enzyme from leafy spurge and small everlasting which glucosylates hydroquinone to arbutin. The detoxification enzyme, a UDPG-dependent glucosyltransferase, was induced in leafy spurge suspension cells in the presence of hydroquinone, however, the activity was lower than that measured in small everlasting. This differential ability to detoxify hydroquinone provides the basis for the observed allelopathic interaction between these two weeds.

The Future

The potential of using allelopathic interactions to control noxious range weeds is on less solid ground than using the same principles in cultivated crops. The real potential lies in the investigation of allelopathic mechanisms and their biochemical reactions. This kind of information will thus lead to strategies to selectively inhibit the growth of unwanted forbs while maintaining the integrity of desirable plants and grasses. For instance, in the case of leafy spurge, the knowledge that it is less efficient in detoxifying hydroquinone than small everlasting may lead to the application of hydroquinone (or some derivative) which will selectively hinder the growth of leafy spurge. The application of hydroquinone need not be direct but may be indirect, such as finding or developing a plant which exudes the toxic substance. It may be possible to transfer the detoxifying genes from the small everlasting to plants which are more vigorous and suitable for propagation for range forage.

References

1. Theophrastus. *Enquiry into Plants and Minor Works on Odors and Weather Signs*, Vol. 2, book 8, pp. 183-185. W. Heinemann, London, 1916.
2. Plinius Secundus C. *Natural History*, (trans. H. Rackman, WHS Jones, and DE Eichhloz), p. 459. Harvard University Press, Cambridge, Mass., 1938-63.
3. DeCandolle MA-P. *Physiologie Vegetale*, Vol. III. Bechet Jeune, Lib. Fac. Med., Paris, 1832.
4. Moses. *Bible*, Exodus 23:11.
5. Molisch H. *Der Einfluss einer Pflanze auf die andere-Allelopathie*, Fisher, Jena, 1937.
6. Massy AB. Phytopathology 15:177 (1952).
7. Funk DT. In: *Proc. First North Central Tree Improv. Conf.*, pp. 28-32. Madison, Wis, 1979.

8. Funk DT, RC Schlesinger, and F Ponder Jr. Bot. Gaz. 140 (suppl.):S110-S114 (1979).

9. Rietveld WJ, RC Schlesinger, and KJ Kessler. J. Chem. Ecol. 9:1119 (1983).

10. Fletcher RA and AJ Renney. Canad. J. Plant Sci. 43:475 (1963).

11. Stevens KL and GB Merrill. In: *Symposium on the Chemistry of Allelopathy, Biochemical Interactions among Plants*, ACS symposium series no. 268, pp. 83-98. American Chemical Society, Washington D.C., 1985.

12. Stevens KL. J. Chem. Ecol. 12:1205 (1986).

13. Stevens KL. In: *The Science of Allelopathy*, (AR Putnam and C-S Tang, Eds.), pp. 219-228. John Wiley & Sons, New York, NY, 1986.

14. Riopelle RJ, RJ Boegman, PB Little, and KL Stevens. Neurosciences Letters, in press.

15. Muir AD, W Majak, F Balza, and GHN Towers. In: *Allelochemicals: Role in Agriculture and Forestry*, pp. 238-246. American Chemical Society, Washington, DC, 1987.

16. Platt KB. J. Range Manage. 12:64 (1959).

17. Ueckert DN. J. Range Manage. 32:216 (1979).

18. McDaniel KC, RD Pieper, and GB Donart. J. Range Manage. 35:219 (1982).

19. Sperry OE, GO Dollahite, GO Hoffman, and BJ Camp. *Texas Plants Poisonous to Livestock*. Texas Agr. Exp. Sta. Bull. 1028 (1964).

20. McDaniel KC. In: *Proc.-Poison Plant Manage in the Trans-Pecos*, pp. 10-121. Texas A&M Univ., Ft. Stockton, Texas, 1981.

21. Ueckert DN. J. Range Manage. 32:216 (1979).

22. Rhodes DF. Biochem. Syst. Ecol. 5:281 (1977).

23. Gergel M. Chemtech. 12:140 (1980).

24. Elkovich SD and KL Stevens. J. Chem. Ecol. 11:27 (1985).

25. Mabry TJ, DR Di Feo Jr., M Sakakibara, CF Bohnstedt Jr, and D Siegler. In: *Creosote Bush: Biology and Chemistry of Larrea in New World Deserts* (TJ Mabry, JH Hunziker, and DR Di Feo, Eds.), pp.115-133. Dowden, Hutchinson and Ross, Stroudsburg, Penn., 1977.

26. Oliveto EP. Chem. Ind. 17:677 (1972).

27. Bendall GM. Weed Research 15:77 (1975).

28. Stachon WJ and RL Zimdahl. Weed Science 28:83 (1980).

29. Amor RL and RV Harris. Weed Research 15:407 (1975).

30. Helgeson EA and R Konzak. North Dakota Agric. Exp. Sta. Bull. 12:71 (1950).

31. Selleck GW. Weed Sci. 20:189 (1972).

32. Manners GD and DS Galitz. Weed Sci. 34:8 (1985).

33. Hogan ME and GD Manners. J. Chem. Ecol. 16:931 (1990).

34. Hogan ME and GD Manners. J. Chem. Ecol., submitted.

13

Improved Grasses and Forbs for Intermountain Rangelands

K. H. Asay, M. D. Rumbaugh, and T. A. Jones

Abstract

Revegetation with improved cultivars of grasses and forbs represents an economical and long-lasting means of improving depleted rangelands. Plant germplasm generated by plant breeding programs has had a major impact on rangeland improvement and excellent opportunities exist for continued progress. Interspecific hybridization is a valuable breeding tool for combining the genetic resources of related grass species. The crested wheatgrass cultivar Hycrest, which is a hybrid between induced tetraploid *Agropyron cristatum* and natural tetraploid *A. desertorum*, along with 'Bozoisky-Select' Russian wildrye (*Psathyrostachys juncea*), are notable examples of recent genetic progress in range grasses. The new cultivar NewHy, an advanced-generation hybrid between quackgrass (*Elytrigia repens*) and bluebunch wheatgrass (*Pseudoroegneria spicata*), has demonstrated exceptional potential as a quality forage grass for saline sites. Wide hybridization is being used to improve important native grasses, including bluebunch wheatgrass, thickspike wheatgrass (*Elymus lanceolatus ssp. lanceolatus*), and Snake River wheatgrass (another subspecies of *E. lanceolatus*). Superior populations of dryland alfalfas (*Medicago sativa*) and kura clover (*Trifolium ambiguum*) have been widely distributed. Elite germplasms of other leguminous and nonleguminous rangeland forbs are being developed and evaluated.

Introduction

Rangelands of the Intermountain region are an important source of red meat, wool, water resources, wildlife habitat, and recreation. Unfortunately, even though many areas of rangeland have improved, mismanagement has left much of this valuable resource short of the desired condition. In general, rangelands can be upgraded by better management or replacement of existing vegetation with improved grasses, forbs, and shrubs. Revegetation is a relatively economical and long-lasting means of improving depleted rangelands. Compared to crop species, little breeding work has been done to improve forage cultivars for rangelands, and most of the early releases were developed for use in the Northern Great Plains. We tend to underestimate the value of improved plant germplasm. As we expand our understanding of genetic variability within and among species, the contributions of plant breeding range improvement and management will increase.

Breeding New Grasses for Intermountain Range

Research is in progress by the Agricultural Research Service of the U.S. Department of Agriculture (USDA-ARS) in cooperation with the Utah Agricultural Experiment Station to develop improved cultivars of important native and introduced grasses. Species of major interest include: crested wheatgrass (*Agropyron cristatum, A. desertorum,* and *A. fragile*); Russian wildrye (*Psathyrostachys juncea*); Altai wildrye (*Leymus angustus* and *L. karelinii*); bluebunch wheatgrass (*Pseudoroegneria spicata*); thickspike wheatgrass (*Elymus lanceolatus* ssp. *lanceolatus*); Snake River wheatgrass (another ssp. of *Elymus lanceolatus*); Indian ricegrass (*Oryzopsis hymenoides*); and interspecific hybrids involving wheatgrasses, wildryes, and related species.

Crested Wheatgrass

Since its successful introduction from Eurasia in 1906 (1), crested wheatgrass has had more impact on revegetation of western rangelands than any other grass. As a cool-season grass, it is particularly valued for its productivity and nutritional value during early spring. The quality of its forage declines rapidly during the late summer. Primarily a bunchgrass, it is resistant to drought and cold and is best adapted to areas that receive 20 to 40 cm annual precipitation at altitudes below 2,400 m. This long-lived perennial has moderate tolerance to excess soil salinity, produces excellent seed yields, and is relatively easy to establish on semiarid range sites (2).

Crested wheatgrass is actually a complex of diploid (2n=14), tetraploid (2n=28), and hexaploid (2n=42) species. The most widely used diploid cultivar is Fairway, which was developed by Agriculture Canada at Saskatoon, Saskatchewan, and released in 1932. It is leafier and more decumbent, but less drought tolerant than the major tetraploid cultivars. Agriculture Canada subsequently released another diploid cultivar, Parkway, which was selected from Fairway. A relatively new cultivar, Ruff, was developed from Fairway-type germplasm by the USDA-ARS in cooperation with the Nebraska Agricultural Experiment Station. It has a spreading (broad-bunch) growth habit and is recommended for grazing and revegetation of problem sites in the low precipitation zones of the Great Plains (3,4).

The most commonly used tetraploid cultivars have been the Standard types, Nordan and Summit, along with the Siberian cultivar, P-27, all released in 1953. Nordan was developed by the USDA-ARS Northern Great Plains Research Center at Mandan, North Dakota, in cooperation with the North Dakota Agricultural Experiment Station. It was derived from collections made in the U.S.S.R. and is noted for its upright growth habit, relatively large seeds, and good seedling vigor (5). The Siberian cultivar, P-27, was released by the USDA-Soil Conservation Service (USDA-SCS). It has narrow, awnless spikes and relatively fine leaves and stems. P-27 is reported to be well adapted on light droughty soils (3) and it matures later and remains green longer than typical Standard types. The tetraploid cultivar, Ephraim, was released in 1983 by the USDA-Forest Service, USDA-SCS, Utah Division of Wildlife Resources, and several state agricultural experiment stations. It is reported to be persistent and drought resistant; however, it is noted primarily for its sod-forming characteristics. The cultivar is slightly shorter than Fairway, but they are similar in biomass production (6).

Dewey (7) concluded that the same basic genome, modified by structural rearrangements, occurred at the three ploidy levels in crested wheatgrass. He advised breeders to treat the species in the complex as a single gene pool. Although interspecific crosses are often difficult to make and sterility problems are encountered in hybrid progenies, all possible crosses have been made among the three ploidy levels and several schemes have been devised and tested to effect interploidy genetic transfer (8).

The greatest progress from interploidy breeding in the crested wheatgrass complex has been achieved at the tetraploid level. Tai and Dewey (9) obtained fertile tetraploid plants by treating diploid Fairway clones with colchicine. Dewey and Pendse (10) later crossed these induced tetraploids with natural (Standard) tetraploids. The hybrids, which represented a genetic combination of Fairway and Standard, were reasonably fertile. Moreover, their data indicated that higher fertility levels could be achieved through selection and many of the hybrid plants were substantially more vigorous than the parental lines.

This germplasm was included in the USDA-ARS breeding program at Logan in 1974. A breeding nursery consisting of 20,000 spaced plants was evaluated for several characteristics, including vegetative vigor, disease and insect resistance, seed yield potential, leafiness, and drought resistance. Selections from this nursery were subjected to more intense progeny testing in the laboratory and field. Eighteen clones were ultimately selected and intermated to produce an experimental strain. The new hybrid strain was evaluated on several range sites in the Intermountain West and was subsequently released as the cultivar, Hycrest. It is the first interspecific hybrid of crested wheatgrass to be released (11).

Hycrest has been consistently more vigorous and productive than Nordan and Fairway on Intermountain range sites, particularly during stand establishment. This is exemplified in trials conducted at Lakeside, UT, near the Bonneville salt flats. This location receives an average of 15.5 cm annual precipitation and the soils are moderately saline. In addition, the experimental area was infested with halogeton (*Halogeton glomeratus*) and cheatgrass (*Bromus tectorum*). Seedings were made in early April without herbicides. Excellent stands of Hycrest were obtained, and after two years the halogeton and cheatgrass were virtually eliminated. Forage yields averaged over two trials at Lakeside during the first year after stand establishment were 2610, 1550, and 960 kg per ha for Hycrest, Nordan, and Fairway respectively. During the subsequent season, the three cultivars produced 2050, 1500, and 1410 kg of forage per ha, respectively. Similar results were obtained in evaluation trials in northwest Utah where Hycrest produced as much forage as Nordan and Fairway combined during the establishment year. On sites where soil moisture was less limiting than at Lakeside, the advantages of Hycrest were less pronounced. Results from cytological studies have shown the new cultivar to be genetically stable and an excellent seed producer (12).

Hexaploid (2n=42) forms of crested wheatgrass have received very little attention in North American range improvement or breeding programs. Accessions recently received from the Soviet Union and Iran, however, show excellent potential, particularly for seedling vigor and leafiness (W.H. Horton, unpublished). Higher chromosome numbers are associated with heavier seeds and seedling vigor in other species. Limited data indicate that hexaploid crested wheatgrass accessions have heavier seeds than diploid or tetraploid cultivars.

Hexaploid germplasm also may provide a gene pool for improving the forage quality of crested wheatgrass. Hexaploid introductions from the Soviet Union with more and broader leaves have been identified in breeding nurseries at Logan. Selection and hybridization programs have been initiated to develop leafier hexaploid cultivars and to transfer this character and associated nutritional benefits into tetraploid cultivars.

Demand for drought resistant, low maintenance grasses for arid turf or soil stabilization has increased in recent years. Crested wheatgrass introductions from Turkey, Iran, and China have shown exceptional adaptation for these purposes. These accessions are shorter in stature, have finer leaves, and are more rhizomatous than typical crested wheatgrass cultivars. Selected lines from these plant materials were included in the USDA-ARS breeding program at Logan and three experimental strains have been developed. Release of a turf-type crested wheatgrass cultivar from the this breeding program is anticipated within three years.

Russian Wildrye

Russian wildrye is an excellent source of livestock forage on semiarid rangelands of the Intermountain West and the Northern Great Plains. This cool-season perennial bunchgrass is native to the steppe and desert regions of the U.S.S.R. and China. Although it was introduced into the U.S.A. in 1927, its value in reseeding depleted rangelands was not fully recognized until the 1950s (3). Russian wildrye is drought resistant and has dense basal leaves that are palatable and high in nutritive value. Although noted for its productivity during the early spring, its nutritive value is retained better during the late summer and fall than many other grasses, including crested and intermediate wheatgrass (*Thinopyrum intermedium*). General acceptance of Russian wildrye on western rangelands has been impeded by difficulty in seedling establishment, particularly on harsh range sites, and tendency of its seed to shatter soon after maturity. Breeding programs in the U.S.A. and Canada have emphasized the improvement of seedling vigor (13-15). Selection to reduce seed shattering has been limited largely to an Agriculture Canada breeding program at Lethbridge, Alberta.

The Russian wildrye cultivars 'Vinall' and 'Sawki' were released in the 1960s. Vinall was developed by the USDA-ARS at Mandan, North Dakota primarily on the basis of large seed, seed yield, and forage yield. Sawki, a 10-clone synthetic cultivar with improved forage and seed yield, was released by Agriculture Canada at Swift Current, Saskatchewan. The cultivar 'Mayak' was released in 1971 from the breeding program at Swift Current. It was reported to be more productive than Sawki in terms of forage and seed (3).

Russian wildrye cultivars released more recently include 'Swift', 'Cabree', 'Bozoisky-Select', and 'Tetracan'. Swift was developed by Agriculture Canada, Swift Current, Saskatchewan, primarily on the basis of improved seedling vigor. This cultivar has demonstrated excellent stand establishment vigor in Canadian trials (15). Cabree was developed by Agriculture Canada at Lethbridge with emphasis on improved seed retention (16).

Bozoisky-Select, derived from an introduction from the USSR (PI 406468, Bozoisky), was released by the USDA-ARS in cooperation with the Utah Agricultural Experiment Station and the USDA-SCS (17). During its development, the breeding population was subjected to recurrent screening for improved seedling vigor. Bozoisky-Select has been easier to establish and more productive than Vinall on semiarid range sites. Plant exploration and breeding programs are continuing to improve the stand establishment of this and related populations. An experimental strain, 'Syn-A', which is easier to establish than Bozoisky-Select, will be released from the USDA-ARS grass breeding program at Logan within two years. Tetracan, a tetraploid cultivar released by Agriculture Canada, Swift Current in 1988, is reported to have larger seeds and wider leaves than its diploid counterparts. Tetracan has demonstrated superiority to the diploid cultivar Swift in stand establishment vigor and is equivalent to Swift in terms of forage and seed yield (18).

Bluebunch and Snake River Wheatgrass

Bluebunch wheatgrass (*Pseudoroegneria spicata*) is a widely adapted and valuable native grass in western North America. This species is resistant to drought and has excellent forage quality; however, because of its high palatability and poor grazing tolerance, its dominance has declined since the introduction of domestic livestock. Potential of bluebunch wheatgrass in rangeland seeding programs is also limited by poor seedling vigor. Snake River wheatgrass, including the cultivar Secar, was not originally distinguished from bluebunch wheatgrass, but is now recognized as a separate entity (19). This species has a more limited distribution than bluebunch wheatgrass and is found primarily in the Salmon, Snake, and Columbia River drainages.

Snake River wheatgrass is morphologically similar to bluebunch wheatgrass, but is generally more productive and more easily established (20). It can be distinguished from bluebunch wheatgrass on the basis of floral characteristics. Snake River wheatgrass spikelets are shorter, wider, and separated by shorter internodes than spikelets of bluebunch wheatgrass. Snake River wheatgrass glumes are narrow and tapered in contrast to the wide and blunt glumes of bluebunch wheatgrass.

An objective of the USDA-ARS breeding program at Logan is to combine selected characters of Snake River wheatgrass and thickspike wheatgrass. An awnless thickspike wheatgrass accession, T21076, with excellent seedling vigor and extensive rhizomes was collected by the USDA-SCS near The Dalles, OR. This accession was crossed by USDA-ARS to Secar Snake River wheatgrass and other accessions from sites receiving 30 to 35 cm annual precipitation. Snake River wheatgrass is awned and leafier and more rust-resistant than T21076. Results indicate that an awnless, leafy cultivar with improved seedling and mature plant vigor, moderate rhizome development, and rust resistance will eventually be released.

Indian Ricegrass

Indian ricegrass (*Oryzopsis hymenoides*) is a highly desirable native species on winter range sites. Seed is generally highly dormant. Mechanical and sulfuric acid scarification reduce mechanical dormancy and stratification, gibberellic acid, and kinetin reduce physiological dormancy. A better understanding of the relationship between mechanical and physiological dormancy may lead to a practical procedure for breaking dormancy in harvested seed. Alternatively, a better understanding of environmental factors influencing seed production may lead to production of seed already low in dormancy. Establishment success will depend on development of appropriate seedbed management practices for various soils. Reduction of currently large shattering losses would have a favorable impact on the economics of seed production. Introduction of additional cultivars will have little favorable impact until progress is made on seed dormancy and seed retention. An interdisciplinary approach including seed physiology, seedbed ecology, seed technology, and plant breeding can potentially solve these problems. The potential of seeding Indian ricegrass for improving rangelands can only be realized after low dormancy seed becomes available, appropriate seedbed management practices are developed, and seed shattering losses are reduced.

Interspecific Hybridization in Triticeae Grasses

Plant breeders have long been intrigued with the potential of using wide hybridization to transfer genes among species or to combine the best traits from two or more species into a "new species". Perennial Triticeae grasses are particularly well suited to this method of genetic improvement. Hybridization occurs frequently among these species and many are themselves natural polyploids of hybrid origin. For example, Dewey (21) concluded that the S, H, J, and N genomes from *Pseudoroegneria, Hordeum, Thinopyrum,* and *Psathyrostachys,* respectively, are all represented in western wheatgrass (*Pascopyrum smithii*).

Formidable barriers prevent breeders from making desired genetic exchanges among species. Most F_1 interspecific hybrids are partially to completely sterile and agronomically inferior to the parental species. Fertility can be achieved by doubling the chromosome number of the sterile hybrids, but sterility problems often intensify in subsequent generations (22). In some hybrids, vegetative vigor declines after chromosome doubling and segregation for deleterious characters such as chlorophyll deficiencies may occur for several generations. Modern techniques involving embryo rescue, protoplast fusion, and other aspects of biotechnology offer considerable optimism; however, it is apparent that genetic progress through interspecific hybridization is resource intensive.

D.R. Dewey, USDA-ARS cytogeneticist at Utah State University, has assembled the world's largest living collection of perennial triticeae grasses, from which over 250 hybrid combinations have been derived. Promising hybrids included in the USDA-ARS breeding program at Logan, UT are:

- Quackgrass (*Elytrigia repens*) X bluebunch wheatgrass.
- A natural and artificial hybrid between a form of quackgrass and an Asian relative of bluebunch wheatgrass.
- Quackgrass X thickspike wheatgrass.
- Thickspike wheatgrass X *E. caninus* (an introduced relative of slender wheatgrass).
- Thickspike wheatgrass X Bluebunch wheatgrass.
- Intermediate wheatgrass X *T. acutum.*
- Altai wildrye (*Leymus angustus*) X Great Basin wildrye (*L. cinereus*) and Mammoth wildrye (*L. racemosus*).
- A natural hybrid between a species of crested wheatgrass and an Asian relative of bluebunch wheatgrass.

The quackgrass X bluebunch wheatgrass hybrid is particularly promising. The objectives in this hybridization program were to combine the drought resistance, caespitose growth habit, and forage quality of bluebunch wheatgrass with the salinity tolerance, productivity, and persistency of quackgrass. Quackgrass, because of its overly aggressive growth habit, is a noxious and troublesome weed in many areas, but may serve as a valuable source of forage in many temperate regions (2).

The initial cross between quackgrass and bluebunch wheatgrass made in 1962 (23) was a major disappointment. It was a pentaploid (2n=35), meiotically irregular, morphologically variable, and in general had poor vegetative vigor and various "offtypes" were common. Because the hybrid was partially fertile, generations were advanced without chromosome doubling (24). Eight cycles of selection yielded a breeding population with relatively good fertility and a stable chromosome number of 2n=42. Selected characteristics of both parental species were represented in the population, which provided the basis for the release of the RS-1 and RS-2 germplasm pools in 1980 (25), and for the cultivar 'NewHy' in 1990 (26).

NewHy has been evaluated on several range sites in the Intermountain West and to a lesser extent in the Great Plains. It has excellent salinity tolerance and is best adapted to range sites receiving at least 33 cm annual precipitation. With adequate moisture, it is more persistent and withstands grazing pressure better than the bluebunch wheatgrass parent. Results from the Utah State University Experiment Station in northwestern Utah are typical. Although two years were required to obtain satisfactory stands, NewHy produced more forage than any of the other 16 entries included in the trial after the second growing season. It is noteworthy that NewHy, unlike its quackgrass parent, did not spread into adjacent plots. Similar trends were observed in a trial established in the foothills of the LaSal Mountains in southern Utah, at an altitude of 1,900 m and with an annual average of 33 cm precipitation. After stand establishment, NewHy was equivalent to the most productive entries in the trial.

Because of the aggressive growth habit of the quackgrass parent, the degree of rhizome development in the hybrid cultivar is of major concern. Growth habit ranges from caespitose (bunch types) to moderately rhizomatous. On sites receiving from 33 to 38 cm annual precipitation, most of the plants develop rhizomes of less than 0.5 m during the season. Two cycles of selection in the breeding population have yielded plants that are essentially true-breeding for the caespitose growth habit (27).

A breeding program is in progress with the quackgrass X *P. stipifolia* hybrid. The *P. stipifolia* parent is an Asian relative of bluebunch wheatgrass. This hybrid is similar to the RS hybrid, but has fewer deleterious off-types and appears to be more productive. The hybrid between quackgrass and crested wheatgrass is another notable example. Although the F_1 generation of this hybrid is sterile, it has shown exceptional potential for revegetating disturbed areas. Chromosome doubling improves the fertility of the hybrid, but apparently at the expense of vegetative vigor.

A germplasm pool from the hybrid between bluebunch wheatgrass and thickspike wheatgrass (designated SL hybrid) was recently released from the USDA-ARS breeding program at Logan. Because the F_1 of this hybrid was sterile, chromosome doubling was necessary to achieve the fertility necessary to initiate a breeding program (28). After six cycles of selection, a fully fertile population that is more upright than thickspike wheatgrass and more rhizomatous than bluebunch wheatgrass has been obtained. It has larger culms, more leaves per culm, and longer and wider leaves than either parent. In seeded and space planted trials on three semiarid range sites in northern Utah, forage yield and forage quality (neutral detergent fiber and crude protein) of the hybrid was equivalent or superior to cultivars of the parental species. A cultivar release from this germplasm pool is expected within three years.

Breeding New Forbs for Intermountain Range

A diverse vegetation consisting of well-balanced mixtures of grasses and forbs, and on some rangelands, shrubs, is usually more productive over time than grass monocultures. Forbs may be present as either dominants or subdominants in the climax vegetation of most grassland range types. Many forbs are relished by livestock and contribute substantially to animal nutrition. Forbs often remain relatively high in phosphorus and crude protein content late into the fall. In contrast, nutrient content in most grasses drops significantly after flowering. Forbs thus extend the grazing season and increase total productivity of pastures. Cook (29) stated, "Thus, for no other reason than to meet protein and phosphorus requirements of grazing animals, grassland ranges should be managed for a mixture of forbs and grasses in the stand."

Leguminous forbs interact symbiotically with bacteria (*Rhizobium* spp.) to fix atmospheric N that plants can utilize for growth and development. The fixed N ultimately is made available to associated grasses and nonleguminous forbs. Total forage production and livestock-carrying capacity can be increased two- to three-fold by inclusion of adapted legumes in seeding mixtures or by sod-seeding of legumes into existing swards (30). Forbs benefit wildlife, contribute to soil and water conservation, and are aesthetically pleasing. The agronomic and pasture traits of several forbs are being improved by the USDA-ARS at Logan.

Alfalfa

Alfalfa (*Medicago sativa*) is the most widely used legume in semiarid pastures. A number of cultivars have been bred specifically for grazing rather than for hay production (31). The USDA-ARS, Logan, UT, developed and released two germplasm pools, GP52-III and BC-79, which have been widely distributed and used in proprietary breeding programs.

The use of alfalfa for grazing could be extended if it were better adapted to saline soils. Research at the University of Arizona has led to the development of salt-resistant alfalfa germplasm adapted to the desert southwest (32). A similar program to develop a salt-resistant, cold-hardy, northern-adapted alfalfa has been initiated by the USDA-ARS at Logan. More than 1100 alfalfa populations from throughout the world have been screened for germination in saline environments (33). The second stage of selection, evaluation of seedling establishment in saline soils, has been initiated (34).

A major concern of livestock producers when pasturing alfalfa is that alfalfa causes bloat-induced mortality in approximately 1% of all ruminants that graze it (35). This loss amounts to more than $300 million per year in the U.S.A. Reduced animal productivity from avoidance or limited use of bloat-inducing legumes, treatment costs of animals, and labor required to treat or prevent the disease probably exceed losses due to mortality and morbidity (35-37).

Other legumes such as sainfoin (*Onobrychis viciifolia*) do not incite blot. Sainfoin contains condensed tannins, polymeric phenylpropanoid compounds that complex with soluble proteins to prevent pasture bloat. These condensed tannins also reduce rumen microbial activity to slow the release of soluble proteins. More than 100,000 plants within the genus *Medicago* (alfalfa and its relatives) have been examined but none produced tannin in the foliage. Production of foliar tannin is believed to be controlled by dominant gene(s) in sainfoin. A cooperative program has been initiated to move that gene into alfalfa. Procedures being explored to induce somatic hybridization and genetic transformation include electroporation and particle acceleration. Progress has been made in development of the appropriate biotechnology (38).

Sweetclover

Sweetclover (*Melilotus alba* and *M. officinalis*) is the second most prevalent leguminous crop seeded in semiarid pastures (30). Two areas of research are being investigated with sweetclovers. The first area relates to our basic understanding of the evolution and speciation in sweetclover. The genetic distances among species have not been quantified. An examination of the phyletic relationships using chemotaxonomy procedures has been initiated. Seed proteins of sweetclover accessions representing the 18 species within the genus were isolated by electrophoresis. Relationships among the populations are being assessed and verified by multivariate statistical procedures. These results will be related to conventional taxonomic designations based on morphology. Quantification of the relationships among sweetclover species and among accessions within those species is expected to lead to an increased understanding of the phylogeny within the genus, the potential for interspecific genetic transformations, and the risk of genetic erosion of germplasm resources.

One of the major failings of sweetclover is its susceptibility to predation by the sweetclover weevil (*Sitona cylindricollis*). The USDA-ARS sweetclover collection (484 accessions) was scored for weevil feeding and sampled for chemical analysis at two northern Utah locations in 1990. The relationship between foliar nitrate concentration and weevil predation in Utah and Minnesota will be determined (39,40).

Sicklepod Milkvetch

Native legumes indigenous to arid ranges often are members of the genus *Astragalus*. Some are potentially useful for rangeland and pasture seedings if they can be domesticated. Sicklepod milkvetch (*A. falcatus*), for example, is adapted to the more favorable sites on pinyon-juniper and big sagebrush ranges (41), and ranks relatively high for seedling vigor, ease of stand establishment, and persistence (42,43). When grown in a mixture with crested wheatgrass, sicklepod milkvetch increased the yield and protein content of the grass (44).

Astragalus species are numerous and distributed worldwide, but often are toxic to livestock (45). Sicklepod milkvetch synthesizes aliphatic nitro compounds that catabolize to toxic 3-nitropropionic acid (3-NPA) in the digestive tract of ruminants (46). Rumbaugh et al. (47) found that near infrared reflectance spectroscopy (NIRS) was moderately successful for initial screening of *Astragalus* species for such nitrotoxins. This technique is now being applied in an intensive breeding effort to develop a sicklepod milkvetch with reduced levels of toxins. One thousand plants were established in a field nursery and tested for toxins in each of two years. Ten of the least toxic plants were selected as parents of an experimental synthetic for which seed was produced in 1990. Genetic gain and heritability will be evaluated in 1991 to estimate the amount of progress made and to determine if additional selection will be required prior to germplasm release.

Kura Clover

Kura clover (*Trifolium ambiguum*) is one of the most drought resistant of the true clovers, yet it grows very well in mesic sites (48). Sparse nodulation, low nitrogen-fixation activity, and limited genetic diversity have restricted the development of productive populations of kura clover. A cooperative introduction and selection program initiated in 1978 led to the release of ARS-2678 kura clover (49). This species will find increasing acceptance in a variety of habitats, including the more favorable foothill ranges as well as in riparian sites and meadows.

Globemallows

There are few leguminous forbs adapted for seeding of rangelands that receive less than 30 cm of precipitation annually. The genus *Sphaeralcea*, a nonleguminous forb, has been considered as a candidate for such programs (50,51). These drought-tolerant native perennials grow mixed with other species from salt desert sites to dry foothills in sagebrush and pinyon-juniper ranges (41). Globemallows are indigenous in many of the same areas in which crested wheatgrass is adapted and are logical candidates for inclusion as a forb component in seed mixtures (52). Past use of these plants has been restricted by the high cost of seed collected by hand from native sites. Research to determine the agronomic suitability of these plants for profitable seed production was required before they could be used extensively.

The available germplasm base was extended by a major domestic collection effort in 1986. Fifty accessions representing six globemallow species were transplanted into replicated nurseries at two locations in 1987. Data were acquired on agronomic attributes in 1987 and 1988. Significant variation in important attributes was noted among species as well as among populations within species (53). An accession of *S. coccinea* was observed to spread by means of lateral roots. This may be an attribute of value in conservation plantings. Indeterminate flowering and seed shattering were identified as major agronomic problems within this genus. *Sphaeralcea parvifolia* appeared to be best adapted to the harsher environment (Curlew Valley, ID) and *S. munroana* appeared to be best adapted to the more mesic environment (Cache Valley, UT).

Although globemallows have been considered moderately palatable for sheep but less palatable for cattle (50,51,54,55), rancher experience has been variable. A cooperative experiment to determine sheep preference, intake, and nutritional value of 14 globemallow populations representing four species when interplanted with crested wheatgrass has been initiated.

References

1. Dillman AC. J. Amer. Soc. Agron. 38:237 (1946).
2. Asay KH and RP Knowles. In: *Forages: The Science of Grassland Agriculture, Ed. 4* (RF Barnes, DS Metcalfe, and ME Heath, Eds.), pp. 166-176. Iowa

State Univ. Press, Ames, 1985.

3. Hanson AA. *Grass Varieties in the United States*. Agric. Handb. 170, U.S. Govt. Printing Off., Washington, DC, 1972.

4. USDA Extension Service. ESC 584 (No. 13):209-211. U.S. Dept. Agric. Ext. Serv., Washington, DC, 1978.

5. Rogler GA. N. Dak. Agric. Exp. Sta. Bull. 375, 16:150-152 (1954).

6. Stevens R, SB Monsen, N Shaw, ED McArthur, G James, G Davis, KR Jorgensen, and JN Davis. Notice of naming and release of Ephraim crested wheatgrass, 1983.

7. Dewey DR. Crop Sci. 14:867 (1974).

8. Asay KH and DR Dewey. In: *Proc. Int. Grassl. Cong.* (JA Smith and VW Hays, Eds.), 14:124-127. Lexington, KY, 1983.

9. Tai W and DR Dewey. Crop Sci. 6:223 (1966).

10. Dewey DR and PC Pendse. Crop Sci. 8:607 (1968).

11. Asay KH, DR Dewey, FB Gomm, DA Johnson, and JR Carlson. Crop Sci. 25:368 (1985).

12. Asay KH, DR Dewey, FB Gomm, WH Horton, and KB Jensen. J. Range Manage. 39:261 (1986).

13. Asay KH and DA Johnson. Canad. J. Plant Sci. 60:1171 (1980).

14. Berdahl JD and RE Barker. Canad. J. Plant Sci. 64:131 (1984).

15. Lawrence T. Canad. J. Plant Sci. 59:515 (1979).

16. Smoliak S. Canad J. Plant Sci. 56:993 (1976).

17. Asay KH, DR Dewey, FB Gomm, DA Johnson, and JR Carlson. Crop Sci. 25:575 (1985).

18. Lawrence T, AE Slinkard, CD Ratzlaff, NW Holt, and PG Jefferson. Canad. J. Plant Sci. 70:311 (1990).

19. Carlson JR. MS Thesis, Oregon State University, Corvallis, 1986.

20. Carlson JR and DR Dewey. Agronomy Abstracts, p. 59, ASA, Madison, WI, 1987.

21. Dewey DR. Amer. J. Bot. 62:524 (1975).

22. Asay KH and DR Dewey. Crop Sci. 16:508 (1976).

23. Dewey DR. Amer. J. Bot. 54:93 (1967).

24. Dewey DR. Crop Sci. 16:175 (1976).

25. Asay KH and DR Dewey. Crop Sci. 21:351 (1981).

26. Asay KH, DR Dewey, WH Horton, KB Jensen, PO Currie, NJ Chatterton, WT Hansen II, and JR Carlson. Crop Sci. (in press) (1991).

27. Asay KH and WT Hansen II. Crop Sci. 24:743 (1984).

28. Dewey DR. Bot. Gaz. 126:269 (1965).

29. Cook CW. Rangelands 5:217 (1983).

30. Rumbaugh MD and CE Townsend. In *Proc. Selected Papers Presented at the 30th Annual Meeting of The Society for Range Management*, pp.137-147. Salt Lake City, UT, 1985.

31. Rumbaugh MD. *In* "Alfalfa for Dryland Grazing" (AC Wilton, Ed.), pp. 15-19. USDA Agric. Info. Rpt. 444, US Govt. Printing Office, Washington, DC, 1982.

32. Dobrenz AK, Robinson DL, Smith SE, and Poteet DC. Crop Sci. 29:493 (1989).

33. Rumbaugh MD and Pendery BM. Plant and Soil 124:47 (1990).
34. Al-Niemi TS, Campbell WF, and Rumbaugh MD. Agronomy Abstracts, p. 118, ASA, Madison, WI, 1990.
35. Rumbaugh MD. *In* "Forage Legumes for Energy-Efficient Animal Production" (RF Barnes, PR Ball, and RW Brougham, Eds.), pp. 238-245. Proc. Tri-National Workshop, Palmerston North, New Zealand, April 30-May 4, 1984 (1985).
36. Essig HW. *In* "Clover Science and Technology" (NL Taylor, Ed.), pp. 309-324. Agronomy 25, 1985.
37. Jacobson NL. *In* "Proc. New Horizons in Legume Bloat Control," pp. 10-28. A.E. Staley Manufacturing Co., Schiller Park, IL, 1967.
38. Bishop DL, Rumbaugh MD, Campbell WF, and Wang RR-C. Agronomy Abstracts, p. 194, ASA, Madison, WI, 1990.
39. Akeson WR, Gorz HJ, and Haskins FA. Science 163:293 (1969).
40. Spanarkel RW, Rumbaugh MD, and Clark DH. Agronomy Abstracts, p. 131, ASA, Madison, WI, 1990.
41. Plummer AP, Christensen DR, and Monsen SB. Pub. 68-3, Utah Div. Fish and Game, Salt Lake City, 1968.
42. Townsend CE and McGinnies WJ. Agron. J. 64:699 (1972).
43. Townsend CE, Hinze GO, Ackerman WD, and Remmenga EE. Colorado State Univ. Exp. Sta. Gen. Series 942, Ft. Collins, CO, 1975.
44. Rumbaugh MD, Johnson DA, and Van Epps GA. J. Range Manage. 35:604 (1982).
45. James LF, Keeler RF, Johnson AE, Williams MC, Cronin EH, and Olsen JD. USDA-SEA, Agric. Info. Bull. 415, p. 90, 1980.
46. Williams MC. Canad. J. Bot. 60:1956 (1982).
47. Rumbaugh MD, Clark DH, and Lamb RC. *In* "Proc. XV Intl. Grasslands Congress, Kyoto, Japan," pp. 315-316. The Science Council of Japan and The Japanese Society of Grassland Science, Nishi-nasuno, Japan, 1985.
48. Rumbaugh MD. *In* "Advances in New Crops: Proc. First National Symposium on New Crops Research, Development and Economics" (J Janick and JE Simon, Eds.), pp.183-190. Timber Press, Inc., Portland, OR, 1990.
49. Rumbaugh MD, Johnson DA, and Carlson JR. Crop Sci. 31: (in press) (1991).
50. Shaw N and Monsen SB. USDA Forest Service Gen. Tech. Rpt. 157, pp. 123-131, 1983.
51. Stevens R, Shaw N, and Howard CG. Proc. 38th Ann. Mtg. Soc. Range Manage., p.220, Salt Lake City, UT, 1985.
52. Pendery BM and Rumbaugh MD. Utah Sci. 47 (1986).
53. Pendery BM and Rumbaugh MD. J. Range Manage. 43:428 (1990).
54. Parker KG. Utah Agric. Ext. Service EC 383, 1983.
55. Wasser CH. USDI Fish Wildl. Serv. FSW/OBS-82/56, p. 347, 1982.

14

Impact of Weeds on Herbage Production

Bill E. Dahl and Ronald E. Sosebee

Abstract

This paper reviews the impact of a select group of undesirable range plants on forage yield. Annual weeds that suppress warm season grasses typically start growth in winter while the forage grasses are dormant. Thus, grass growth can be greatly enhanced, sometimes over 200%, by suppressing dense winter populations of weeds. Suffrutescent plants like broom snakeweed also start growth before warm season grasses, adversely impacting grass yields. Woody species, such as oaks and mesquite, compete strongly with herbaceous forages, are sometimes poisonous, and dense stands restrict livestock management, particularly the working and gathering of livestock. While mesquite is considered the major undesirable woody plant in the southwest, the oaks restrict forage production more severely. Mesquite is more discriminated against because it makes livestock handling difficult. Numerous examples are given illustrating the degree of forage reduction caused by different degrees of undesirable plant infestation.

Introduction

Many plants are undesirable because they compete with more useful species, they make desirable plants unavailable, or they are poisonous to grazing and browsing animals (1). While it is common knowledge that annual and perennial weeds suppress herbage yield of nonweed species, the amount and timing of this suppression is neither obvious nor documented. Therefore, the purpose of this paper is to review the impact of a select group of noxious or otherwise undesirable plants on forage yield.

A major reason for suppressed growth of desirable plants is that competing plants have growth peaks at different seasons. For example, winter growing annual and perennial plants (cool season plants) usually exhaust soil moisture before spring and summer growing (warm season) plants initiate growth in the spring. Consequently, little, if any, soil moisture remains to allow for growth of warm season forages. When adequate soil moisture does exist, cool season plants have a competitive advantage because they have a high proportion of mature leaves, i.e., large photosynthetic capacity, whereas warm season plants, just initiating growth, must rely heavily for new growth on carbohydrates from stem bases and roots—which are quickly exhausted. Contrarily, warm season plants have a competitive advantage during years with little fall and winter soil moisture.

In the southern Great Plains and the southwestern U.S., where dominant forage species are warm season grasses, forage yield declines most from competing species that initiate growth well ahead of the warm season dominants.

Herbaceous Broadleafed Weeds

Annual Weeds

Control of annual weeds on rangeland has been much debated because of their erratic occurrence, their desirability as forage, ineffectiveness of herbicidal control, and questionable cost effectiveness. The question is— do they suppress desirable forage enough to warrant control? Also, will good grazing management keep infestations to an acceptable level? When perennial vegetation density on rangeland becomes low due to drought injury, winter mortality, or abusive grazing, plants with high fecundity invade and retard recovery of resident perennials. In addition, many annuals, biennials (including winter annuals), and short-lived perennials respond to favorable climatic patterns and occur in high densities during years that favor their establishment regardless of grazing management or stand thinning of perennial plants.

Common broomweed (*Amphiochyris dracunculoides*), Texas thistle (*Cirsium texensis*)—on clayey soils—and camphor weed (*Heterotheca latifolia*) —on sandy soils—are southern Great Plains winter annual weeds that compete for spring moisture with the resident perennial forage grasses.

TABLE 14.1 Herbaceous standing crop (kg/ha) following weed[1] control with herbicides on sandy rangeland near Post, Texas (3).

	1985[2]		1986[2]	
	Site 1	Site 2	Site 3	Site 4
Grass Control				
Untreated	359	572	802	1065
Herbicide treated	1222	1939	1900	1872
Forbs Control				
Untreated	359	790	353	695
Herbicide treated	62	202	73	90

[1] Weeds were primarily annual with some perennials in 1985; they were primarily perennial ragweed in 1986. Yields were taken in the treatment year.
[2] The 1985 studies were conducted on the Post-Montgomery Estate Ranch and the 1986 studies were on the adjoining Middleton Ranch.

Because dense infestations are difficult to anticipate, few have studied the impact of such plants on yield of resident perennial forage. Sosebee and Gordon (2) reported that in 1977 common broomweed produced 730 kg/ha and grass produced 1065 kg/ha. Where the broomweed was controlled, grass produced 1905 kg/ha and the broomweed only 28 kg/ha. On sandy rangeland infested with sand shinnery oak (*Quercus havardii*) and camphor weed, grass production increased up to 240% by controlling the weeds (Table 14.1). Herbicide treatment also suppressed the sand shinnery oak in this study, so the increased grass production cannot all be attributed to decrease of herbaceous weed competition.

Bitterweed (*Hymenoxys odorata*) is a winter annual commonly found in high densities on overgrazed ranges. It is apparently more of a threat from poisoning than from reduced grass production. Nevertheless, Chambers (4) showed that only 164 kg/ha of bitterweed could reduce grass yields 26% (from 383 to 304 kg/ha) by mid-May. Also, Bunting and Wright (5) found that controlling bitterweed increased grass cover from 58 to 78% by late May. Fortunately, kind of seedling infestation to be expected for most problem annual weeds can be determined by late March or early April, which gives ample time in the Southern Plains to decide whether control should be considered.

Perennial Forbs

Dwyer (6) reported that taprooted forbs do not significantly decrease prairie tall grass yield because they use moisture and nutrients below

TABLE 14.2 Impact of forbs on prairie tall grasses (6).

Rhizomatous Forbs	Grass reduction due to Forbs	Taprooted Forbs	Grass reduction due to Forbs
Heath aster	68	Black sampson	19
Stiff goldenrod	53	Scurfpea	13
Western ragweed	56	Broom snakeweed	12
Aromatic aster	51	Sensitive briar	11
Velvety goldenrod	46	Prairie clover	10

grass roots. However, rhizomatous forbs reduced big bluestem (*Andropogon gerardi*) production over 50% (Table 14.2) and decreased grass roots and rhizomes as much as 61%.

Also, Dahl et al. (3) reported that control of perennial ragweed (*Ambrosia psilostachya*), a rhizomatous species, and partial control of sand shinnery oak increased grass yield 137% (Table 14.1, Site 3). Where the grass competition was mostly from perennial ragweed, grass yield was increased 76% (Table 14.1, Site 4).

Suffrutescent Plants

Broom snakeweed (*Gutierrezia sarothrae*) is probably the most widespread (7) undesirable suffrutescent plant of western rangelands. It not only causes abortion in cattle, but it competes strongly with desirable grasses. During the past 15 years in eastern New Mexico and western Texas, precipitation patterns have favored broom snakeweed invasions to the point that dense stands cover huge expanses of these short grass plains. Broom snakeweed stands yield up to 1500 kg/ha of snakeweed, often with less than 200 kg/ha of grass yield (9). During the 1970s, partial control of moderate infestations of broom snakeweed in Texas and New Mexico yielded moderate increases in grass yield (8) (Table 14.3). As infestations became more overpowering in the 1980s, grass yields were more drastically curtailed (9) (Table 14.3). As with other undesirable plants, partial control yields little extra grass production. Rather, almost complete control of weed species seems necessary to effectively increase yields of desirable species.

Ueckert (10) reported only a 188 kg/ha increase in grass yield with 50% thinning but an 1175 kg/ha increase with 100% of broom snakeweed removed during 1976. In 1977, these plots provided 375 and 2201 kg/ha

TABLE 14.3 Impact of broom snakeweed on grass production in Texas and New Mexico.

	Grass	Forb	Broom Snakeweed
		kg/ha	
Untreated[1]	1547	963	643
Tebuthiuron (1.1 kg/ha)[1]	1842	294	67
Untreated[2]	587	68	705
Tebuthiuron (1.1 kg/ha)[2]	959	55	47
Untreated[3]	968	196	831
Tebuthiuron (1.1 kg/ha)[3]	1321	108	55
Untreated[4]	200	NR	1500
Tordon 22K (0.28 kg/ha)[4]	770	NR	0
Tordon 101 (1.1 kg/ha)[4]	1090	NR	100

NR = Not reported. [1] Treated in 1975; [2,3] same plots sampled in 1976 and 1977 with no further herbicide treatment (8); [4] Sosebee et al. (9).

TABLE 14.4 Impact of various broom snakeweed thinning treatments on grass yield (10). (Plots were also treated to remove desert termites.)

Year	Treatment	Perennial Grasses	Broom Snakeweed
	% thinned		kg/ha
August 1976	0	1094	2377
	25	1094	2736
	50	906	1686
	100	2269	0
July 1977	0	679	2473
	25	704	3084
	50	1054	2200
	100	2880	0

Perennial grass production/cm of water (kg/ha/cm)

	1976	1977
With snakeweed	24	11
Without snakeweed	49	45

TABLE 14.5 Impact of broom snakeweed on grass production in Wyoming after five years (11).

| | Picloram Treatment (kg/ha a.i.) | | |
	0.0	0.56	1.12
Control after 5 years (%)	0	95	85
Blue grama (kg/ha)	123	617	762
Needle-and-thread (kg/ha)	62	560	191
Misc. grasses (kg/ha)	17	140	67
Total grasses	202	1317	1020

TABLE 14.6 Impact of common goldenweed on forage yield in May 1973 in Zapata, Texas (12).

| | Season of treatment | | | |
	Fall	Spring	Summer	Ave.
No treatment	869	379	896	715
2,4-D + dicamba	1270	1717	840	1276

more grass under 50 and 100% broom snakeweed removal (Table 14.4). Research on broom snakeweed from Wyoming provides similar results to the southwestern U.S. studies.   Gesink et al. (11) reported grass increases as much as 1115 kg/ha with 95% control of broom snakeweed (Table 14.5).

Common goldenweed (*Isocoma coronopifolia*), an aggressive half-shrub, is rapidly increasing as a management problem on south Texas rangeland. Spring herbicide treatment provided over 350% increase in forage (12) (Table 14.6). Thus, undesirable suffrutescent range plants reduce grass production as severely as any other kind of range weed, woody or herbaceous.

Mesquite

More control efforts are directed toward mesquite than any other woody plant in the southwest. This is due more to inability to work and observe livestock (13) than from reduced forage production because herbaceous desirable understory forages commonly coexist with mesquite. The relatively sparse foliage of mesquite provides only light shade and during hot summers, grasses will often yield more in moderate stands of mesquite compared to mesquite-cleared areas (14-16). Nevertheless, mesquite competes for habitat necessities and significantly reduces herbaceous yield most years (14,17-20).

Mesquite reduces herbaceous yield more in the drier climates, such as Arizona, as compared to most of Texas. However, simply thinning a mesquite stand does not always provide a proportional increase in forage production. At the Santa Rita Experimental Range in Arizona, mesquite (*P. velutina*) trees were thinned to various stand densities. Density and yield of perennial grasses was double that of untreated sites (339 vs 161 kg/ha) within three years. Annual grasses produced five times (24 vs 117 kg/ha) that of untreated areas. Velvet mesquite densities of 37 plants/ha seriously competed with grass. Effective control would require a mesquite density of less than 37 plants/ha (17).

On another Arizona study, velvet mesquite was thinned in 1945 to leave 0, 22, 40, and 62 trees/ha compared to an unthinned stand (Figure 14.1). Similar plots were established at four elevations from 960 to 1250 m. Yields on plots with 40 and 62 trees/ha were about half as great as on those with no mesquite. In 1958, 14 years after treatment, plots without mesquite yielded from four to ten times more grass than

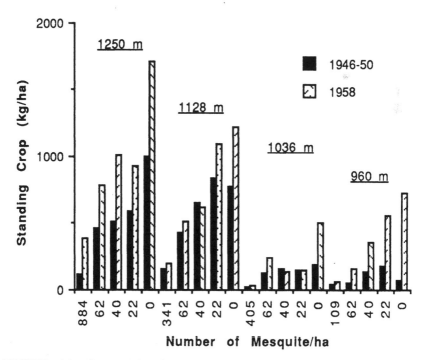

FIGURE 14.1 Grass yields following mesquite thinning treatments on four elevations in Arizona (17,18).

unthinned plots (17,18). Relative forage increases were greater from mesquite thinning at the lower elevations, i.e., the drier areas. However, the increase at the drier areas was mostly from annuals (18).

Cable and Martin (21) found that perennial grass production on the Santa Rita Experimental Range increased over twice as much from 1958 to 1963 on mesquite-killed units as on untreated units (497 vs 247 kg/ha). Although substantial, this increased yield was less than from earlier studies from the same area. They accounted for the different responses by indicating that their study was conducted with a more sparse mesquite stand which had a better stand of perennial grass at the start of the study. This implies that forage production is enhanced more by mesquite control if the mesquite stand is relatively dense and the understory vegetation relatively sparse. Work in Texas with honey mesquite control provides a similar conclusion (14).

Texas research to evaluate benefits of mesquite removal to forage and animal production has been conducted since the mid 1940s. On the Texas High and Rolling Plains, annual precipitation made a large difference in mesquite's impact on herbaceous forage. Studies by Dahl et al. (22) showed that during dry years mesquite control increased herbaceous yields from 70 to 220%. However, during above average soil moisture conditions, mesquite competition did not retard grass yields. During normal years, mesquite control increased herbaceous yields from 0% to a maximum of 50%. An attempt to show differences in grass yield response to different densities of mesquite showed that a 30% canopy of mesquite reduced grass yields little more than a 7.5% canopy, thus confirming results from Arizona studies (18).

Another Texas study relating grass yield to honey mesquite (*P. glandulosa*) density or canopy cover was conducted by Dahl et al. (14) from 1970 to 1975 in west Texas. Forage increases from mesquite control were considerably less than reported from Arizona. We believe this is due to (1) the wetter climate (510 mm per year vs less than 360 mm), (2) a more dense understory of grasses, and (3) the fact that fewer mesquite plants were actually killed. Increased forage was again seemingly related to climate during the year. For example, grass yield increased only 2% the second year after spraying compared to 26% increased grass yields the third year after herbicide treatment. Grass yields increased dramatically with reduced mesquite canopy and with increased proportion of plants apparently killed the year of treatment. However, mesquite control in this study increased perennial grass yields only an average of 22% for the five years included. From a similar study with mesquite sprayed in 1969 or 1970 at four ranch locations (Table 14.7), Fisher et al. (19) reported grass yields in 1971 were 63% greater on treated pastures. Total forage (grass and forbs) increases were only 29% in their study.

TABLE 14.7 Grass and forb yields (kg/ha) following honey mesquite spraying in western Texas (19).

Location	Aerial sprayed		Untreated	
	Grass	Forbs	Grass	Forbs
Vernon	1175	157	1090	99
Garden City	3600	179	1184	1031
Guthrie	1767	336	1112	664
Menard	1125	135	1330	81
Average	1917	202	1179	469

In south Texas, treating mixed brush with herbicides increased grass yield from 717 kg/ha to as much as 1818 kg/ha the year after treatment (20). However, this additional response was expected to be short-lived because of rapid brush recovery.

Oak Species

Oaks (*Quercus* sp.) are major causes of livestock poisoning during spring green-up. Also, they have broader leaves than mesquite, providing more dense shade, and they apparently compete more severely with herbaceous forage. Also, shrubby oaks such as sand shinnery oak (*Quercus havardii*) leaf out in the spring three to six weeks ahead of the dominant warm season grasses of the area, providing a strong competitive advantage to the oak. Contrary to findings with broom snakeweed and mesquite, oak control apparently does not need to be complete to get a major forage response. Elwell (23) reported that a 50% kill of post (*Q. stellata*), blackjack (*Q. marilandica*), and chinquapin (*Q. prinoides*) oaks increased native grass production up to 10-fold.

Dense stands of post oak and blackjack oak in Texas, Oklahoma, Arkansas, and Missouri suppress herbaceous forage enough that land managers have sought to reduce or eliminate its impact for many years. Scifres et al. (24) give three examples of the kind of forage increases one can expect from suppression of these trees (Table 14.8). Depending on the year, pattern of rainfall, and available species, it is obvious that grasses can be increased two to nine times by suppressing these woody plants.

Marquiss (25) obtained nearly 100% forage increase with only a 47% stem-kill of Gambel oak (*Q. gambellii*). Following blue oak (*Q. douglasii*) control in California, Johnson et al. (26) reported a 500% forage increase.

TABLE 14.8 Herbaceous forage (kg/ha) response from oak control by tebuthiuron application to Post Oak Savannah in Texas (24).

	Months after treatment	Grasses		Forbs	
		Untreated	Treated	Untreated	Treated
Study 1	4	1140	950	125	130
	13	670	2590	1380	575
	25	1260	2150	1120	885
	40	1380	1645	350	290
Study 2	4	0	0	0	0
	12	0	70	0	0
	27	192	1330	0	135
	36	250	1640	0	0
Study 3	4	200	133	21	16
	13	105	401	124	84
	25	300	963	143	228
	30	71	627	24	155

Oklahoma researchers consistently doubled grass production with single applications of phenoxy herbicides to sand shinnery oak (27-30). Consecutive yearly herbicide applications provided three-fold increases in grass yields. Forage production soon declined and stabilized at about two times that of controls (28). Robison and Fisher (31) obtained three to five times more forage by applying 2,4,5-T for three consecutive years in the Texas Rolling Plains. Also, Scifres (32) increased grass yield three- to five-fold the year after herbicide treatment and up to six times two years after. Tebuthiuron and picloram suppression of sand shinnery oak has also allowed for several-fold increases in peak standing crop of herbaceous plants in west Texas (33). Jones (34) and Plumb (35) obtained impressive herbaceous yield increases in follow-up west Texas studies using tebuthiuron to suppress sand shinnery oak (Table 14.9). Note that in Plumb's study considerable recovery of the oak in 1983 still provided 603 extra kg/ha of grass, a 147% increase over no oak control.

These studies indicate that increases in yield of herbaceous plants by oak control approximately equal the loss in yield of the shrub. Also, Galbraith (36) physically removed sand shinnery oak in one set of plots, herbaceous plants in another set, and left everything intact in another set. His data showed higher herbaceous production with oak removed the first year than total production on intact plots. The reverse was true the second year. He concluded that a particular site is capable of producing at a certain level. And, unless water or fertilizer is added, one is not

TABLE 14.9 Plant response of tebuthiuron treatment on sand shinnery oak-infested rangeland in Cochran County, Texas (34,35).

| | Sand shinnery oak | | Grass | | Forbs | |
	Untreated	Treated	Untreated	Treated	Untreated	Treated
Jones (33)						
1978*	458	458	214	298	30	7
1979	906	1	280	1600	50	100
1980	781	24	130	680	31	30
1981	1149	0	100	690	55	260
Plumb (34)						
1982	1396	392	197	882	51	248
1983	2488	1433	365	903	71	63

*Year herbicide applied. Herbicide was applied in 1980 for Plumb's study.

likely to increase total biomass production on a site; we merely replace one kind of plant with another. It is also important to remember that different precipitation patterns from year to year favor one kind of plant one year and another the next.

Other Woody Plants

Other woody species besides the ones described above tend to become so dense that they also suppress herbaceous forage. Blackbrush (*Acacia rigidula*), huisache (*Acacia farnesiana*), sagebrush (*Artemisia* sp.), Junipers (*Juniperus* sp.), and salt cedar (*Tamarix* sp.) are species in western rangelands that also occur in stands dense enough to greatly reduce herbaceous forage production. We believe the examples already provided suffice to illustrate that dense stands of woody range plants reduce forage for wildlife and livestock.

References

1. Platt KB. J. Range Manage. 12:194 (1959).
2. Sosebee RE and RA Gordon. Texas Tech Univ. Range and Wildlife Dept. Manage. Note 1. Lubbock, TX, 1983.
3. Dahl BE, JC Mosley, PF Cotter, and RL Dickerson Jr. J. Range Manage. 42:400 (1989).
4. Chambers RC. MS Thesis, Texas Tech Univ., 1971.
5. Bunting SC and HA Wright. J. Range Manage. 27:281 (1974).
6. Dwyer DD. J. Range Manage. 11:115 (1958).

7. Platt KB. J. Range Manage. 12:64 (1959).
8. Sosebee RE, WE Boyd, and CS Brumley. J. Range Manage. 32:177 (1979).
9. Sosebee RE, DJ Bedunah, WW Seipp, GL Thompson, and R Henard. Down to Earth 39(2):17 (1981).
10. Ueckert DN. J. Range Manage. 32:216 (1979).
11. Gesink RW, HP Alley, and GA Lee. J. Range Manage. 26:139 (1973).
12. Mayeaux HS Jr, DL Drawe, and CJ Scifres. J. Range Manage. 32:271 (1979).
13. Hoffman GO. Texas Agr. Ext. Ser. Misc. Publ. MP-386 (1975).
14. Dahl BE, RE Sosebee, JP Goen, and CS Brumley. J. Range Manage. 31:129 (1978).
15. Heitschmidt RK, RD Schultz, and CJ Scifres. J. Range Manage. 39:67 (1986).
16. Scifres CJ and DB Polk Jr. J. Range Manage. 27:462 (1974).
17. Parker KW and SC Martin. U.S.D.A. Agric. Circ. No. 908 (1950).
18. Martin SC. Prog. Agr. in Ariz. 15(4):15 (1963).
19. Fisher CE, HT Wiedemann, JP Walter, CH Meadors, JH Brock, and BT Cross. Texas Agr. Ext. Serv. Misc. Publ. MP-1043 (1972).
20. Scifres CJ, GP Durham, and JL Mutz. Weed Sci. 25:48 (1977).
21. Cable DR and SC Martin. USDA Rocky Mt. Forest and Range Expt. Sta. Res. Note RM-30 (1964).
22. Dahl BE, RE Sosebee, and JP Goen. Texas Tech Univ. Noxious Brush and Weed Control Research Highlights. 5:10-12 (1974).
23. Elwell HM. Agron. J. 56:411 (1964).
24. Scifres CJ, JW Stuth, DR Kirby, and RF Angell. Weed Sci. 9:535 (1981).
25. Marquis RW. J. Range Manage. 26:57 (1973).
26. Johnson WC, CM McKell, RA Evans, and LJ Berry. J. Range Manage. 12:18 (1959).
27. McIlvain EH and CG Armstrong. Proc. 12th Meeting South. Weed Sci. Soc. 12:134 (1959).
28. McIlvain EH and CG Armstrong. U.S. Southern Great Plains Field Sta. Prog. Rept. 6301 (1963).
29. McIlvain EH and CG Armstrong. U.S. Southern Great Plains Field Sta. Prog. in Range Res. Woodward Brief 535 (1965).
30. Greer HAL, EH McIlvain, and CG Armstrong. Okla. State Univ. Ext. Facts No. 2765 (1968).
31. Robison ED and CE Fisher. In: *Brush Research in Texas: 1968*, pp. 5-8. Texas Agr. Expt. Sta. Cons. Prog. Rept. 2583-2609, 1968.
32. Scifres CJ. J. Range Manage. 25:386 (1972).
33. Pettit RD. J. Range Manage 32:196 (1979).
34. Jones VE. PhD Diss. Texas Tech Univ., Lubbock, TX, 1982.
35. Plumb GE. MS Thesis, Texas Tech Univ., Lubbock, TX, 1984.
36. Galbraith J. MS Thesis, Texas Tech Univ., Lubbock, TX, 1983.

15

Consequences of Shrub Die-Off

James A. Young and Lynn F. James

Abstract

All or portions of several recent symposiums have been devoted to potential causes of widespread deaths of native shrubs on rangelands in western North America. Just as the causes of shrub mortality may be varied, certainly the consequences will be highly variable and site-specific. The relation of shrub mortality to this conference on noxious weeds is the fear that widespread shrub mortality will lead to subsequent increases in opportunistic weeds on rangelands. We are not sufficiently presumptive to offer a detailed or even vague prescription for what is going to happen and specific range sites where extensive shrub mortality may occur, but offer this brief review to encourage ranchers, land managers, and scientists who work in the affected environments to be keen observers as the natural drama of post shrub mortality succession unfolds.

Stand Renewal

Successional patterns are preconditioned by how the previous community that occupied the site was destroyed. This ecologic principal is so simple it can be overlooked. Forest succession is obviously different following catastrophic wildfires versus individual trees sporadically becoming senescent over time, toppling, and creating sunspots through the forest overstory.

Following four years of record drought in southern Utah, we can expect some level of shrub mortality and subsequent succession. One would expect that successional trend following mortality from drought would be different in terms of native versus exotic species, annual versus perennial species, than the post mortality succession in the northern Intermountain Area that occurred after several seasons of near record precipitation.

Seedbanks

The senior author happened to have a long-term study of seedbanks on the annual-dominated ranges of California when a two-year severe drought occurred (1). The study encompassed one year of relatively normal available moisture for plant growth, two years of drought, and a following season of moderate precipitation. The plant communities in the study were almost entirely composed of annual species so the reaction to extreme drought was readily apparent. The first year of the drought, the species composition and relative abundance closely resembled those of predrought conditions, but the actual plant density was greatly reduced. Many of the dormant grasses in these communities did not produce seeds that persisted in the seedbank. Therefore, there was a decided shift in species composition the second year of drought. Legume species with hard seed coats dominated the seedbank and increased in relative density as the drought continued. The other group of increasers was *Erodium* spp. that through the process of self burial were ideally suited for the earliest germination (2).

Based on this annual plant model, it would appear that natural disasters such as droughts can be escaped through dormant seeds in seedbanks or exploited by species whose survival strategies outlast the disaster.

Poststand Renewal Dynamics

Green rabbitbrush (*Chrysothamnus viscidiflorus*) provides a good model for this concept. The aerial portion of green rabbitbrush plants is consumed in wildfires, but about 80% of the plants sprout from crown beds (3). The widespread examples of dominance of green rabbitbrush after big sagebrush (*Artemisia tridentata*) plant communities are burned are often attributed to this sprouting characteristic. Dominance in post burn conditions actually results from abundant production of achenes by the sprouting plants and subsequent germination and abundant seedling production. Plants with inherent potential to persist through periods of stress and then respond with enhanced reproductive potential can assume dominance following stand renewal.

Demography

Some forms of stand renewal such as intense wildfires are relatively broad scale in their destructions of the age classes that compose the community being destroyed. Senescent plants with a load of insect and disease pests also can form a disproportionately susceptible age class to stress. Demography has two major facets in terms of stand renewal. Obviously, some age classes may be more susceptible to stress than others, but the relative distribution of age classes within a community also governs the resiliency of the community to stress. Even-aged stands, especially of high risk age classes such as senescent plants or seedlings, may be at the greatest risk. Even a short period of drought during the germination-establishment period can prove fatal to seedlings. Catastrophic stand renewal processes such as intense wildfires may produce relative even-aged stands.

Herbivory

Many of the environmental disasters that occurred on the western range in the late 19th century were the result of overstocking during times of drought. Currently, stocking rates of domestic livestock are often dramatically reduced on rangelands during periods of drought. The net effect of these two extremes (overstocking versus livestock removal) could have enormous ecological implications on post stress plant communities.

It is more difficult to rapidly reduce the populations of native large herbivores. The lack of snow in the mountains of southern Utah during the current drought may have allowed over-utilization of browsing resources on sites where snow cover would have normally seasonally limited such use (4). Jackrabbit or grasshopper populations may interact with environmental stress such as prolonged drought. A certain population density of grasshoppers may have little prolonged influence on normal years but could produce critical results on plants already stressed by drought.

Livestock Preference

During the 19th and early 20th centuries, livestock losses from poisonous plants were often more severe during periods of drought. This was attributed to changes in livestock preference as overgrazing depleted the available forage. Currently, we are much more sophisticated in our approaches to livestock behavior and especially preference for poisonous species.

Refugia

The changes in plant community structure and distribution that occur subsequent to periods of prolonged environmental stress offer opportunities to further our understanding of the ecological fabric of rangeland ecosystems. Who survives drought and where plants find refuge from environmental stress is basic to understanding relative adaptation to specific environments.

The reactions of clonal species, especially these that may have existed since the Pleistocene, to prolonged drought is especially interesting.

References

1. Young JA, RA Evans, CA Raguse, and JR Larson. Hilgardia 49:1-37 (1981).
2. Young JA, RA Evans, and BL Kay. Agron. J. 67:54-57 (1975).
3. Young JA and RA Evans. J. Range Manage. 27:127-132 (1974).
4. Richard Madril, Dixie National Forest, personal communication.

16

Economic Impact, Classification, Distribution, and Ecology of Leafy Spurge

Rodney G. Lym

Abstract

Leafy spurge (*Euphorbia esula*) is a long-lived perennial plant that was introduced into the United States in 1827. The weed currently infests over 1.1 million ha in the Northern Great Plains of the U.S. and Prairie Provinces of Canada. The plant emerges in early spring, producing showy, yellow bracts which appear in late May with the true flower emerging in mid-June. The plant spreads by both seed and roots and contains a white sticky latex that prevents grazing by many animals. Leafy spurge is found on a variety of terrain from flood plains to grasslands and mountain slopes. The plant reduces the carrying capacity of rangeland to near zero as cattle will not graze in areas with a 10 to 20% leafy spurge cover. Over $14.4 million is lost each year in North Dakota alone due to reduced forage production and utilization.

Distribution

Leafy spurge is a dicotyledonous herbaceous deep-rooted perennial of the spurge (Euphorbiaceae) family. The plant is difficult to control and spreads by both seed and roots. It has become particularly troublesome in the Upper Great Plains of North America.

Leafy spurge is widespread in continental Europe as far south as central Spain, Italy, and the Balkans, extending eastward through central Russia into Siberia (1). Leafy spurge apparently is not a weed of economic or agronomic significance in its area of origin. The earliest report of the weed in the United States is a herbarium specimen collected in 1827 in Newbury, Massachusetts. Leafy spurge has been known in North Dakota since 1909 (2). The weed was not recognized in western Canada until the early part of the 20th century (3). Early reports in Manitoba were from Russian Mennonite settlements where seed may have been in crop seed brought from Russia (2). Dunn (4) conducted a survey in 1979 and found the greatest abundance of the weed now is in the Northern Great Plains of the United States and the Prairie Provinces of Canada where it infests over 1.1 million ha.

Leafy spurge is found in six Canadian provinces and at least 26 states of the United States where 144 counties have infestations greater than 200 ha (4). In Canada, leafy spurge is most important in the Prairie Provinces of Manitoba, Saskatchewan, and Alberta. However, Canada has less than 5% of the total number of hectares infested in North America. Probably an active federal program for research and control in Canada for the past 30 yr has limited the spread of the weed. Leafy spurge has been reported as far south as Ecatapec in the state of Mexico (D.F.), Mexico as a minor roadside weed (5).

Leafy spurge was introduced into North America in several ways. The infestation along the east coast probably resulted from multiple introductions in ballast soil that was carried in ships from Europe to North America and dumped on the shore (5). Leafy spurge probably was introduced into the prairie states by Mennonites as a contaminant in seed grain. The original infestation found in Canada may have come from contaminated smooth bromegrass (*Bromus inermis*) brought by immigrants from eastern Europe and Russia.

Morphology

Leafy spurge is one of the first plants to emerge in the spring. It generally emerges during March in Iowa (6) and Wisconsin (3), in early April in North Dakota (2), and between mid-April and May 1 in Saskatchewan (3,7).

The plants grow rapidly and the first yellowish-green bracts appear from early to late May across the region, depending on the geography and temperature for that growing season (8). The yellow bracts are visible from late May through June and the plants are from 0.75 to 1.25 m tall.

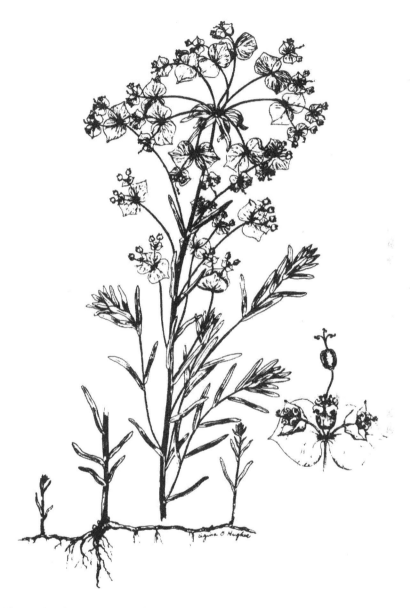

FIGURE 16.1 Leafy spurge plant during the flowering growth stage. Features "shown" include the spreading root system, with another shoot that developed from a root bud, the stem with minimal branching, the terminal cluster of flowers (inflorescence), the bracts that surround the true female and male flowers (cyathium) with the three-chambered seed capsule (inset).

Flowers

The yellowish-green flower-like clusters are the most conspicuous feature of leafy spurge. The flowers (inflorescence) are borne on an umbel at the tip of the stem or on lateral branches near the top of the stem (Figure 16.1). Each stalk (pedicel) of the umbel bears the unique inflorescence of the spurge family known as a cyathium. Each cyathium produces one female (pistillate) and 11 to 20 male (staminate) flowers. Two opposite branches arise from the base of each cyathium and each branch produces a cyathium. The process is repeated until the single original floral branch has divided into 8 to 16 branchlets (3).

The true flower parts are surrounded by cup-like bracts (involucre) of the cyathium. The yellowish-green bracts are the most colorful parts of the plant but are not true male and female flower parts. The bracts generally appear 2 to 3 weeks before the true flower parts are functional. The sepals (calyx) and petals (corolla) are absent (2). The involucre surrounds a cluster of 11 to 20 stalked staminate flowers, each consisting of a single stamen, and one single-stalked pistillate flower (3,7). The stigma of the pistil opens and becomes receptive to pollen 1 to 3 days after emergence of the female flower. The glands of the cyathium begin secreting nectar 3 to 8 days later. Then the pistil inverts by elongation of the pedicel, and one or two male flowers appear. On the day following inversion, 2 to 5 male flowers appear, and the total of 11 to 20 male flowers appear within 3 days.

Mature pollen is found on the anthers 24 to 48 hr after emergence of the male flower and is orange and sticky when ripe. Pollination apparently is almost entirely by insects. Ants, bees, flies, and mosquitoes have been observed feeding on the nectar and may be responsible for much of the pollination (3,7). The pollination mechanism in leafy spurge minimizes self-pollination.

Commercial honey producers utilize leafy spurge as an early-season food source (8). The plant flowers prior to the prime honey producing months. Also, leafy spurge honey does not granulate quickly in cold weather, so it makes good honey to feed bee colonies in the winter.

The leafy spurge fruit develops from a three-celled ovary. Nearly 50% of the fruits produce only one mature seed, about 35% produce two seeds, and only 15 to 20% produce three seeds (2). The individual seed stalks produce 10 to 50 fruits under normal conditions (6). Selleck et al. (3) reported up to 252 seeds/shoot of leafy spurge in competition with native grass. The ripe fruit dehisces explosively along the seam of the union of the carpels beginning in mid-July and continues through fall (2).

The average seed production in the center of leafy spurge patches was 2500 seeds/m^2, with a range from 790 to 8020 seeds/m^2 (9). Flowering generally ceases in terminal inflorescences between late June and mid-July. The development and maturation of seed extends about 30 days after appearance of the last flower. Generally, the first viable seed is produced by late June and most of the seed matures during July, but small amounts of seed can be produced in August and until frost.

Seeds

Seeds are silvery gray to light gray tinged with purple (2,10). A prominent yellow protrusion, called the caruncle, is located at the narrow end of the seed and a brown line extends from the caruncle to the opposite end. Seeds go through several color changes during development from yellow through orange and brown to gray at maturity (11,12). Seeds are viable by the brown stage, which occurs 10 to 13 days after fertilization, depending upon environmental conditions. Leafy spurge control to prevent viable seed production must occur before the seeds have turned brown. The approximate seed maturity can be determined by breaking the fruit open to observe the seedcoat color.

Seed germination in the laboratory was most rapid at alternating 20 to 30 C, intermediate at constant 30 C, and slowest at constant 20 C (2). Field germination of seed in the spring follows several days of 25 to 28 C air temperature (3). Most seeds germinate in the spring, but some seeds will germinate through the growing season. Under favorable conditions, 99% of the seeds will germinate or be destroyed within 2 yr, and the remainder either germinate or deteriorate within 8 yr.

Seed dispersal begins when the mature capsule dehisces, hurling the seed up to 4.5 m and distributing them uniformly (2). Seeds can float for several hours on runoff water or streams which facilitates spread.

Wild and domestic animals, birds, and insects are agents of dispersal. Birds as primary disseminators of leafy spurge seed have been suggested because of frequent feeding on seed and frequent occurrence of new patches under trees and fences. Viable seeds have been found in the droppings of some birds, such as sharptail grouse. Mourning doves (Zenaida macroura) may spread seed especially when ground-nesting, but less than one intact leafy spurge seed/g was found in fecal materials (13). Seeds probably move with mud on animal feet or hair. Some leafy spurge seeds can occur in sheep manure, and probably can occur in the manure of other animals. Seeds also move on machinery and in hay.

Seedlings

The plant emerges by elongation of the hypocotyl, and the seed leaves (cotyledons) function as the first leaves. The first true leaves generally develop 6 to 10 days after emergence (3). The first true leaves are opposite, but the subsequent leaves are distinctly alternate. Most seedling emergence occurs from 1.3 to 5 cm soil depth, although germination can occur from the surface to 15 cm deep.

Seedlings rapidly develop perennial characteristics. Of 22 seedlings in the 6-leaf stage which were cut 1.2 cm below the soil surface, 11 produced new shoots (3). All of the seedlings in the 10-leaf stage survived when severed at the same depth. Seedling roots can be 60 cm long and the stem 13 cm tall within two months after cotyledons expand (2).

Seedling survival generally is low, especially with competition (2,3). Selleck et al. (3) reported 2800 seedlings/m² on May 27, 1951, in a natural infestation, but 82% of the seedlings had died by September 12. Seedling survival was 5.1, 1.5, and 0.4% of the seed sown on cultivated land, western snowberry, and native grassland habitats, respectively (9).

Stems and Leaves

Stems usually are erect, 40 to 80 cm high, and unbranched except for the inflorescence. However, axillary buds frequently develop into branches when the stem tip has been injured. The stems become woody toward the base, so the dead stem remains standing for a year or more. Stems are pale green in summer but become yellow to red in the fall.

The density of leafy spurge stems varies greatly among stands and can fluctuate greatly from year to year with changing environmental conditions. Usually the number of stems does not increase when a density around 200 shoots/m² is reached (3). Densities over 980 shoots/m² have been recorded, however.

Leaves are linear-lanceolate to ovate, broader above the middle, tapering to the base, sessile, margins entire or slightly sinuate, leaves thin, and not glossy (14). Leaves are bluish-green in color and weakly veined except for the midrib. The epidermis has a fairly thick layer of cutin, especially toward the margins of the leaf (2).

Roots

The most important part of the plant for leafy spurge control is the root system and associated vegetative buds. There are numerous coarse roots that occupy a large volume of soil. The roots are woody and tough in structure with numerous buds. New shoots are produced readily from small pieces of roots. Roots remain viable within the first 6 yr of plant development (3).

The horizontal underground structures are roots and not underground stems (rhizomes) (15). All underground stem tissue examined in leafy

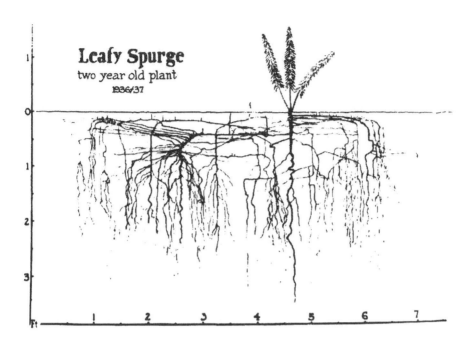

Leafy Spurge
two year old plant
1936/37

FIGURE 16.2 Leafy spurge root system [from Hanson and Rudd (2)].

spurge was vertical and a product of adventitious bud development from horizontal, vertical, and oblique roots. There are two types of roots produced on plants (16,17) (Figure 16.2). "Long" roots arise from the vigorous primary root of the seedling and have extensive longitudinal growth and considerable cambial activity. "Short" roots arise from long roots but have limited growth and no cambial activity. Long roots give rise to new lateral long roots and shoot buds that account for the vegetative spread of leafy spurge patches. Short roots do not produce shoot buds but are important for water and nutrient uptake.

Roots of leafy spurge have great capacity to produce new shoots from various depths. Coupland et al. (18) covered leafy spurge roots with 0.25, 0.5, or 1 m of weed-free soil using two methods; (a) piling and tamping soil to the desired depth on top of a leafy spurge patch and (b) excavating all the soil in a leafy spurge patch to the desired depth and tamping clean soil over the remaining roots. Shoots at all locations penetrated 0.5 m of soil within 12 months, and some shoots emerged through 1 m of soil. Following regrowth of shoots from 0.25 and 0.5 m deep, new crowns were formed near the soil surface from which regrowth occurred the following year. Roots had the ability to produce vegetative shoots for five successive years from a depth of 1 m after most of the root system had been removed by excavation.

Hanson and Rudd (2) observed emergence of new shoots after roots had been killed to a depth of 30 cm or more by sodium chlorate. Leafy spurge control treatments such as herbicides or tillage must kill the root system at least 1 m deep to prevent emergence of established plants, or treatments must be repeated until deep root reserves are depleted.

Vegetative Buds

Buds occur on both the roots and underground portion of shoots. The maximum depth at which buds occur on plants varies from 30 to 175 cm (19). The average number of buds per root was 35 to 272 in different plant groups. The number of vegetative buds tends to be greatest just below the soil surface and decreases with increasing depth. The number of vegetative buds at each depth seems to be directly related to the weight of root material in each layer.

Roots have vegetative buds on both the horizontal and vertical portions of the long roots. New long roots do not develop buds until cambial activity has begun, while older roots develop additional buds without any particular pattern (16). Only a few of the root buds develop into shoots. Shoots first develop on horizontal roots of an intact root system at some distance from the parent shoots, commonly at or near the point of vertical turning. Additional shoots may arise later on the horizontal root closer to the parent root.

The underground portion of leafy spurge is composed predominately of roots, but buds which develop into shoots have an underground portion which is persistent after the aerial portion has died back (16). Buds are formed on these underground stems both in the axils of the scale-like leaves and adventitiously elsewhere. Several stem buds usually develop into shoots, forming a crown of shoots after several years. The underground stems are vertical and do not contribute directly to the spread of the plant, but they do serve an important function in perennial survival of leafy spurge. Moreover, they give rise to adventitious roots, some of which are spreading long roots that contribute to vegetative spread of the plant.

Latex and Chemical Properties

Latex is present throughout the plant (6). Injury to any part of the plant will result in immediate flow of the white, sticky latex to seal the wound. Apparently the latex tubes originate as one cell beginning in the embryo of the seed (2). The latex has been evaluated as a potential rubber or energy source, but these alternative uses seem impractical for the foreseeable future. Extracts of leafy spurge have been shown to stimulate hair growth on rabbits (20).

Leafy spurge contains a toxic substance that, when taken internally, is an irritant, emetic and purgative. It causes scours and weakness in cattle and may result in death (3). The toxin has produced inflammation and loss of hair on the feet of horses from freshly mowed stubble during haying (21) and has caused mortality of sheep in Alberta (11). Animals will eat the dried plant in hay, but livestock, particularly cattle, avoid eating growing plants. Sheep and goats are less affected by the toxic principle in the latex and will graze young plants. Thus, sheep and goats have been used in management programs for leafy spurge control.

There is indirect evidence that leafy spurge has allelopathic properties, i.e., the weed releases chemicals that inhibit the growth of other plants in the same area (22). For example, the small number of forbs in patches of leafy spurge, even when bare ground is visible between shoots, suggests that this species exerts inhibitory effects on other plants (7). However, specific chemicals have not been identified to verify the occurrence of allelopathy.

Genetic Diversity

Leafy spurge generally is identified as *Euphorbia esula*, but its correct botanical identity is uncertain. Richardson (23) in 1968 identified leafy spurge as *Euphorbia podperae* Croiz. Dunn and Radcliffe-Smith (24) in 1980 examined 28 herbaria specimens of leafy spurge from 12 states and identified five morphologically separable taxa. However, Ebke and McCarty (25) in 1980 established leafy spurge from 36 locations across the northern United States and Canada and reported that most of the collections separate into a single taxa: *Euphorbia x pseudovirgata*, a hybrid of *E. esula* and *E. virgata*.

Regardless of the correct taxonomic identification, the genetic diversity in leafy spurge may limit the suitability of some diseases and insects as potential biological control agents (26). Galitz (unpublished data) at North Dakota State University observed a highly virulent fungal growth on one leafy spurge biotype among approximately 15 selections in the greenhouse; the fungus apparently is a powdery mildew. Dunn and Radcliffe-Smith (24) reported that a clear-winged moth (*Chamaesphecia tenthrediniformis*) collected on leafy spurge in Switzerland and Austria failed to become established on leafy spurge in either Canada or the United States.

Leafy Spurge Ecology

Leafy spurge grows on a wide variety of terrain from flood plains to river banks, grasslands, ridges, and mountain slopes (6). It is primarily found in untilled non-cropland habitats such as abandoned cropland, pastures, rangeland, woodland, roadsides, and waste areas (27). The plant grows in diverse environments from dry to subhumid and from subtropic to subartic (15). It occurs on many topographic positions from the flat bottom of glacial lakes to the slopes of sand dunes and glacial moraines. After leafy spurge is introduced into an area, there does not seem to be any topographic limits to its invasion of new areas.

Leafy spurge tends to occupy sites having high sand content, at least as the site for initial infestation (28). Leafy spurge often is the dominant species in bottomland positions, with less on the toeslope, summit, and shoulder slope, respectively. The favored site associations seem more related to moisture and fertility conditions favorable for plant growth than to edaphic factors.

About 95% of leafy spurge infestations within a 374 ha native prairie area were associated with soil disturbances such as vehicle tracks or road construction and fireguards which removed native plant cover and exposed mineral soil (29). After leafy spurge invasion, plant diversity declined from 11 species outside the infestation to three species at the center. The only species that were positively correlated with leafy spurge were smooth brome and Kentucky bluegrass (*Poa pratensis*), both of Eurasian origin. This correlation may have occurred because Eurasian agricultural species readily invade disturbed soil.

Alfalfa (*Medicago sativa*) and leafy spurge occurred together in only 8% of over 700 areas sampled during three surveys (30). A parasitic rust fungus, *Uromyces striatus* Schroet., infects both leafy spurge and alfalfa as alternating hosts. It was hypothesized that nonconcurrence of alfalfa and leafy spurge may be due to naturally occurring biocontrol.

Soil disturbance by humans promotes the establishment of leafy spurge. Over 45 times more seeds established on bare soil than in undisturbed vegetation (9). In non-cultivated areas leafy spurge patches increased in radius by 0.3 to 0.9 m/yr, with a median of 0.612 m (3). Spread is potentially much greater in cultivated habitats because of reduced competition and movement of root fragments (2,3).

Many plant population models have been developed to predict the rate of expansion for leafy spurge patches (31-33). These models include many environmental and physiological variables that simulate leafy spurge communities. However, these models are difficult to use in applied situations.

Stroh et al. (34) have proposed a simple formula to estimate leafy spurge patch expansion. The formula is based on a review of the literature and research on native grasslands in the upper Great Plains.

Leafy Spurge Patch Expansion Formula

$X = \pi * [(Y-4) * 0.61m]^2$
$Z = X * (100 \text{ stems}/m^2)$
where Y = years
   M = meters
   X = area of patch in m^2
   Z = total stems in patch

The formula is based on the premise that more than 4 yr are required before a seedling will start to spread vegetatively. Thus a single leafy spurge seedling could infest 0.5 ha in 80 yr. However, the actual rate of increase would be faster since the formula does not generate information on new patches formed from seed dispersal.

Economic Impact

The influence of leafy spurge on long-term land value is difficult to assess (35,36). However, short-term return can be estimated by measuring changes in forage production and use by livestock following leafy spurge control (36,37).

Leafy spurge reduces the livestock carrying capacity 50 to 75% (38,39). In North Dakota, cattle used 20 and 2% of the forage available in zero- and low- (<20% cover) density leafy spurge infestations by mid-season (40). Moderate- and high-density infestations were avoided until early fall when the milky latex in leafy spurge disappeared. Leafy spurge canopy cover of 10% or less and shoot control of 90% or more was necessary to achieve 50% forage use by cattle in Montana (41).

Moderate and high leafy spurge densities reduced long-term herbage production approximately 16.5 to 33% in North Dakota on land that was 50 to 100% infested, respectively (37). A ranching enterprise would lose approximately 17.5% of its net income from cattle refusing to graze herbage in moderate to heavy leafy spurge infestations and an additional 17.5% from lost annual production (40). Besides production losses, control costs to manage infested sites and potential for increased infestation each year must be included in assessing the economic impact of leafy spurge.

The first attempt to estimate direct economic losses due to leafy spurge was made in North Dakota in 1985. Messersmith and Lym (36) estimated a direct annual loss to the state of nearly $7 million of hay and beef cattle production plus $5.5 million from expenditures for treatment with herbicides. This $12.9 million/yr loss was occurring on 348,000 ha or only 6% of the land most likely to be infested.

Thompson et al. (42) estimated both the direct and secondary effects of reduced livestock carrying capacity due to leafy spurge in North Dakota in 1990. They found the reduction in carrying capacity is best approximated by the equation:

C.C. = 100 - 1.25 (P.I.)
P.I. = Percent land area covered by leafy spurge.

Thus a leafy spurge infestation covering 80% of the land area would reduce the carrying capacity to zero from a practical management standpoint. They estimated a direct reduction of 577,000 animal unit months or $8.6 million annually, similar to the earlier report (36). The decreased production due to the lost carrying capacity was $14.4 million (42). The secondary impacts were estimated at $25 million in lost personal income. Substantial impacts were also shown for the retail trade sector ($19.3 million) and the agriculture-crop sector ($10.7 million). The total estimated annual loss was $75 million. They further estimated a reduction in grazing land value of $137 million.

The economic impact of leafy spurge on wildlife habitat is currently being evaluated at North Dakota State University.

References

1. Bybee TA. MS Thesis, North Dakota State Univ., Fargo (1976).
2. Hanson HC and VE Rudd. N.D. Agric. Exp. Stn. Bull. 266 (1933).
3. Selleck GW, RT Coupland, and C Frankton. Ecol. Monographs 32:1-29 (1962).
4. Dunn PH. Weed Sci. 27:509-516 (1979).
5. Dunn PH. In: *Leafy Spurge Monogr. 3* (AK Watson, Ed.), pp. 7-13. Weed Sci. Soc. Amer., Champaign, IL, 1985.
6. Bakke AL. Exp. Stn. Res. Bull. 198 (1936).
7. Selleck GW. Weed Sci. 20:189-194 (1972).
8. Messersmith CG, RG Lym, and DS Galitz. In: *Leafy Spurge Monogr. 3* (AK Watson, Ed.), pp. 42-56. Weed Sci. Soc. Amer., Champaign, IL, 1985.
9. Best KF, GG Bowes, AG Thomas, and MG Maw. Can. J. Plant Sci. 61:651-663 (1980).
10. Krockmal A. Weeds 1:243-255 (1952).

11. Johnston A and RW Peake. J. Range Manage. 13:192-195 (1960).
12. Wicks GA and LA Derscheid. Weeds 12:175-176 (1964).
13. Blockstein DE, BD Maxwell, and PK Fay. Weed Sci. 35:160-162 (1987).
14. Moore RJ. Can. J. Bot. 35:547-559 (1958).
15. Meyers GA, CA Beasley, and LA Derscheid. Weeds 12:291-295 (1964).
16. Raju MVS, TA Steeves, and RT Coupland. Can. J. Bot. 41:579-589 (1963).
17. Raju MVS, TA Steeves, and JM Naylor. Can. J. Bot. 42:1615-1628 (1964).
18. Coupland RT, GW Selleck, and JF Alex. Can. J. Agric. Sci. 35:477-484 (1955).
19. Coupland RT and JF Alex. Can. J. Agric. Sci. 35:76-82 (1955).
20. Roslycky EB. Can. J. Plant Sci. 52:844-845 (1972).
21. Kingsbury JM. *Poisonous Plants of the United States and Canada.* Prentice-Hall, Inc., Englewood Cliffs, NJ, 1964.
22. Steenhagen DA and RL Zimdahl. Weed Sci. 27:1-3 (1979).
23. Richardson JW. Univ. of Kansas Sci. Bull. 8:45-112 (1968).
24. Dunn PH and A Radcliffe-Smith. Res. Rep. North Cent. Weed Control Conf. 37:48-53 (1980).
25. Ebke DH and MK McCarty. Proc. North Cent. Weed Control Conf. 35:13-14 (1980).
26. Harvey SJ, RM Nowierski, PG Mahlberg, and JM Story. Weed Sci. 36:726-733 (1988).
27. Watson AK. In: *Leafy Spurge Monogr. 3* (AK Watson, Ed.), pp. 1-6. Weed Sci. Soc. Amer., Champaign, IL, 1985.
28. Bjugstad AJ, R Francis, and PW Skinner. Forestry Committee, Great Plains Agric. Counc., Billings, MT (1983).
29. Belcher JW and SD Wilson. J. Range Manage. 42:172-175 (1989).
30. Stack RW and GD Statler. Proc. N.D. Acad. Sci. 43:86 (1989).
31. Bowes G and AG Thomas. *Proc. 1st Intern. Rangeland Congress*, pp. 254-256 (1978).
32. Maxwell BD, MV Wilson, and SR Radosevich. Weed Technol. 2:132-138 (1988).
33. Watson AK. In: *Leafy Spurge Monogr. 3* (AK Watson, Ed.), pp. 93-104. Weed Sci. Soc. Amer., Champaign, IL, 1985.
34. Stroh RK, JA Leitch, and DA Bagsund. N.D. Farm Res. 47(66):15-17 (1990).
35. Lym RG and CG Messersmith. Weed Technol. 4:635-641 (1990).
36. Messersmith CG and RG Lym. N.D. Farm Res. 40(5):8-13 (1985).
37. Lym RG and CG Messersmith. Res. Prog. Rep. West. Soc. Weed Sci., p. 19-20 (1987).
38. Alley HP, NE Humberg, JK Fornstrom, and M Ferrell. Res. Weed Sci. Wyo. Agric. Exp. Stn. Res. J. 192:90-93 (1984).
39. Reilly W and KR Kaufman. N.D. State Univ. Coop. Ext. Serv., unnumbered publication (1979).
40. Lym RG and DR Kirby. Weed Technol. 1:314-318 (1987).
41. Hein DG. PhD thesis, Univ. of Wyo., Laramie (1988).
42. Thompson F, JA Leitch, and FL Leistritz. N.D. Farm Res. 47:9-11 (1990).

17

Biological Control of Leafy Spurge

Norman E. Rees and Neal R. Spencer

Abstract

Leafy spurge (*Euphorbia esula*) is an aggressive, perennial weed of the northern United States and southern Canada which reproduces by seeds and vegetative root buds. Dense stands of leafy spurge replace grasses and forbs and restrict cattle grazing on rangeland. Eight Eurasian insect species which attack leafy spurge are currently approved for release and have become established in the United States. Eight additional insect species are in various stages of United States clearance procedures. A total of twelve insect species have been approved for release in Canada and many are established there. In 1990, a large scale five-state research project was started to determine methods for establishment, population increase and assessment of impact of the flea beetle, *Aphthona nigriscutis* (COLEOPTERA:CHRYSOMELIDAE) on leafy spurge.

Introduction

Leafy spurge is an aggressive, persistent, deep-rooted perennial plant of Eurasian origin that has become dominant on rangelands and pastures, displacing useful forage plants in North America. First reported in the United States in 1827, leafy spurge now extends throughout much of southern Canada and the northern United States (1-3). Leafy spurge produces a milky latex that is poisonous to some animals and can cause blisters and irritation on contacted skin, and similar results to the digestive tract when ingested by man and some animals. In cattle, it causes scours and weakness, and in larger ingested amounts, can cause death (4,5). Cattle usually refuse leafy spurge as food unless it is given to them in weedy hay or when better forage is not available.

Leafy spurge reproduces both by vegetative regrowth and the production of large quantities of seeds which are often distributed by birds, wildlife, man, water, etc. (6,7). Leafy spurge plants are able to maintain high root reserves through an extensive root system, ranging from a massive network of small roots near the soil surface to deep penetrating tap roots. This ability to maintain high root reserves permits the plant to recover quickly from physical and most chemical damage (8,9).

A conservative estimate of loss in the United States, in terms of expenditures for controlling leafy spurge and loss of productivity, was given by Noble et al. in 1979 as $10.5 million annually (10). Derscheid and Wrage report that the problem is most severe on undisturbed lands, but on cultivated cropland areas, it can reduce crop yields from 10-100% (11). Thompson et al. recently concluded that in North Dakota, 405 thousand hectares of leafy spurge had depreciated land values by $137 million and had a total economic impact of $105 million in that state for 1989 (12).

In Europe, there are 105 native *Euphorbia* species which belong to the subgenus *Esula*, the group to which the target leafy spurge belongs. An additional four species belong to the subgenus *Chamaesyce*. In North America, 21 native species belong to the subgenus *Esula*, 26 species belong to the subgenus *Agaloma*, 3 species to *Poinsettia* and 57 species to *Chamaesyce*. Two species (*E. garberi* and *E. deltoides*), belonging to the subgenus *Chamaesyce*, currently have federal protection under the endangered or threatened species act[1] (13-16).

The taxonomic status of the introduced North American leafy spurge complex is in a state of confusion. Whereas Ratcliffe-Smith lists 79 different species, crosses, and hybrids in Europe (17), some believe that the leafy spurge complex in the United States is one species (18). It is difficult to believe that only one spurge species was introduced into the United States from Eurasia. Either way, more perplexity is added when one considers that this weed may have been introduced from multiple sources throughout Eurasia (19,20). These conditions must all be considered, therefore, when biocontrol agents are being selected for introduction into North America (21).

---

[1]Pemberton RW. Petition by Rees, Pemberton, et al. to clear *Dasineura* sp. nr. *capsulae* submitted to T.A.G. 1989. (This species has not yet been named.)

Established Biocontrol Agents

Since leafy spurge reproduces by both seed and vegetative regrowth, and can maintain high root reserves, eventual management programs against leafy spurge must utilize an assortment of biotic agents which are able to concentrate their action at various, distinct locations on the plant, (e.g. root hairs, roots, crown, stems, leaves, flowers, and seeds).

Variations in environment, plant communities, soil conditions and structure, ground aspect, host biotypes, etc. can influence the suitability of a biocontrol agent at specific locations. For example, a root-attacking insect that is adapted to dry, sandy soils in open sunlight would do well at one location, while one that is adapted to wet, loamy soils in the shade would do well in another.

Eight insect species, which are host specific to leafy spurge, are currently approved for release in the United States by U.S. Department of Agriculture, Animal and Plant Health Inspection Service, Plant Protection and Quarantine (USDA/APHIS/PPQ), and are established in the United States. These include a moth that defoliates the plant, a beetle that burrows in the stems and roots, a fly that galls the growing tips of leafy spurge, a clear wing moth whose larvae attack spurge roots, and four species of flea beetles that attack both the foliage of the plant and its roots.

Leafy Spurge Hawk Moth, Hyles euphorbiae
(LEPIDOPTERA:SPHINGIDAE)

Larvae of the leafy spurge hawk moth are defoliators of leafy spurge. There are generally two generations per year, with pupae overwintering in the soil. Adults rest during most of the day but will fly, mate, and lay eggs in the daylight and twilight hours. Eggs are generally deposited on the underside of leafy spurge leaves and resultant dark, hungry larvae hatch and devour the leaves and flower parts. Each instar is somewhat different in color beginning with a dark grey caterpillar and ending in the last larval instar with one which is brightly colored. The bright coloring serves to identify the larvae as being toxic to predators through its sequestering of leafy spurge toxins (22,23)

The hawk moth was first collected in Europe on E. sequieriana and cypress spurge (E. cyparissias) and was maintained in Canada by Peter Harris in both laboratory and field colonies (24). Insect releases in the United States between 1966 and 1974 were obtained from these colonies. Although initial establishment was documented on leafy spurge near Bozeman, Montana, after a 1966 release, moths from numerous releases at other sites from Canadian stock of the hawk moth were never recovered.

Batra collected third and fourth instar larvae at Braeside, Ontario, Canada, and released them at Chestertown (Warren Co.), New York, on cypress spurge in 1977. The insect established on cypress spurge and one of the hybrids of leafy spurge. The population increased from the original 180 larvae to an estimated 1 million insects within 5 years with total defoliation of spurge in some areas (25).

Hawk moths were collected in 1974 near Debrecen, Hungary, from *E. esula* and *E. virgata*. As a result of the release of the Hungarian insects, four field colonies were established. Moths from these colonies have dispersed over large areas in Montana. Redistribution of insects from the Bozeman and New York field colonies to other states and other locations within Montana was conducted in 1985 under the direction of Robert Nowierski (Montana State University) with the assistance of Steven Hight (presently with the USDA/ARS).

Two factors have reduced the potential of the leafy spurge hawk moth as a biocontrol agent for leafy spurge. First, it is a defoliator of leafy spurge. Defoliation has little impact as sheep and goats defoliate leafy spurge with little reduction in the reproduction of leafy spurge. Second, several colonies of the hawk moth are known to contain a nuclear polyhedrosis virus which was either introduced with it or subsequently infected from other lepidoptera in the United States. Under proper conditions, the virus is an important mortality factor.

Red Headed Beetle, Oberea erythrocephala
(COLEOPTERA:CERAMBYCIDAE)

Adults of the red headed longhorn beetle girdle and make hollows in the stems of leafy spurge into which they deposit eggs. Larvae feed first in the stems and then migrate into the crown and roots. Feeding causes major damage to the root reserves of the plant. There is only one insect generation per year with larvae overwintering in the roots near the crown. Adults are generally first located in the field about the first of July (26).

Releases of the red headed longhorn beetle began in 1982 from material collected near Debrecen, Hungary. Establishment first occurred near Bozeman, Montana, and later at three other locations in the state (27). It also became established in North Dakota at two or three sites (Robert Carlson, personal communication). Although of great importance in countries such as Yugoslavia, populations in Montana have never increased to large numbers. Therefore, adults have not been relocated to new sites in the United States.

Gall Midge, Spurgia esulae (DIPTERA:CECIDOMYIIDAE)

Larvae of this gall midge, a delicate small fly, cause terminal leaves of leafy spurge to grow together, forming a gall. The leaf gall becomes a protective niche in which the immature stages of the insect develop. Inner leaves of the gall, which are the apical tip of the stem, eventually die, thus reducing the plant's ability to produce flowers and seeds.

The gall midge has multiple generations per year. Adults generally hide in shaded areas during the day. There they will mate and lay eggs on the terminal spurge leaves. An adult gall midge may live for only 24 to 36 hours. Mass adult emergence is often associated with storm systems which last for four to five days. Mature larvae of the autumn generation overwinter in the soil.

The gall midge was originally thought to be *Bayeria capitigena* when it was screened and much of the life history information on *B. capitigena* from the literature was applied to it. Some differences in the supposed life cycle were detected, requiring new life history studies in Montana. This midge was first released in Montana in 1985 and one site was established in North Dakota by Robert Carlson of NDSU. Three of six Montana releases were established by 1987, but the insect population declined until galls were found only at the original release site in 1989. The size and density at this release site, however, has been increasing each year and is the focus of life history and habit studies. Releases in North Dakota have also been successful (Robert Carlson, personal communication).

Chamaesphecia empiformis (LEPIDOPTERA:AEGERIIDAE)

Larvae of this clear wing moth attack the roots of leafy spurge. The moth was cleared for introduction into the United States in 1977 to assist in the control of cypress spurge. Although there have been several attempts to establish this species in the western part of the United States, especially in California, Oregon, and Washington, no records of recoveries are known. A second, closely related species, *C. tenthrediniformis*, may have been introduced accidentally, but also is thought not to have estab-lished. Although there are numerous species in this genus, each species is extremely narrow in its host range. The host range of *C. empiformis* is restricted to cypress spurge, and *C. tenthrediniformis* is restricted to leafy spurge. A permit was issued to Robert Pemberton (USDA/ARS) in 1985 to work on *C. tenthrediniformis* in quarantine, but none survived on the spurge offered (Jack Coulson, personal communication).

Flea beetles, Aphthona spp. (COLEOPTERA:CHRYSOMELIDAE)

Flea beetles: black spurge flea beetle, black dot spurge flea beetle, brown dot spurge flea beetle and copper spurge flea beetle.

Adult flea beetles, *Aphthona* spp., feed on leafy spurge leaves and bracts, while the larvae feed on the root hairs and yearling roots. Larval feeding damages the roots and reduces the plant's ability to take up nutrients and moisture. Moderately attacked plants show retarded flowering periods. Canadian releases in the early 1980s have shown that continued pressure by the flea beetles first reduces the average plant height and then, as the insect population increases further, plant density drops to acceptable levels and native vegetation returns (28).

Most *Aphthona* flea beetle species have one generation per year on leafy spurge; adults are present between late June and early September when eggs are laid in the soil near leafy spurge roots. Larvae hatch and immediately begin to feed on leafy spurge roots. As they grow, they move to larger roots where they feed both externally and internally. A portion of the leafy spurge control that has already been documented from flea beetle release sites may be due to the secondary invasion of plant pathogens since they are present in all soils. Mature larvae overwinter in the soil and pupate in late spring or early summer.

Although these four species of flea beetles are similar in action, there are some characteristics which tend to separate them. The black spurge flea beetle (*A. czwalinae*) is totally black in color and resides in areas of higher moisture and relative humidity than are acceptable by the other three species. The copper spurge flea beetle (*A. flava*) is amber colored, while the brown dot spurge flea beetle (*A. cyparissiae*) and the black dot spurge flea beetle (*A. nigriscutis*) are brown and can be separated by a black dot between the forward section of the two elytra on the latter species (29). From work by Peter Harris in Canada, it appears that the black dot flea beetle prefers sandy soils with low humus content, while the brown dot flea beetle does better in higher humus and moisture content soils (Peter Harris, personal communication).

The black spurge flea beetle has not been recovered in Montana from a 1987 release, although several establishments occurred from releases by Robert Carlson (NDSU) and Robert Richard with the USDA/APHIS (Robert Carlson and Robert Richard, personal communications). The copper flea beetle was first released near Bozeman in 1985 and in North Dakota and Idaho in 1986. It established at four of the eight Montana sites and at a single North Dakota site. The brown dot flea beetle was first released in Montana in 1987 and established at two sites. Most 1989 releases of the black dot flea beetle appear to have established in 1990 in Idaho, Montana, Nebraska, and North Dakota.

Additional Canadian Releases

Canada has approved 12 insect species for the biological control of leafy spurge. In addition to the agents cleared and released in the United States, Canada has released a clear wing moth (*C. empiformis*), a second clear wing moth (*C. tenthrediniformis*), a leaf tying tortricid moth (*Lobesia euphorbiana*), a geometrid moth (*Minou murinata*), and two gall flies (*Pegomya curticornis* and *P. euphorbiae*).

Biocontrol Agents Presently Being Screened in the United States

The flea beetle *A. abdominalis* is multivoltine (more than one generation a year). The head and most of the body is reddish-yellow, with the elytra and legs a light, bright yellow. Adults feed on the leaves of leafy spurge as do all the other introduced flea beetles. However, the larvae feed on stem buds of adventitious roots instead of directly on the roots as do the other *Aphthona* species. The USDA/ARS laboratory in Rome, Italy, conducted host specificity testing of this species, and the target date for its release into the United States is the summer of 1991.

Aphthona chinchihi is a flea beetle which was collected in China and is being screened by the USDA/ARS Rome laboratory. It is hoped that testing of this species can also be done at the biological quarantine facility at Bozeman. Target date to complete screening of this species is probably three years away.

A third flea beetle, *A. seriata*, was also collected in China and is under testing at the USDA/ARS Rome laboratory. Once the relationship of *A. seriata* to leafy spurge is established at the overseas lab, and if the data show clear host specificity, a request will be made to import the species into the biological quarantine facility at Bozeman for final testing. Target date to complete screening of this species is probably 1993.

Chamaesphecia crassicornis is a clear winged moth similar to *C. empiformis* and *C. tenthrediniformis*. Research already completed by ARS at their Rome laboratory indicates that this species appears to be restricted to *E. virgata* type leafy spurge. It has one generation a year with emergence occurring from May through July and oviposition thereafter. Eggs are deposited on leafy spurge stems. Hatching larvae migrate down the plant to the crown where they penetrate the plant to feed in the roots. Mature larvae bore upwards to the stem where pupation occurs. Final testing will be conducted at Bozeman with a target release date of summer 1992.

Dasineura sp. nr. *capsulae* is another species of gall fly similar to *S. esula*, except that the gall is an encapsulate gall and the species has only one generation per year. This species is still awaiting a proper name. The petition to introduce this species into the United States was approved last September by the Technical Advisory Group (TAG) of Plant Pest & Quarantine (PPQ), but issuance of permits is awaiting approval of the Environmental Assessment.

Eurytoma euphorbiana is a small wasp that attacks the seeds of leafy spurge. Testing of this species is just beginning and will be conducted at both the Rome and Bozeman quarantine facilities from specimens obtained in the USSR.

Oxicesta geographica is a multivoltine noctuid moth whose larvae defoliate leafy spurge. The early instars are gregarious, with the last two being solitary. Most of the host specificity testing was conducted in Europe and will be concluded in Bozeman this year.

Simyra dentinosa is a univoltine moth whose larvae feed in silk webs on the apex of leafy spurge. There are six instars requiring a total of 30 days, with all instars being hairy. The first instar is yellow and black, the second black, the third to sixth are dark brown with light brown banding. The first four instars are gregarious, while the last two are solitary. Most of the host specificity testing has been done in Europe, but should be concluded in Bozeman this year.

Additional Canadian Screening

Aphthona lacertosa is a black flea beetle species, with a light to dark brown hind femur, which separates it from *A. czwalinae* with a black hind femur. Each female can deposit two to three hundred eggs in small groupings under the soil surface near the stem of their hosts, leafy and cypress spurge. Mature larvae overwinter in the roots. In Hungary, this species occurs on a range of mesic-dry to moist sites (28).

Large Scale Testing of the Flea Beetle, *Aphthona nigriscutis*

A large scale experiment was begun in 1990 to observe the various conditions which affect the efficiency of the black dot flea beetle to suppress leafy spurge. Whenever possible, seven treatments with nearly matching soil, plant, physical and environmental conditions were established at six sites in five states. Each treatment/check area

190

measured ca 500 m X 100 m, with a 100 m dia. treatment at one end and its equal check at the other. This experiment has the following objectives:

1. To study the efficacy of various numbers and patterns of releases of *Aphthona nigriscutis* as to its effects on populations of the leafy spurge plant, flowering periods, growth patterns, and spurge density.
2. To study the effects and interactions of different associated plant communities on the efficacy of the beetle in suppressing populations of leafy spurge.
3. To study the effects of different leafy spurge densities on the efficacy of the beetle.
4. To study the direct and indirect effects of different soil types and composition on the efficacy of *A. nigriscutis*.
5. To determine factors that affect how fast *A. nigriscutis* populations can increase and expand.
6. To monitor the changes in plant community constituents, density, and biomass production as leafy spurge competition is reduced.
7. To determine the optimum size of release for *A. nigriscutis*. From this information, protocols for the sizes of future release for various *Aphthona* species may be established.
8. To determine the optimum period for releasing adults.
9. To locate detrimental factors for *A. nigriscutis* establishment and population dynamics and to investigate how best to avoid or manipulate such factors.
10. To compare efficiency of sweep sampling of the flea beetle on leafy spurge with D-VAC sampling.
11. To determine effectiveness of the beetle in relation to weather conditions, elevation, site conditions, etc.
12. To investigate the effect of host-plant genetics (biotype) on the efficacy of *A. nigriscutis* against leafy spurge.

This research is being conducted by the USDA/ARS in cooperation with Bureau of Land Management (BLM), U.S. Forest Service (USFS), USDA/APHIS, Resource Conservation and Development (RC&D), Agriculture Canada, and Montana State University (MSU). Six sites were selected in five states, Colorado, Idaho, Montana, Nebraska, and North Dakota.

From three to seven treatments and duplicate checks were randomly assigned at each site, such that topography, vegetation, soil conditions, etc. were similar throughout the study area, and particularly between each treatment and its check. Where possible, a minimum of one kilometer exists between treatments, and 200 to 300 meters separates a treatment and its check. Transects radiate outward from the center in each of eight directions: north, northeast, east, southeast, south, southwest, west, and northwest. Along each transect, marked wooden stakes identify sampling loci along the transect. The stakes begin 8 m from the center spike (the point of insect release in the treatment), and continue outward every 4 m for a total of 40 m. Outer ends and the center are demarcated with railroad spikes imprinted with the treatment/check number and direction of the transect.

Plant composition, canopy cover, spurge plant height, spurge density and number of flowering stems are to be sampled and recorded from each of the 0.1 m² north, east, west, and south test loci. The four remaining transects are to be used for clipping samples at each 0.1 m² locus. Plant clippings are taken each year by advancing the sample site outward the width of the Daubenmire frame for each previous year sampled, with this year's sample being clipped on the opposite of the line from that of the last.

Conclusion

Since leafy spurge is not considered a problem in its native lands, there is a very strong likelihood that in time, with the proper understanding and the correct combination of biocontrol agents, we can recreate that balance of nature in North America that endures in Eurasia, and ultimately force leafy spurge to take a role as simply a member of the North American plant community rather than as a noxious weed.

References

1. Britton NL. J. New York Bot. Gard. 22:73-75 (1921).
2. Dunn PH. Weed Sci. 27:509-516 (1979).
3. Hanson HC and VE Rudd. ND Agric. Exp. Sta. Bull. 266, Fargo, 1933.
4. Kingsbury JM. *Poisonous Plants of the United States and Canada.* Prentice-Hall Inc., Englewood Cliffs, NJ, 1964.
5. Messersmith CG and RG Lym. ND Farm Res. 40(5):8-13 (1983).
6. Selleck GW, RT Coupland, and C Framltpm. Ecol. Monogr. 32:1-29 (1962).
7. Best KF, GG Bowes, AG Thomas, and MG Maw. Can. J. Plant Sci. 60:516-663 (1980).
8. Lym RG and CG Messersmith. J. Range Manage. 40(2):139-144 (1987).
9. Muemscher WC. *Poisonous Plants of the United States.* pp. 142-144. Macmillan Co., New York, 1940.
10. Noble D, PH Dunn, and L Andres. ND State Univ. Coop. Ext. Serv., Bismarck, ND, 1979.
11. Derscheid LA and FS Wrage. SD State Univ. Ext. F. S. 449, 1972.
12. Thompson F, JA Leitch, and FL Leistritz. ND Farm Res. 47(6):9-11 (1990).

13. Webster GL. Taxon 24:593-601 (1975).
14. Webster GL. J. Arnold Arboretum 48:363-430 (1967).
15. Wheeler LC. Rhodora 43:97-154; 168-205; 223-268 (1941).
16. Wheeler LC. Amer. Midl. Nat. 30:456-503 (1943).
17. Radcliffe-Smith A. Mono. Ser., Weed. Sci. Soc. Amer. 3:14-25 (1985).
18. Harvey SJ. Weed Sci. 30:726-733 (1988).
19. Dunn PH and A Radcliffe-Smith. Research Report, North Central Weed Cont. Conf. 37:48-53 (1980).
20. Dunn PH. Weed Sci. Soc. Amer. Monograph 8:7-13 (1985).
21. Pemberton RW. In: *Proc. VI Int. Symp. Biol. Contr. Weeds* (ES Delfosse, Ed.), pp. 365-390. Can. Govt. Pub. Centre, Ottawa, 1985.
22. Harris P, PH Dunn, D Schroeder, and R Vonmoss. Weed Sci. Soc. Amer. Monograph 8:79-92 (1985).
23. Marsh N, M Rothschild, and FJ Evans. In: *The Biology of Butterflies*, p. 135. Academic Press, London, 1984.
24. Harris P and J Alex. *Biological Control Programmes in Canada 1959-1968*, 1971.
25. Batra SWT. New York Entomol. Soc. 91(4):304-311 (1983).
26. Schroeder D. Zeitschrift für angewandte Entomologie 90:237-254 (1980).
27. Rees NE, RW Pemberton, A Rizza, and P Pecora. Weed. Sci. 34:395-397 (1986).
28. Maw E. MS Thesis, Univ. of Alberta, Edmonton, 1981.
29. Sommer G and E Maw. Final Report Commonwealth Inst. Biol. Contr., European Station, Delemont, Switzerland, 1982.

18

Controlling Leafy Spurge with Grazing Animals

Peter K. Fay

Abstract

Leafy spurge (*Euphorbia esula*) is a deep rooted perennial weed that is very difficult to control. It is palatable to sheep and goats but not cattle. It is being controlled by sheep grazing on thousands of hectares in Montana. Availability of sheep is a major factor limiting their use as control agents. Timeliness of grazing is also a problem since it is difficult to graze large areas of leafy spurge at the correct time. Overgrazing and animal loss to predators have been problems when producers use grazing as a means of control. Animal containment has been achieved with portable electric fences.

Introduction

Leafy spurge is an introduced perennial weed infesting over 400,000 ha in Montana. Eradication is not possible and even short-term control with herbicides is not cost effective (1).

Sheep Grazing Leafy Spurge

While ranchers have been using sheep to graze leafy spurge for more than 30 yr in Montana, there has been relatively little scientific research conducted. Johnston and Peake (2) examined the effect of selective grazing by sheep on a mixed crested wheatgrass-leafy spurge pasture.

A 12 ha field infested with leafy spurge was seeded with crested wheatgrass (*Agropyron cristatum*). While excellent stands of crested wheatgrass were established, the competition did not control leafy spurge. Twelve yr after seeding, 45 ewes were placed on the pasture from May to September each yr for a period of five yr. At the conclusion of the study, the basal area of leafy spurge was reduced 98% while the basal area of crested wheatgrass increased 20%.

Muenscher (3) stated that large infestations of leafy spurge should be seeded with grass and pastured closely with sheep for at least three or more yr. Bibbey (4) reported that sheep effectively controlled leafy spurge and nearly eradicated the plant if grazing continued over a period of yr.

Helgeson and Thompson (5) grazed sheep on 0.4 ha pastures (mowed and unmowed) infested with leafy spurge. After 2 yr of grazing, the number of leafy spurge stalks was reduced 68 and 31% in the mowed and unmowed pastures, respectively. The total basal area of bluegrass increased 179 and 212% in the mowed and unmowed pastures, respectively. Baker (6) found that grazing leafy spurge with sheep reduced the initial infestation 34% and led to a 10% increase in grass density after six yr. Wood (7) reported several cases where sheep were used to bring heavily infested pastures under control after two to four yr of continuous grazing.

Bowes and Thomas (8) studied the longevity of leafy spurge seeds following various control programs. Picloram (4-amino-3,5,6-trichloro-picolinic acid) was applied at two locations while a third location was grazed by sheep for eight yr. Seeds were recovered from soil and tested for viability. The number of leafy spurge shoots/m^2 was recorded annually. During the first 3 to 4 yr, picloram was as effective as sheep grazing in reducing the density of leafy spurge, but the number of viable seeds in the soil increased nearly three-fold. The number of viable seeds in the grazed pastures decreased from 3,500 to 15 seeds/m^2 in the eight-yr test period.

There is disagreement in the literature concerning the effect of leafy spurge on sheep. Muenscher (3), Christensen et al. (9), Wood (7), and Baker (6) state that sheep suffered no deleterious effects as a result of consuming leafy spurge. Bakke (10) stated that neither sheep nor cattle will eat leafy spurge because of its acrid latex. He did indicate that sheep in an enclosed area will eat leafy spurge if starved and no harmful effects were noted. Helgeson and Thompson (5) found that sheep suffered no ill effects from consuming leafy spurge with the exception of several lambs which scoured.

Alternatively, Johnston and Peake (2) attributed the deaths of an unknown number of sheep to poisoning as a result of consuming relatively large leafy spurge plants. However, no losses were reported when sheep were placed in the pastures at an earlier date. The authors stated that post-mortem examinations indicated that leafy spurge was responsible for the poisoning losses.

Bartik and Piskac (11) stated that leafy spurge and other closely related species produce a poisonous white latex which affects almost all animal species. These authors added that drying does not remove the plant's toxicity. After ingestion, an animal will experience gastroenteritis, violent vomiting, sometimes diarrhea with blood in the feces, feeble heartbeat, and depression of the central nervous system.

Farnsworth et al. (12,13) conducted a preliminary phytochemical and biological evaluation of leafy spurge. An extract of leafy spurge at doses ranging from 5 to 400 mg/kg produced a very weak central nervous system depression in mice. Tertiary and quaternary alkaloids were present in the aerial parts of leafy spurge.

Helgeson and Thompson (5) observed the grazing habits of sheep on spurge. They reported that sheep first ate all the blossoms and seeds, and later stripped the stems of the leaves. They reported that the sheep would alternatively graze leafy spurge and grass, eating the spurge rather "greedily" for a time and then shifting to grass. Johnston and Peake (2) observed that sheep preferred younger leafy spurge plants and avoided more mature plants. They felt that sheep distribution does not need to be controlled for the first three yr of grazing since sheep will naturally congregate in areas infested with leafy spurge. After three yr sheep should be moved to avoid overgrazing.

Baker (6) observed that sheep placed on leafy spurge required an initial adjustment period. Following this adjustment, the sheep appeared to prefer the plant. Christensen et al. (9) reported most of the lambs fed leafy spurge silage did not consume much the first day. After a few days, all but two of the animals were eating enough leafy spurge to continue the test for 34 d without any apparent ill effects. This initial avoidance, called neophobia, is a recognized foraging strategy in ruminant animals and an adjustment period prior to acceptance of a new feedstuff is not uncommon (14). Wood (7) reported that sheep had a decided preference for leafy spurge over other herbage in a pasture dominated by crested wheatgrass.

Helgeson and Thompson (5) found that ewes and lambs grazing leafy spurge-infested pastures gained weight over a three-month period that was comparable to animals in uninfested pastures. Ewes averaged 1.8 and 3.1 kg gains while lambs averaged 12.6 and 18 kg gains on leafy spurge-infested and uninfested pastures, respectively.

Christensen et al. (9) conducted a study which determined the digestibility of leafy spurge silage. Three lots of silage were prepared:

(1) leafy spurge alone; (2) leafy spurge plus phosphoric acid; and (3) leafy spurge plus molasses. The total digestible nutrients (TDN) for each silage were 45.2, 50.5, and 50.3%, respectively. The percent crude protein (CP) was 9.2, 9.5, and 9.4%, respectively. Lambs fed each of the silages lost an average of 1.5 kg per animal during a 10-d feeding trial. They attributed the loss in weight to the fact that the silages did not meet the minimum maintenance requirements for CP or TDN.

Landgraf et al. (15) conducted a feeding experiment for three months during the winter of 1981 to determine if leafy spurge hay would have deleterious effects upon sheep. Three groups of sheep were fed a hay ration of introduced grass species, an increasing level of leafy spurge, or entirely leafy spurge. Blood samples were analyzed throughout the experiment to detect internal physiological effects. Leafy spurge hay did not cause any deleterious effects.

A field grazing study (15) was also conducted during the summer of 1981 where esophageal fistulated ewes were placed in pastures containing light, moderate, or heavy infestations of leafy spurge to determine if leafy spurge intake would be influenced by infestation level. No definite preference for or an avoidance of leafy spurge was observed. There was a one to three week initial adjustment period before sheep selected a significant amount of leafy spurge. Intake of leafy spurge steadily increased during the study and reached 40 to 50% of the diet for sheep on all three infestation levels. Weights of the ewes in leafy spurge-infested pastures were not significantly different from those in control pastures containing no leafy spurge. The authors concluded that leafy spurge is relished by sheep and should be classified as desirable sheep forage.

The Present Situation

The sheep grazing research conducted in Montana (15) received considerable publicity in the state over the years. The failure of herbicides to control leafy spurge, and the absence of success to date of classical biological control, has led to widespread acceptance of sheep grazing as a means of controlling leafy spurge. At the present time, almost every large infestation of leafy spurge in Montana is being grazed by sheep.

The major problem with this technique is the limited availability of sheep. There are simply not enough sheep to address the leafy spurge problem primarily because it is not cost effective to haul sheep very far even if the spurge-infested pastures are provided rent-free. The most successful attempts to incorporate sheep have occurred when the landowner purchased sheep, or a very close neighbor with sheep brought his animals to the infested land.

The second major problem is timeliness of grazing. It is difficult to get sheep across large infestations of spurge in a timely fashion to prevent seed production and reduce competitiveness of spurge.

A third problem that has occurred frequently is overgrazing. There have been instances where sheep have been put on infested land for too long. The result is that overgrazing by sheep becomes far more of a problem than leafy spurge.

Predators

A problem commonly encountered is animal loss to predators. Large sheep producers have historically addressed the problem of predators with herders. However, competent herders are very difficult to find. Many ranches heavily infested with spurge are cattle ranches. When the managers consider the addition of sheep, the issue of predators is a major deterrent to adopting sheep grazing. The unavailability of herders has led to the use of herder substitutes.

One herder substitute method that is being tested is the use of llamas. Normally, a single llama is placed with a herd of sheep. Llamas are both curious and protective. If a disturbance occurs, the sheep will move toward the llama, and the llama will inspect the disturbance. Llamas hate dogs and have been effective in warding off coyotes. It is important that humans leave the llama alone to prevent domestication. Human interference has reduced llamas' attention to sheep.

Containment

Another problem that occurs with grazing animals for weed control is containment on the weed infestation. It is often not feasible to fence the weedy area, which permits animals to wander. There have been successful efforts to use portable electrical fencing systems. One effective system uses plastic twine which is intertwined with small gauge stainless steel wire. The plastic twine is strung on nylon posts and a two-strand fence has contained both cattle and sheep.

Electric net fencing has also been effective for sheep containment on leafy spurge and the net arrangement provides coyote protection. Net fencing systems use a solar-powered transmitter, are easily erected, and very portable. Mr. Reeves Petroff, a weed supervisor in Bozeman, MT, is using net fencing to contain sheep on highly visible 0.4 to 0.8 ha sites infested with spurge that are too sensitive for herbicide use. The project serves as an example to the public that the weed supervisor is using an integrated approach to weed management and not just herbicides.

Goats

Goats eat leafy spurge and in some instances serve as an effective method of control. A major problem faced with goats is containment. We successfully tested a dog containment system (16) where an electric shock collar is placed around the neck of the animal. The area of containment is delineated by a single 14 gauge wire which is either laid on or placed just below the soil surface. A small current is transmitted through the wire which creates a magnetic field. When the animal wearing the electric shock collar enters the magnetic field, the collar emits a tone. If the animal does not withdraw within 2 sec, a shock is received. The system is manufactured by the Invisible Fence Company of Wayne, Pennsylvania. The Invisible Fence System worked well since goats were easily trained to the collar system and stayed in the containment area. Unfortunately, the cost of the commercial dog collars is prohibitive for livestock containment.

With the help of several electrical engineering students at Montana State University, we have improved the transmitter to permit containment to areas as large as 14 ha. At the present time, we are testing an inexpensive collar system which employs current electronic circuitry. It is our intent to promote production of inexpensive collars for livestock containment.

Range Weed Management

Leafy spurge and many other range weeds occur in three basic phases. Phase One occurs when there are 20 plants per ranch. Phase Two infestations consist of 20 patches of leafy spurge per ranch, and Phase Three leafy spurge infestations cover 8,000 ha per ranch. If not addressed in Phase One, spurge will spread and become a Phase Two situation. If left unchecked, the infestation moves to Phase Three. In Montana at the present time, many Phase Three infestations are being addressed with sheep grazing. Many Phase One infestations are being effectively treated with herbicides. The most difficult phase to control is the Phase Two infestation since it's uneconomical to spray herbicides and difficult to manage with sheep because small patches of leafy spurge are spread over very large areas.

Conclusion

Grazing animals can be used to control leafy spurge and other weed species. Unfortunately, each infestation is unique so considerable creativity is needed to make grazing an effective solution. Problems such as predators, animal availability, and animal containment must be overcome to achieve success. Most important, good animal and grazing management techniques must be employed.

References

1. Alley P. In: *Proceedings: Leafy spurge symposium*. North Dakota Coop. Ext. Service, unnumbered publication, pp. 53-69 (1979).
2. Johnston A and PW Peake. J. Range Manage. 13:192-195 (1960).
3. Muenscher WC. New York State Coll. Agri. Ext. Bull. 192, 10 pp. (1930)
4. Bibbey RO. Ontario Dept. Agri. Circ. 125. 5 pp. (1952).
5. Helgeson EA and EJ Thompson. Dakota Agri. Exp. Sta. Bi-monthly Bull. Vol. II:5-9 (1939).
6. Baker LO. Personal communication. Dept. of Plant and Soil Science, Montana State University, Bozeman (1981).
7. Wood HE. Manitoba Dept. Agr. Pub. No. 200. 9 pp. (1945)
8. Bowes GG and AG Thomas. J. Range Manage. 31:137-140 (1978).
9. Christensen FW, TH Hopper, EA Helgeson, and EJ Thompson. Proc. Amer. Soc. Anim. Prod. 31:311-316 (1938).
10. Bakke AL. Iowa Agri. Exp. Sta. Res. Bull. 198:209-245 (1936).
11. Bartik M and A Piskac. *Veterinary Toxicology*. Elsevier Scientific Publishing Co., Amsterdam, 1981.
12. Farnsworth NR, LK Henry, GH Svoboda, RN Blomster, MJ Yates, and KL Euler. Lloydia 29:101-122.
13. Farnsworth NR, H Wager, L Hörhammer, HP Hörhammer, and HHS Fong. J. Pharm. Sci. 57:933-939 (1968).
14. McClymont GL. In: *Handbook of Physiology, Vol. I. Control of Food and Water Intake*, pp 129-137. American Physiological Soc., Washington, D.C., 1967.
15. Landgraf BK, PK Fay, and KM Havstad. Weed Sci. 32:348-352 (1984).
16. Fay PK, VT McElligott, and KM Havstad. Appl. Anim. Behavior Sci. 23:165-171 (1989).

19

Chemical Control of Leafy Spurge

Rodney G. Lym and Thomas D. Whitson

Abstract

Herbicides commonly used to control leafy spurge (*Euphorbia esula*) include 2,4-D, dicamba, glyphosate, and picloram. Picloram is the most effective herbicide while a combination of picloram plus 2,4-D is the most cost-effective treatment. Most herbicides are applied during the leafy spurge true-flower growth stage, but glyphosate is most effective in the fall. Dichlobenil can be used to suppress leafy spurge growth under trees while fosamine, glyphosate, and 2,4-D can be used adjacent to water. Sulfonylurea and imidazoline herbicides control leafy spurge but may cause severe grass injury. Once grasses were established, Russian wildrye, pubescent wheatgrass, big bluegrass, and intermediate wheatgrass were more competitive than other grass species in leafy spurge-infested rangeland and maintained at least a 90% cover for 4 yr.

Introduction

Leafy spurge is difficult to eradicate, but topgrowth control and a gradual decrease in the underground root system is possible with a persistent management program. Nearly all experimental herbicides have been tested on leafy spurge since the introduction of 2,4-D in the 1940s. Most herbicides have little or no activity on leafy spurge.

Herbicides commonly used to control leafy spurge include 2,4-D, dicamba, glyphosate, and picloram (3). Picloram, dicamba, and 2,4-D are selective herbicides that control broadleaf weeds while glyphosate is nonselective and controls both grass and broadleaf weeds. Dichlobenil (2,6-dichlorobenzonitrile) suppresses leafy spurge growth only and can be used under trees (4) and fosamine can be used adjacent to water (5).

Long-term control of leafy spurge is extremely difficult to achieve. The most cost-effective control method depends on the size and location of the infested area. Small patches of leafy spurge can be eliminated with a persistent herbicide program; however, large areas will require continued control measures. A combination of chemical and cultural treatments such as cultivation, cropping, and grazing may be necessary to stop the spread of leafy spurge (1,2).

The key to controlling leafy spurge is early detection and treatment of the initial invading plant. A persistent management program is needed to control topgrowth and to gradually deplete the nutrient reserve in the root system.

In the preface of the WSSA monograph, *Leafy Spurge* (2), Lloyd Andres states "as we review our present knowledge of leafy spurge, we again are reminded of the fact that in our attempts to control weeds we are dealing with a complexity of factors that can stagger the mind. A technique that provides control in one area is quite often unacceptable or ineffective in others. It is becoming obvious that for a weed as widespread and tenacious as spurge a variety and combination of control techniques will be needed." Even so, the research during the last 20 yr has provided us with a data base from which to work (6). This review discusses the most effective herbicides for leafy spurge control in several management situations.

Timing of Application

Timing of herbicide application to leafy spurge is important for maximum control and is based on the morphological and physiological development of the plant (7). Leafy spurge control with 2,4-D, dicamba or picloram is best either when flowers and seed are developing during early summer growth or when fall regrowth has developed but before a killing frost (Figure 19.1). Glyphosate is most effective for leafy spurge control either after seed filling in midsummer or after fall regrowth has begun until a killing frost. Glyphosate generally provides less long-term control than 2,4-D, dicamba, and picloram when applied during spring growth. Split-treatment applications of glyphosate are more effective than single applications and may be used in grasslands.

The yellow bracts (not true flowers) of leafy spurge usually appear in late May and reach maximum color in early June, but this is prior to optimum treatment time (Figure 19.1) and earlier than necessary to prevent seed production (3). Leafy spurge development when control is

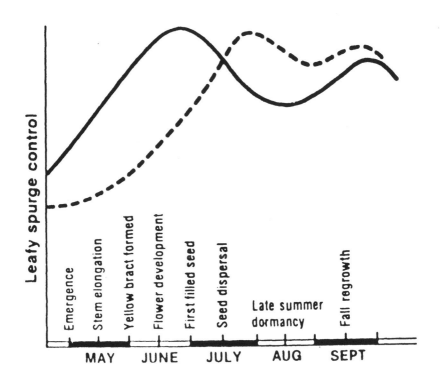

FIGURE 19.1 Susceptibility of leafy spurge to 2,4-D, dicamba, or picloram (——) and glyphosate (- - -) applications made at several times during the growing season.

most effective is characterized by lush green leaves along the entire stem, visible swelling of seed capsules, and vigorous plant growth. This growth stage for optimum treatment with certain herbicides usually begins in mid-June and ends with seed dispersal during hot, dry weather in July. Treatment during the early portion of the optimum period will control established plants and prevent seed production. Herbicide treatment during the latter part of seed set will control established and seedling leafy spurge plants, but viable seed has been produced by this time.

Leafy spurge begins a dormant period after seed dispersal and most leaves fall from the stem (3,7). The dormant stage usually coincides with hot dry weather of mid-summer and continues until new fall regrowth is stimulated by cooler weather and rainfall in August or early September. Herbicides generally are not effective when applied during the dormant period. Fall growth is characterized by a leafless main stem with two or more branches developing below the original flowering branches. The branches usually are 10 to 15 cm long with small leaves, and the leaves often turn red or yellow as the fall progresses. The plants may appear in poor vigor in late fall, but herbicide treatments have been effective at this time (8).

Recent research has shown that picloram does not move with root carbohydrates to the roots in the fall as is widely believed (9). Maximum ^{14}C-picloram translocation to the roots occurred during the flowering and seed set growth stages with nearly five times as much picloram reaching the root at this time compared to any other growth stage.

Leafy spurge control with herbicides generally follows the morphological and physiological development of the plant as described, but plant response to herbicides can vary almost day-by-day with growing conditions. In particular, control declines with low soil moisture and unseasonably high or low temperatures (9).

Control in Pasture and Rangeland

Picloram is the most effective herbicide for controlling leafy spurge (10). Picloram at 2.2 kg ai/ha applied in mid-June or early September will give 90 to 95% leafy spurge control that normally will be maintained for 2 to 4 yr depending upon soil type and precipitation. However, picloram at a 2.2 kg/ha application rate may not be economically feasible if a large area is infested with leafy spurge.

Research at North Dakota State University has shown that a more cost effective option is a tank mix of picloram at 0.28 to 0.56 kg/ha plus 2,4-D at 1 kg ae/ha applied in June and repeated annually. Annual applications of low rates of picloram plus 2,4-D will gradually reduce leafy spurge infestations by at least 90% in 3 to 5 yr (3,11,12). This treatment also increased forage production up to 300% (13). Research at the University of Wyoming has shown that picloram at 0.56 kg/ha applied 4 successive yr was as effective for leafy spurge control as picloram at 1.1 or 2.2 kg/ha (1).

Dicamba is an alternative to picloram for leafy spurge control in pasture and rangeland (3). Dicamba at 6.7 to 9 kg ai/ha will give 80 to 90% leafy spurge control for 1 yr, but usually decreases rapidly during the second yr. A retreatment program is necessary to maintain leafy spurge control. Dicamba at these high rates often decreases native forage production. The bluegrasses (*Poa* spp.) are the most tolerant grasses to dicamba at high rates. Dicamba at 2.2 kg/ha will provide good topgrowth control for 1 yr. Annual application of dicamba at 2.2 kg/ha has given good leafy spurge control with no grass injury in North Dakota (13).

Several cost effective follow-up treatments can be applied after the initial picloram or dicamba application. The most effective herbicide retreatments both for leafy spurge control and forage production in Wyoming have been picloram at 0.25 kg/ha, 2,4-D at 1.1 kg/ha, dicamba at 2.2 kg/ha (12,14), and picloram at 0.5 kg/ha (6).

The 2,4-D low volatile ester, oil soluble amine, or water soluble amine at 1 to 2 kg/ha applied annually gave short-term topgrowth control of leafy spurge (3). Spring-applied 2,4-D controlled the topgrowth, but leafy spurge control declined to 50% or less by fall and less than 20% by the following spring. Fall and spring applications of 2,4-D have shown similar leafy spurge control.

Applications of 2,4-D both spring and fall have prevented the spread of leafy spurge but generally have provided only small reductions of density (12). Long-term control of leafy spurge was not improved when 2,4-D was applied at 10 times the recommended rate (1). However, 2,4-D alone does permit a large increase in forage production compared to no treatment. In the short term, 2,4-D is an economical treatment but annual applications may be required indefinitely.

Glyphosate plus 2,4-D at 0.43 plus 0.73 kg/ha was first labeled for leafy spurge control as a fall treatment in pasture and rangeland in 1990 (13,15). Leafy spurge control averaged 60 to 70% and grass injury 30 to 70% when applied in August and September, respectively, in a five state regional trial (13). Leafy spurge control in Wyoming was much lower than in Colorado, Montana, North Dakota, and South Dakota. September treatments provided better leafy spurge control than treatments applied in August at all locations except Montana.

Control in Trees

Leafy spurge is a frequent invader of shelterbelt and other wooded areas (16). Neither picloram nor dicamba can be used under trees because these herbicides have a long soil residual and are toxic to trees when leached to the roots.

The 2,4-D amine or oil soluble amine at 1.1 to 2.2 kg/ha gives leafy spurge topgrowth control in trees. The 2,4-D must not contact the tree foliage either by direct spray or drift to prevent tree injury. The 2,4-D ester formulations are volatile, so they are not recommended in trees. Application of 2,4-D must be made annually to prevent the spread of leafy spurge (4,5).

Glyphosate at 0.84 kg/ha generally gives 80 to 90% control of leafy spurge if applied during fall growth until frost kills the leafy spurge stems (3). A follow-up treatment of 2,4-D is required for control of seedlings the following spring (5,10).

Dichlobenil suppresses leafy spurge for about one season. Dichlobenil at 6.7 to 9 kg ai/ha must be applied before leafy spurge emerges in early spring; either in early to mid-April, as early as possible after snow melt, or in late November when above-freezing temperatures are no longer expected. Dichlobenil applied at 9 kg ai/ha in November provided 80% suppression of leafy spurge the following June but control declined to 20% by September (4). Dichlobenil must be applied very early in the spring because it will prevent leafy spurge emergence only and does not affect emerged plants. Dichlobenil is an expensive treatment but is safe under trees and may be useful in small shelterbelts or under fruit and shade trees.

Control for Infestations Which are Small, Near Water, or in Noncropland

Leafy spurge confined to small, well-defined areas should be treated immediately to avoid spread. Application 3 to 5 m around leafy spurge patches will help control spreading roots and seedlings around the established stand. A persistent follow-up program is necessary for several years to control surviving stems and seedlings. Many attempts to control leafy spurge have failed because follow-up treatments were not applied (7,16). All of the herbicide treatments mentioned previously can be used on noncropland.

Leafy spurge control along open water is desirable to prevent further spread by seed. However, most herbicides used for leafy spurge control cannot be used near water. Fosamine, glyphosate (Rodeo formulation only), and 2,4-D can be used safely near water. Fosamine should be applied at 6.7 to 9 kg/ha during the true flower growth stage. Control with fosamine has been inconsistent (5). Fosamine application is best or recommended when soil moisture is abundant and the relative humidity is high.

Fosamine or glyphosate require a follow-up treatment the next yr to control seedlings (7,16). A 2,4-D formulation labeled for use near water at 0.56 to 1.1 kg/ha will prevent seedling establishment if applied from June to mid-July.

Economics and Longevity of Leafy Spurge Control

The influence of leafy spurge control on long-term land value is difficult to assess (17). However, short-term returns can be estimated by measuring changes in forage production and utilization by livestock following leafy spurge control (16,18).

TABLE 19.1 Leafy spurge control, forage production, and estimated net return from several herbicide treatments in eastern and western North Dakota, during a 5-yr management program.

Original treatment date and herbicide	Rate	Retreatment applied	Cost	Control[a] Aug. 1988	Total Yield Forage	Leafy spurge	Total net return
	(kg/ha)	(Year)	($/ha)	(%)	---(kg/ha)---		($/ha)
Spring 1984				**Eastern North Dakota**			
2,4-D[b]	2.2	85-88	75	30	10,780	4170	356
Picloram + 2,4-D[b]	0.28+1.1	85-88	175	70	11,480	2210	284
Picloram[c]	2.2	1988	405	100	12,770	1760	105
Dicamba[c]	9	85-87	1010	90	12,180	2230	-523
Fall 1983							
2,4-D[b]	2.2	84-87	75	0	8,320	7390	258
Picloram + 2,4-D[b]	0.28+1.1	84-87	175	20	10,890	3830	261
Picloram[c]	2.2	1985	405	90	12,310	330	87
Dicamba[c]	9	1986	505	70	12,080	860	-20
Control	--	--	0	0	10,480	8630	
LSD (0.05)				15	1,600	850	60
Spring 1984				**Western North Dakota**			
2,4-D[b]	2.2	85-88	75	40	4,780	590	116
Picloram + 2,4-D[b]	0.28+1.1	85-88	175	90	7,070	180	108
Picloram[c]	2.2	86,87	610	100	6,920	140	-333
Dicamba[c]	9	85-87	1010	100	5,670	390	-783
Fall 1983							
2,4-D[b]	2.2	84-87	75	10	5,520	1550	146
Picloram + 2,4-D[b]	0.28+1.1	84-87	175	20	5,110	1420	29
Picloram[c]	2.2	1986	405	70	6,690	50	-137
Dicamba[c]	9	85,86	755	60	6,280	120	-504
Control	--	--	0	0	4,610	3230	
LSD (0.05)				20	850	450	35

[a] Control 12 months after last treatment.
[b] Annual retreatment.
[c] Retreated when control declined to less than 70%.

Leafy spurge reduces the livestock carrying capacity of pasture and rangeland 50 to 75% (1,19). In North Dakota, cattle used 20 and 2% of the forage available in zero- and low- (<20% cover) density leafy spurge infestations respectively, by midseason. Moderate and high density

TABLE 19.2 Longevity of leafy spurge control.

Original	Years without treatment		
Control	1	2	3
95 or more	85	70	<20
80	60	<20	0
70	<30	0	0
60	20	0	0

Values given in % control; compiled from Lym and Messersmith (3,11).

infestations were avoided until early fall when the milky latex in leafy spurge disappeared. Leafy spurge canopy cover of 10% or less and shoot control of 90% or more were necessary to achieve 50% forage utilization by cattle in Montana (20).

Four herbicide treatments (Table 19.1) applied in the spring or fall were evaluated for leafy spurge control and forage production in eastern and western North Dakota during a 5-yr experiment (10). All treatments gradually reduced the leafy spurge infestation at both locations except 2,4-D at 2.2 kg/ha applied annually in the spring or fall and picloram plus 2,4-D at 0.28 plus 1.1 kg/ha applied annually in the fall (Table 19.1).

The most cost-effective treatment was picloram plus 2,4-D applied annually in the spring which resulted in a net return of $284 and $108/ha in eastern and western North Dakota, respectively, and averaged 80% leafy spurge control (10). Picloram at 2.2 kg/ha was reapplied only when leafy spurge control declined to less than 70% and provided an annual net return of $96/ha in eastern North Dakota but a net loss of $235/ha in western North Dakota. Dicamba at 9 kg/ha was applied up to four times to maintain 70% control with an average annual net loss of $458/ha. Total potential forage production value varies considerably with location and rainfall and will impact the cost-effectiveness of these treatments.

Regardless of the herbicide applied, once leafy spurge control reaches 95% or more, control will gradually decline to approximately 70% in 2 to 3 yr if a retreatment program is not maintained (Table 19.2) (3). Once control declines to less than 70% leafy spurge will reinfest an area rapidly.

Grass Competition

Some perennial grass species can effectively compete and provide leafy spurge control. Several grass species in a leafy spurge-infested area have been evaluated for establishment and productive capabilities under a tilled or nontilled program (21). Established grasses included: Luna pubescent wheatgrass (*Agropyron trichophorum*), Ephraim crested wheatgrass (*Agropyron cristatum*) mountain rye (*Secale montanum*), Sherman big bluegrass (*Poa ampula*), RS1 hybrid wheatgrass (*Agropyron repens* x *A. spicatum*), Lincoln smooth bromegrass (*Bromus inermis*), Oahe intermediate wheatgrass (*Agropyron intermedium*), Secar bluebunch wheatgrass (*Agropyron spicatum*), Rosana western wheatgrass (*Agropyron smithii*), Bozoisky Russian wildrye (*Elymus cinereus*) and Critana thickspike wheatgrass (*Agropyron dasystachyum*) (6).

Four years after seeding, areas tilled before seeding and then established to Russian wildrye, pubescent wheatgrass, big bluegrass, and intermediate wheatgrass maintained greater than 90% leafy spurge control, with dry matter yields of 1411, 2281, 3297, and 3490 kg/ha, respectively (21). In no-tilled areas, big bluegrass, and pubescent wheatgrass maintained leafy spurge control with dry matter yields of 2330 and 1168 kg/ha, respectively.

Future Research

Recent field research has shown that certain sulfonylurea and imidazolinone herbicides control leafy spurge. However, many also caused severe grass injury. Compounds currently being field tested include CGA-136,872 (Beacon){2-[[[[[4,6-bis(difluoromethoxy)-2-pyrimidinyl]amino]carbonyl]amino] sulfonyl]benzoic acid methyl ester}, DPX-V9360 (Accent){2-([[[[4,6-dimethoxypyrimidin-2-yl]aminocarbonyl]] amino sulfonyl])-N-N-dimethyl-3-pyridinecarboxamide monohydrate}, imazethapyr (Pursuit),{(±)-2-[4,5-dihydro-4-methyl-4-(1-methylethyl)-5-oxo-1H-imidazol-2yl]-5-ethyl-3-pyridinecarboxylic acid}, imazaquin (Sceptor),{2-[4,5-dihydro-4-methyl-4-(1-methylethyl)-5-oxo-1H-imidazol-2yl]-3-quinolinecarboxylic acid}, and quizalofop (Assure),{±-2-[4[(6-chloro-2-quinoxalinyl)oxy]phenoxy]propanoic acid} (13,22).

Biological control is viewed as the most cost-effective long-term control method. A successful control program will require integration of chemicals, land management, and biological control agents (6). Research is in progress at North Dakota State University to incorporate biological control agents with herbicides. Initial data indicate insects establish better in leafy spurge recently treated with picloram plus 2,4-D than in untreated leafy spurge.

Public comments and frustrations heard for not controlling leafy spurge center on two themes: it doesn't pay to control leafy spurge and leafy spurge can't be controlled (16). Leafy spurge can be controlled and research has shown that some herbicide treatments on rangeland and pastures are cost-effective. The economic return for control on nonagricultural land is the nondirect benefit of minimizing future losses and reducing spread onto larger areas.

References

1. Alley HP, NE Humburg, JK Fornstrom, and M Ferrell. Res. Weed Sci., Wyo. Agric. Exp. Sta. Res. J. 192:90-93 (1984).
2. Dersheid LA, LJ Wrage, and WE Arnold. In: *Leafy Spurge Monogr. 3* (AK Watson, Ed.), pp 57-64. Weed Sci. Soc. Amer., Champaign, IL, 1985.
3. Lym RG and CG Messersmith. J. Range Manage. 38:149-154. 20-year summary (1985).
4. Lym RG and CG Messersmith. Proc. North Cent. Weed Control Conf. 39:66-67 (1982).
5. Lym RG and CG Messersmith. Proc. North Cent. Weed Control Conf. 45:86-87 (1988).
6. Alley HP. *Proc. Range Weeds Revisited.* Washington State Univ. Coop. Ext. Unnumbered. p. 53-58 (1990).
7. Lym RG, CG Messersmith, and DE Peterson. N.D. Ag. Exp. Stn. Circ. W-765R. 4 p. (1988).
8. Lym RG and CG Messersmith. Res. Progress Rep. West. Soc. Weed Sci. p. 13-14 (1989).
9. Lym RG and CG Messersmith. J. Range Manage. 44:254-258 (1991).
10. Lym RG and CG Messersmith. Weed Technol. 4:635-641 (1990).
11. Lym RG and CG Messersmith. N.D. Farm Res. 43(1):7-9, 14 (1985).
12. Lym RG and CG Messersmith. J. Range Manage. 40:194-198 (1987).
13. Lym RG and CG Messersmith. *Proc. Great Plains Ag. Council. Leafy spurge control in the Northern Great Plains*, Gillette, WY. Univ. of Wyo., Laramie, 82071. p. 21-22 (1990).
14. Lym RG and CG Messersmith. J. Range Manage. 38:386-391 (1985).
15. Beck KG. *Proc. Great Plains Ag. Council. Leafy Spurge Control in the Northern Great Plains*, Gillette, WY. Univ. of Wyo., Laramie, 82071. p. 17. (1990).
16. Lym RG and CG Messersmith. N.D. Farm Res. 43(1):3-6 (1985).
17. Messersmith CG and RG Lym. N.D. Farm Res. 40(5):8-13 (1983).
18. Lym RG and DR Kirby. Weed Technol. 1:314-318 (1987).
19. Reilly W and KR Kaufman. In: *Proc. Leafy Spurge Symposium*, pp. 21-24. N.D. State Univ. Coop. Ext. Serv., Unnumbered publication (1979).
20. Hein DG. PhD Thesis, University of Wyoming, Laramie (1988).
21. Whitson TD, DW Koch, AE Gade, and MA Ferrell. Abstr. Weed Sci. Soc. Amer. 30:31 (1990).
22. Masters R, RN Stougaard, and SJ Nissen. *Proc. Great Plains Ag. Council. Leafy spurge control in the Northern Great Plains*, Gillette, WY. Univ. of Wyo., Laramie, 82071. p. 16. (1990)

20

Classification, Ecology, and Economics of Perennial Snakeweed

K.C. McDaniel, J.W. Briede, and L. Allen Torell

Abstract

Gutierrezia sarothrae (broom snakeweed) and *Gutierrezia microcephala* (threadleaf snakeweed) occur widely on rangelands throughout western United States, Canada, and northern Mexico. This paper presents an overview on these and some ecologically allied species. Broom snakeweed and threadleaf snakeweed are drought tolerant and have a high photosynthetic capacity, which may give them a competitive edge over associated plants during stressful periods. On rangelands, broom snakeweed and threadleaf snakeweed are extremely competitive for early spring moisture and nutrients, and they maintain this competitiveness throughout the year. Both species produce numerous secondary plant products such as saponins which produce toxic effects in livestock and some wildlife. Control measures have been studied, and herbicidal control has the highest efficacy, whereas biological control and prescribed burning are receiving increased attention.

Introduction

The genus *Gutierrezia* is composed of low-growing woody and herbaceous plants native to the more arid regions of North and South America. *Gutierrezia* is a member of the family Asteraceae and contains annual and perennial species (Table 20.1). The genus in general, but in particular broom snakeweed, is regarded as a serious pest on western

TABLE 20.1 Different species in the genus *Gutierrezia*, their distribution, growth habit, chromosome number, and date of discovery.

Distribution		Habit		Chromosome #	Date/Reference
alamanii	Mexico	P	HRo	4,8	1852 (2)
ameghinoi	Argentina	P	SGl	-	- (1)
argirocarpa	Mexico	P	Op	4	1902 (2)
arizonica	Arizona	A	SGl	4	1877 (2)
baccharoides	Southern Andes	P	Cu	-	- (1)
braceata	California	P	SGl	8,12	1907 (1)
californica	California	P	Op	12	1836 (1,2)
conoidea	Mexico	A	SGl	4	1881 (2)
espinosae	Northern Chile	P	SGl	-	- (1)
gayana	Northern Chile	P	Gl	16	- (1)
glutinosa	Southern Texas/ Northern Mexico	A	Gl	4	1847 (1)
grandis	Northern Mexico	P	SGl	-	1907 (1,2)
mandonii	Argentina	P	Gl	12	- (1)
microcephala	South west U.S./ Northern Mexico	P	Gl	8,16	1836 (1,2)
neaeana	Northern Chile	P	SGl	-	- (1)
petradoria	Western U.S.	P	HRo	4	1980 (2)
ramulosa	California/Baja	P	Gl	4	1889 (2)
repens	Northern Argentina	P	Cr	-	- (1)
resinosa	Chile	P	Op	28	- (1)
ruiz-lealii	Northern Argentina	P	Cu	16	- (1)
sarothrae	West U.S.A./Canada/ Northern Mexico	P	GL	4,8	1804 (1,2)
sericocarpa	Mexico	P	HRo	4,8,16	1880 (2)
serotina	Arizona/Northern Mexico	P	GL	4	1935 (1,2)
sphaerocephala	South west U.S.A./ Northern Mexico	A	Op	4,5	1847 (2)
taltalensis	Northern Chile	P	Op	-	- (1)
texana	Texas	A	Gl	4	1828 (1,2)
triflora	Texas	A/P	Gl	5	1895 (2)
wrightii	South west U.S.A./ Northern Mexico	A	SGl	4	1873 (2)

A - Annual, P - Perennial, GL - Globose, SGl - Semi-globose, Op - Open, HRo - Herbaceous rosette, Cr - Creeping, Cu - Cushion plant

rangelands and has received a great deal of attention from researchers (3,4). A closely allied species to broom snakeweed is threadleaf snakeweed and collectively these two species are commonly referred to as perennial snakeweeds or simply, snakeweeds. Other common names often used to describe these species include turpentine weed, rubberweed, rockweed, stinkweed, yellowtop, matchweed, and perennial broomweed.

Taxonomy

Gutierrezia was first recognized as a genus containing one species in 1816 by Lagasca (2,5-7). Lagasca eponymously dedicated the genus to a Spanish noble family named Gutierrez. The specimen was lost, and Lane (6,8) selected a neotype (a "new original" herbarium specimen) for the name. Gradually, several other species were added (Table 20.1, Figure 20.1). During his voyage on the Beagle (1831-1836), Darwin collected the genus in South America (9).

Shinners (10,11) placed *Gutierrezia* in the genus *Xanthocephalum*, which was based mainly on morphological characteristics. Since that time, the generic alignment has vacillated between the two genera, and publications using either name can be found between 1950 and 1982. Ruffin (12) argued for the name *Xanthocephalum*, whereas Solbrig considered *Gutierrezia* a distinct genus (7). Lane (13) championed the use of the name *Gutierrezia*, and concluded that mistakes made by Bentham (14) and Hemsley (15) led to the original confusion (2). Bentham and Hemsley included several taxa from *Gutierrezia* in *Xanthocephalum*, creating a false impression that there was an alliance between the two genera. Lane authoritatively resolved the issue when she provided evidence that the name *Gutierrezia* was taxonomically more appropriate (6,8,15).

Authors of contemporary floras separate the two species taxonomically using flower morphology (13,16-18). Broom snakeweed usually has more than three ray flowers per capitulum, whereas threadleaf snakeweed has one or two ray flowers (16,18,19). Ray flowers in broom snakeweed range from two to seven, while zero to nine disk flowers can be found per capitulum (16,18-20). Threadleaf snakeweed usually has two or less disk flowers, which abort and do not produce viable seed (7,16). Another difference between the two species is the involucre size. Threadleaf snakeweed has a slender involucre that is turbinate to cylindrical, whereas broom snakeweed has a larger and definitely turbinate involucre (17,19). These differences result in flower heads that are bigger in both length and width for broom snakeweed. A gross morphological feature, useful for differentiating between the two species under field conditions, is that threadleaf snakeweed flowers are more loosely clustered at the end of branches, while the slightly smaller broom snakeweed has a relatively compact flower arrangement about the canopy perimeter.

Distribution and Ecology

Snakeweeds are common throughout western rangelands from Canada to Mexico (Figure 20.1) (2). In 1959, it was estimated that *Gutierrezia* covered 57 million hectare of rangeland (21). No new estimates are available, but according to reports (Dr. Chris Call, Utah State University, 1989 personal observation), snakeweed is increasing throughout the West, invading many native vegetation types (salt desert shrub, pinon-juniper, and sagebrush types) in Utah, Nevada, and Idaho.

Snakeweed has long been native to North America. Studies of fossilized pack rat middens showed *Gutierrezia* species to have been present in southwestern United States in excess of 15,000 years (22,23). Campbell and Bomberger (24) suggested, on pristine sites, snakeweed may have contributed up to 15% of the plant composition on Chihuahuan desert grasslands. Wooton and Standley (25) considered snakeweeds an extremely common species on disturbed and undisturbed western range lands.

Snakeweeds are considered to be aggressive invaders into many plant communities, although they are not considered to be pioneering species (26,27). The growth habit and physiology of snakeweeds may contribute to this aggressive behavior. Snakeweeds are prolific seed producers (28), although initial germination is low and ripening is necessary (29). Six months of storage at room temperature, followed by a constant temperature between 15C and 25C, with an 8-hr light period is required to produce high germination percentages (29,30). Alternating temperatures result in only a slight increase over these conditions (29,31). Broom snakeweed and threadleaf snakeweed have slightly different germination requirements, which partially explains the relative distribution of the species. Broom snakeweed, with its more northern distribution, has a higher temperature requirement for germination than threadleaf snakeweed (29). At northern latitudes, these higher temperature optima would prohibit germination on warmer days in winter, which could be followed by a killing frost. Seeds remain viable for several years and are able to germinate when conditions are optimal. This is similar to the escape-in-time mechanism found in many arid land species, safeguarding survivorship and fecundity (32-34).

Seedlings produce a taproot that remains intact for the first year and extends to about 27 cm depth (35). Near the end of the first year, plants develop a more extensive root system that is more diffuse, eventually losing their pronounced taproot. Dwyer and DeGarmo (36) found that broom snakeweed produced portionally more root biomass than creosotebush (*Larrea tridentata*) (under four different soil-moisture conditions) and mesquite (*Prosopis juliflora*) (when water-stressed). Osman and Pieper (35) reported root:shoot ratios of about 0.5 for developing broom snakeweed seedlings, which is similar to many arid land species (37).

During the first year of growth, plants usually develop a single upright stem with few flowers. Spring regrowth after the first year is primarily

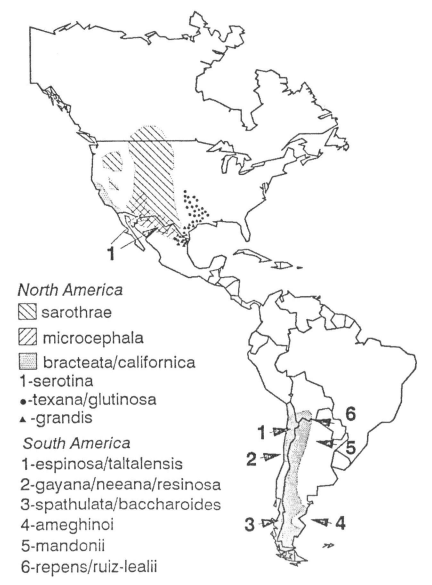

North America
⊠ sarothrae
▨ microcephala
▦ bracteata/californica
1-serotina
•-texana/glutinosa
▲-grandis
South America
1-espinosa/taltalensis
2-gayana/neeana/resinosa
3-spathulata/baccharoides
4-ameghinoi
5-mandonii
6-repens/ruiz-lealii

FIGURE 20.1 Species distribution from the genera *Gutierrezia* in North and South America [adapted from Solbrig (1)].

from vegetative buds developed on the lower parts of the stems the year before. As plants mature, they develop multiple branching from lower stems, resulting in the characteristic compact spheroid canopy (38). Plants attain maximum canopy development after about 3 yr. Ludwig et al. (39), Nadabo et al. (40), Osman and Pieper (35), and Briede (41) discuss some of the growth patterns and size-biomass relations of broom snakeweed. Plants usually flower in August or September, with seed set approximately three weeks post-flowering. Plants may go dormant with the first frost; however, vegetative buds stay green and are capable of photosynthesis when winter temperatures are favorable.

Snakeweeds have a suffrutescent habit, shedding leaves and branches when stressed, thus the plants need to maintain a high gas exchange in spring to compensate for the yearly loss of these photosynthetically active parts. Ehleringer et al. (42) and Briede (41) found spring photosynthesis rates three times higher than summer and fall rates. In addition, Ehleringer et al. (12) found threadleaf snakeweed to have higher stem assimilation rates than broom snakeweed. Typically, threadleaf snakeweed occurs in more southern (warmer) regions and sheds its leaves more frequently, being more dependent on stem photosynthesis. Broom snakeweed, on the other hand, has a broader distribution, covering seemingly less stressful environments, which may result in less evolutionary pressure towards stem photosynthesis.

Control and Management

Control of snakeweed has been attempted in a number of ways; however, herbicidal treatment is the most common method (43-51). Burning and mowing have met with mixed success and need further study (27,47,52). Biological control measures are in the pioneering stage, although more then 300 insect species have been found to be associated with snakeweed (53-55).

Although the fate and mode of action of herbicides in snakeweed have not yet been studied (56), successful use of herbicide depends mainly on rate and time of application (4,46,57). Snakeweed should be photosynthetically active when herbicide is applied. Currently, picloram is the most commonly used herbicide for snakeweed control and metsulfuron is also efficient (45). Other phenoxy-like chemicals that have been investigated, but are usually less effective than picloram, include dicamba, silvex, 2,4-D, and 2,4,5-T (45,50). Tebuthiuron, both as pellet or as wettable powder has also been found effective for snakeweed control (49).

Most brush and weed species invading rangeland continue to be a persistent problem unless controlled by artificial means and only

TABLE 20.2 Probability of snakeweed invasion or die-off in New Mexico.

Snakeweed Classification at t	Snakeweed Classification at $t+1$	
	Non-economic	Economic
Non-economic	0.75	0.25
Economic	0.27	0.73

Economic: Snakeweed production > 300 kg/ha.
Non-economic: Snakeweed production ≤ 300 kg/ha.

occasionally will wildfire, insects, drought, old age, or other events cause a natural die-off of these plants. Snakeweed is unique in this respect. Snakeweed populations have been described as cyclic and relatively short-lived (19). Propagation occurs under favorable environmental conditions and plants survive until conditions become unfavorable. Snakeweed can be a problem one year and not the next; or the range can remain covered with the weed for a number of years if environmental conditions remain favorable.

Areas infested with snakeweed can be classified as economic or non-economic, based on the relative forage suppression and livestock poisoning potential from different densities of snakeweed. Using data collected from nine study sites in New Mexico over an 11-year period, as reported by McDaniel (58) and Torell et al. (59), the probabilities of a snakeweed invasion or natural die-off was estimated for rangelands where infestation by snakeweed is a continuing problem (Table 20.2). As the matrix indicates, if a particular parcel of rangeland has either no snakeweed or a light infestation of the weed (non-economic infestation), there is a 75% probability it will remain at this non-economic level into the next year. Similarly, if an economic infestation of the weed is present, without artificial control, the probability the infestation will remain at economically damaging levels the next year is 73%.

The probability that snakeweed numbers in a given area will either increase or decrease from one year to the next is significant. For example, the probability that an economic infestation of snakeweed will die-back to a non-economic level within any given year is estimated to be 27% in New Mexico (Table 20.2). Similarly, snakeweed-free rangeland in New Mexico can be expected to become infested to economically damaging levels with a 25% probability.

The cyclic nature of snakeweed populations makes management and control decisions extremely difficult. Uncertainty about how long the weed problem will persist both with and without control prevents an accurate estimate of what benefits from control might be. A rancher cannot estimate with a high degree of certainty that if the plant were controlled that the area would remain snakeweed-free for a significant period of time. Just as important, the snakeweed stand may die-off from natural causes if left untreated.

Torell et al. (60) estimated that a four-year treatment life was necessary to break even on an investment to control the weed with picloram provided the herbicide cost $9.50 per acre to apply. If a five-year treatment life is realized, then the range improvement practice was estimated to be very economical with an estimated rate of return on investment of 9.4% with relatively low 1985 beef prices, and 17.9% with record high 1979 beef prices (60). However, if uncertainty about future snakeweed infestation levels is incorporated into the economic analysis and an expected payoff from treatment is computed, then the economics of snakeweed control are variable. As shown by Torell et al. (59), on range sites where snakeweed has been a longlasting and a persistent problem, a positive return from chemical control can be expected. On other sites where snakeweed infestations have been variable and relatively short-lived, then an economic payoff could not be expected.

References

1. Solbrig OT. Brittonia 22:217-229 (1970).
2. Lane MA. Syst. Bot. 10:7-28 (1985).
3. Huddleston EW and RD Pieper (Eds). *Snakeweed: Problems and perspectives*. New Mexico State Agric. Exp. Stat. Bull. 751 (1989).
4. McDaniel KC and RE Sosebee. In: *The Ecology and Economic Impact of Poisonous Plants on Livestock Production* (LF James, MH Ralphs, and DB Nielsen, Eds), pp:43-56. Westview Press, Boulder, CO, 1988.
5. Lagasca M. *Genera et species plantarum quae aut novae sunt.* 35 pp., Madrid (1816).
6. Lane MA. Syst. Bot. 7:405-416 (1982).
7. Solbrig OT. Contr. Gray Herb. 188:1-63 (1960).
8. Lane MA. Taxon 31:330-333 (1982).
9. Hooker WJ and GA Arnott. Hooker, Companion to Bot. Mag. 2:41-53 (1835).
10. Shinners LH. Field & Lab. 18:25-32 (1950).
11. Shinners LH. Field & Lab. 19:74-82 (1951).
12. Ruffin J. Sida 5:301-333 (1974).
13. Lane MA. Sida 8:313-314 (1980).
14. Bentham G. In: *Genera Plantarum, Vol. 2)* (G Bentham and JD Hooker, Eds.), p. 249. Reeve and Co. and Williams and Norgate, London, 1873.
15. Hemsley WB. In: *Biologia Centrali-Americana, Botany, Vol.2* (R Ducane and O Salvin, Eds.), pp. 109-112. RH Porter and Dulun and Co., London, 1881.
16. Lane MA. PhD Diss., Univ. Texas, Austin (1980).

218

17. Kearney TH and RH Peebles. *Arizona Flora*. Univ. of Cal. Press., Berkeley, 1960.
18. Welsh SL, ND Atwood, S Goodrich, and LC Higgins. *A Utah Flora*. BYU Univ. Press, Provo, UT, 1987.
19. McDaniel KC, RD Pieper, LE Loomis, and AA Osman. New Mex. State Univ. Agr. Exp. Sta. Bull. 711 (1984).
20. Solbrig OT. Contr. Gray Herb. 193:67-115 (1964).
21. Platt KB. J. Range Manage. 12:64-68 (1959).
22. VanDevender TR and BL Everitt. Southwest. Nat. 22:337-352 (1977).
23. VanDevender TR, CF Freeman, and RD Worthington. Southwest. Nat. 23:289-302 (1978).
24. Campbell RS and EH Bomberger. Ecology 15:49-51 (1934).
25. Wooton EO and PC Standley. *Flora of New Mexico*, Contrib. from the US Natl. Herbarium, Govt. Printing Off., Washington, DC, 1915.
26. Judd BI. Southwest. Nat. 19:227-239 (1974).
27. Neuenschwander LF, HA Wright, and SC Bunting. Southwest. Nat. 23:315-338 (1978).
28. Ragsdale BJ. PhD Diss., Texas A&M Univ., College Station, TX (1969).
29. Mayeux HS Jr and L Leotta. Weed Sci. 29:530-534 (1981).
30. Kruse WH. J. Range Manage. 23:143-144 (1970).
31. Mayeux HS Jr. In: *Snakeweed: Problems and perspectives* (EW Huddleston and RD Pieper, Eds.), pp. 39-50. New Mexico State Agric. Exp. Sta. Bull. 751, 1989.
32. Angevine MW and BF Chabot. In *Topics in Plant Population Biology* (OT Solbrig, S Jain, GB Johnson, and PH Raven, Eds.), pp. 188-207. Columbia Univ. Press, New York, 1979.
33. Fitter AH and RKM Hay. *Environmental Physiology of Plants*. Academic Press, New York, 1981.
34. Venable DL and L Lawlor. Oecologia 46:272-282 (1980).
35. Osman A and RD Pieper. J. Range Manage. 41:92-93 (1988).
36. Dwyer DD and HC DeGarmo. New Mexico State Univ. Agric. Exp. Sta. Bull. 570, 1970.
37. Barbour MG. Amer. Midl. Natur. 89:41-57 (1973).
38. Briede JW and KC McDaniel. 41st Ann. Soc. Range Manage. Mtg., Corpus Christy, TX, Abstract No. 250 (1988).
39. Ludwig JA, Reynolds JF, and Whitson TD. Amer. Midl. Natur. 94:451-461 (1975).
40. Nadabo S, RD Pieper, and RF Beck. J. Range Manage. 33:394-397 (1980).
41. Briede JW. PhD Diss., New Mexico State Univ., Las Cruces, NM (1990).
42. Ehleringer JR, JP Comstock, and TA Cooper. Oecologia 71:318-320 (1987).
43. Gesink RW, HP Alley, and GE Lee. J. Range Manage. 26:139-143 (1973).
44. McDaniel KC. New Mexico State Univ. Agr. Exp. Sta. Bull. 706 (1985).
45. McDaniel KC. In: *Snakeweed: Problems and Perspectives* (EW Huddleston and RD Pieper, Eds.), pp. 85-100. New Mexico State Agric. Exp. Sta. Bull. 751, 1989.
46. McDaniel KC and KW Duncan. Weed Sci. 35:837-841 (1987).

47. McDaniel KC and LA Torell. In: *Integrated Pest Management on Rangeland* (JL Capinera, Ed.). Westview Press, Boulder, CO, 1987.
48. Schmutz EM and DE Little. J. Range Manage. 23:354-357 (1970).
49. Sosebee RE, WE Boyd, and CS Brumley. J. Range Manage. 32:179-182 (1979).
50. Sperry OE and ED Robinson. Texas Agric. Exp. Sta. Prog. Rep. 2273, 1963.
51. Whitson TD and MA Ferrell. Res. Prog. Rep. West. Soc. Weed Sci. p. 47 (1988).
52. Dwyer DD. J. Animal Sci. 26:934 (1967).
53. Foster DE, DN Ueckert, and CJ Loach. J. Range Manage. 34:446-454 (1981).
54. Liddell CM, DC Thompson, TM Sterling, DB Richman, and CJ DeLoach. New Mexico State Agric. Exp. Sta. Bull. 751, pp. 163-168 (1989).
55. Wangberg JK. J. Range Manage. 35:235-238 (1982).
56. Sterling TM. New Mexico State Agric. Exp. Sta. Bull. 751, pp. 51-60 (1989).
57. Sosebee RE. Texas Tech. Management Note 6, Texas Tech. Univ. (1985).
58. McDaniel KC. New Mexico State Agric. Exp. Stat. Bull. 751, pp. 13-26 (1989).
59. Torell LA, K Williams, and KC McDaniel. New Mexico State Agric. Exp. Sta. Bull. 751, pp. 71-84 (1990).
60. Torell LA, HW Gordon, KC McDaniel, and A McGinty. In: *The Ecology and Economic Impact of Poisonous Plants on Livestock Production* (LF James, MH Ralphs, and DB Nielsen, Eds), pp. 57-69. Westview Press, Boulder, CO 1988.

21

Biological Control of Snakeweeds

C. Jack DeLoach

Abstract

Snakeweeds (*Gutierrezia* spp.) are good targets for biological control because they cause great damage to western rangelands and they have very little beneficial or ecological value. Insects have been identified in southern South America that are good candidates for introduction. Several of these insect species have been tested at the USDA-ARS Biological Control Laboratory in Argentina, others are being tested, and three have been tested in quarantine at Temple, TX. The first releases of one insect, a root-boring weevil, have been made in the field in Texas and New Mexico but the insect is not yet established.

Introduction

The project on biological control of snakeweeds at Temple, TX, is the first attempt worldwide to control a native weed in a continental area by the introduction of foreign natural enemies. Since it is a native plant (1), its control may be more difficult than that of introduced weeds (2,3).

We selected snakeweed as a primary target for biocontrol for three reasons: it causes widespread and heavy damage on western rangelands (4-7); it has little beneficial value in the ecosystem (wildlife food and shelter) (8) or for direct human usage (ornamentals, grazing, honey, etc.); and the possibilities for successful biological control appear good because species of *Gutierrezia* closely related to our snakeweeds are native in semiarid areas of southern South America (9). Several insects in

Argentina damage closely related snakeweeds and might control the North American weedy species if introduced (10).

Our research on snakeweeds has progressed through most of the steps in the accepted protocol (11) of a biological control of weeds program. This has included a careful review of the literature on taxonomy and worldwide distribution, the amount of damage caused, the beneficial and ecological values of snakeweeds, and the native natural enemies in North America that attack them. The taxonomy, damage caused, methods of control other than biological, and economic impacts are discussed by other speakers at this symposium and in the recent snakeweed symposium at Las Cruces, NM (12). We obtained a resolution of those minor conflicts of interest between beneficial and harmful values. Approval to begin a control program, according to established procedures (13), has been given by the Technical Advisory Group for the Introduction of Biological Control Agents of Weeds (TAGIBCAW) of the U.S. Department of Agriculture's Animal and Plant Health Inspection Service (APHIS).

North American Insects That Attack Snakeweeds

In Texas and New Mexico, 247 species of phytophagous insects attack the two major species of snakeweeds, broom snakeweed (*Gutierrezia sarothrae*) and threadleaf snakeweed (*G. microcephala*); these include 149 species of beetles (including 32 species of leaf beetles and 35 species of weevils) and 76 species of bugs and leaf hoppers (14).

The native insect causing the most widespread damage was the cerambycid root borer *Crossidius pulchellus*. At times, it appeared to kill most plants over hundreds of acres (15); however, the damage occurs only sporadically. *Crossidius pulchellus* attacked 60% of the roots of larger plants and dead plants were closely correlated with bored plants (16). Four other species of borers attacked snakeweeds in Arizona (17). A weevil, *Myrmex lineolata*, infested 47 to 94% of the plants and three buprestid beetles each attacked from 1-17% of the plants. Mealybugs on the roots were often associated with large-scale dieback of the plants. A number of other insects were observed attacking the foliage (18). These included a grasshopper, *Hesperotettix viridis*, that feeds almost exclusively on snakeweeds and causes high mortality of young plants (19). Also, leaf tying moths (*Synnoma* sp. and *Synalocha* sp.) damaged plants (15,20,21) and larvae of the noctuid moth *Narraga fimetaria* fed on the foliage of snakeweeds and annual broomweeds (22).

South American Insects That Attack Snakeweed

Our research in Argentina is in close cooperation with the ARS Biological Control Laboratory in Hurlingham, near Buenos Aires, Argentina. Since 1977, we have made 26 exploratory trips from the Hurlingham Laboratory in Argentina, covering more than 60,000 miles, searching for natural enemies of snakeweeds. We have found 67 species of insects, one mite, and one plant pathogen on the four species of snakeweed most similar to the North American snakeweeds: 23 of these species were observed feeding or had a high probability of feeding on snakeweed. Of these, 16 species of insects probably have a narrow host range and are possible candidates for biological control of snakeweed (10).

Argentine Stem and Root Borers

We observed three locations in Argentina where the combined attack of root-boring insects (*Heilipodus ventralis*, *Carmenta haematica*, and three species of buprestid beetles) had killed nearly all snakeweed plants over an area of several hectares (10). Plant death was apparently caused by a combination of drought and heavy damage to the roots by insect feeding. The large stem-boring weevil *H. ventralis* was the first control agent tested at Hurlingham (23). We collected it throughout most of the range of snakeweed in Argentina in areas that correspond to latitudes from Monterrey, Mexico, to northern Wyoming in North America. The adults of *H. ventralis* feed on leaf petioles and stem terminals and females lay eggs in the lower part of the stems. The newly hatched larvae tunnel downward inside the stem and developing larvae feed in the crown and taproot which they damage extensively. This feeding does not cause immediate obvious damage to the plant, but many of the plants die during the following dry season. The weevil typically has one generation per year, and females lay an average of 100 eggs during their life. The adults emerge in the spring and oviposit through most of the summer. The larvae feed through the summer and fall, overwinter in the plant roots, and pupate the following spring.

To determine host range, we examined plants in the field and conducted laboratory tests in Argentina and in quarantine at Temple, TX. In the field in Argentina, workers examined over 4000 plants of 45 species of Compositae (23). They found larvae only in three closely related genera, mostly in *Grindelia* and *Gutierrezia* and a few in *Baccharis*. The laboratory tests using 66 plant species confirmed this host range.

At Temple, TX, we conducted multiple-choice tests of adult feeding and oviposition and no-choice tests of larval development in our quarantine facility. We exposed over 600 adults (50% males and 50% females) to 22 species of Astereae and 29 species of six other tribes of Compositae for periods ranging from one week to one month. These tests demonstrate that *Gutierrezia*, *Grindelia*, and *Gymnosperma* are the only major host genera in North America, though occasional insect development may occur on other closely related genera of Astereae such as *Chrysothamnus*, *Isocoma*, and *Baccharis*.

We released both adults and eggs (which we placed in holes drilled in the stems) at several locations in Texas and New Mexico and at several times during the growing season. These releases were made in cooperation with Dr. Jim Richerson of Sul Ross University, Alpine, TX, and Dr. David Richman of New Mexico State University, Las Cruces. So far, we have no evidence that a self-sustaining population has become established in the field. The reasons, though not well established, appear to be two-fold. At the first release site, the snakeweed plants were in a declining phase and died before the weevils could complete their development. In the second releases, the exceptionally cold winter of 1989-90 either prevented hatching or killed the young larvae. We expect that the finding of healthy snakeweed stands, changes in the release dates, and modification of the release techniques will lead to successful establishment in the near future.

The second insect we investigated in Argentina was the day-flying, clear-winged sesiid moth, *Carmenta haematica*. It also occurs throughout the range of snakeweed, and the larvae also feed in the roots of snakeweed. In the field in Argentina, *C. haematica* attacks species of only two plant genera—gumweed (*Grindelia*) and snakeweed. These observations were confirmed by laboratory host-range tests.

The behavior of *C. haematica* is very special, and considerable basic research was required by the personnel at the Hurlingham Laboratory to discover how to obtain mating, where and how to place the eggs on the plants, and how to obtain normal entry by neonate larvae into the host plant. In quarantine at Temple, our tests indicate that *C. haematica* is sufficiently host-specific to release in the field; however, the final assessment has not been completed.

Three species of flat-headed wood borers (Buprestidae)—*Dactylozodes okea*, *D. alternans*, and *Agrilus leucostictus*—frequently damaged the roots of snakeweed in Argentina. All appear to have narrow host ranges and probably have one generation per year. These species attack smaller plants than do *H. ventralis* or *C. haematica* and would be very valuable in a control program.

Other Insects Attacking Snakeweed in Argentina

Several other insects attack snakeweeds in Argentina (10). Larvae and adults of the leaf beetle *Stolas ingrata* fed on the foliage of snakeweed, gumweed, and baccharis at several sites in southern Argentina. The insect is currently being tested at Hurlingham and at Temple. A variety of insects fed in the flowers, fruits, and seeds of snakeweed. A small phycitid moth, *Homoeosoma* n. sp., and a tephritid fly, *Trypanaresta difficilis*, occurred in a high percentage of the flower heads at several locations in the southern area of Argentina. These insects are now being tested at Hurlingham. Also, a tiny weevil, *Apion* sp., fed in flower heads in the south, and an undetermined species of Hemiptera was associated with flower heads in some areas. Larvae of the tephritid fly, (*Strobelia baccharidis*) made conspicuous white froth balls on the small branches of snakeweed, and larvae of an unidentified species of Gelechiidae made small galls in the stems. Four species of scale insects and a mealybug (Homoptera) also attacked snakeweed species in Argentina. These were common in some areas but seldom caused much damage. We have not yet evaluated their potential for biocontrol.

Plant Pathogens. The plant pathogenic rust *Puccinia grindeliae* infested nearly all snakeweed plants at Neuquen and Peninsula Valdez in southern Argentina. Infection of plants was often heavy. This rust is also widespread in North America, occurring from Wisconsin to Alberta and south to central Mexico. It is known to have hosts in 24 genera of Compositae, including snakeweed (24).

Conclusions

The root borers would be effective control agents only for the perennial snakeweeds in North America because the larvae require living roots in which to overwinter. However, the other types of insects would be able to complete their life cycle on both snakeweeds and the annual broomweeds.

Exciting recent theory proposes that control agents obtained from plant species closely related but different from the target weed have the potential to be more damaging than those obtained from the weed species itself (25). Out of necessity, this is our approach with the snakeweeds. The arguments put forth by Hyrum Johnson (26) indicate that biological control of native weeds would not lead to ecosystem collapse. We propose that control of a weed such as snakeweed, that has little known beneficial value in the ecosystem, will have little adverse effects and that unknown reactions through the food chains of various organisms will be so masked as to be of little or no consequence. In fact, Peter Harris (27) proposes that (in the case of introduced weeds) biocontrol would be less harmful than the herbicides currently used to control them or than doing nothing and allowing the weed to remain overabundant with the attendant reduction in plant diversity. The same argument would appear to apply to overabundant native weeds. If the hypothesis of Johnson and Mayeux discussed at this symposium proves correct (that the observed global increase of woody plants in former grasslands is caused at least in part by the global increase in CO_2), then a "global" or area-wide method, as is biocontrol, would be the most effective control method.

We expect that within the next few years we can demonstrate that a native weed in a continental area can be safely controlled with insects introduced from another part of the world and that these new theories have validity in the control of serious native weeds, both in rangelands and in other areas.

References

1. Lane MA. Systematic Bot. 10:7-28 (1985).
2. DeLoach CJ. In: *Proc. V Internat. Symp. Biol. Control Weeds* (ES DelFosse, Ed.), pp. 175-199. Brisbane, Australia, 1981.
3. DeLoach CJ, PE Boldt, HA Cordo, HB Johnson, and JP Cuda. In: *Management and Utilization of Arid Land Plants* (DR Patton, CE Gonzales V, AL Medina, LA Segura T, and RH Hamre, Eds.), pp. 49-68. USDA-Forest Serv. Gen. Tech. Rep. RM-135, 1986.
4. Torrell AL, HW Gordon, KC McDaniel, and A McGinty. In: *The Ecology and Economic Impact of Poisonous Plants on Livestock Production* (LF James, MH Ralphs, and DB Nielsen, Eds.), pp. 57-69. Westview Press, Boulder, CO, 1988.
5. Carpenter BD, DE Ethridge, and RE Sosebee. Rangelands 12:206-208 (1990).
6. Rittenhouse LR, JS Bluntzer, and MM Kothmann. Texas Agric. Expt. Sta. Res. Cen. Tech. Rep. No. 77-1 (1977).
7. McDaniel KC. In: *Snakeweed: Problems and Perspectives* (EW Huddleston and RD Pieper, Eds.), pp. 13-25. Proc. Symp. Nov. 9-10, 1988, New Mexico State Univ., Las Cruces, 1990.
8. Martin AC, HS Zim, and AL Nelson. *American Wildlife and Plants*. Dover Pub., Inc., New York, 1951.
9. Solbrig OT. Contr. Gray Herb. 197:3-42 (1966).
10. Cordo HA and CJ DeLoach. J. Range Manage. (in press).
11. Huffaker CB. Hilgradia 27:101-157 (1957).
12. Huddleston EW and RD Pieper. Proc. Symp., New Mexico State Univ., Las Cruces, 1990.

13. Coulson JR and Soper RS. In: *Plant Protection and Quarantine, Vol III Special Topics* (RP Kahn, Ed.), pp. 1-35. CRC Press, Inc., Boca Raton, FL, 1989.

14. Foster DE, DN Ueckert, and CJ DeLoach. J. Range Manage. 34:446-454 (1981).

15. Ueckert DN and DE Foster. *Noxious Brush and Weed Control Research Highlights*, Texas Tech. Univ. 7:24-25 (1976).

16. Richman DB and EW Huddleston. Environ. Entomol. 10:53-57 (1981).

17. Falkenhagen TJ. MS Thesis, Univ. Arizona, Tucson, 1978.

18. Thompson DC and DB Richman. In: *Snakeweed: Problems and Perspectives* (EW Huddleston and RD Pieper, Eds.), pp. 179-187. Proc. Symp. Nov. 9-10, 1988, New Mexico State Univ., Las Cruces, 1990.

19. Parker MA. Ecology 66:850-860 (1985).

20. Wangberg JK. J. Range Manage. 35:235-238 (1981).

21. Wisdon CS, CS Crawford, and EF Aldon. J. Ecology 77:685-692 (1989).

22. DeLoach CJ and RE Psencik. Ann. Entomol. Soc. Amer. 75:631-635 (1982).

23. Cordo HA. In *Proc. VI Internat. Symp. Biol. Control Weeds* (ES DelFosse, Ed.), pp. 709-720. Vancouver, Canada, 1985.

24. Cummins GR. *Rust Fungi on Legumes and Composites in North America*. Univ. Ariz. Press, Tucson, 1978.

25. Hokkanen H and D Pimentel. Canad. Entomol. 116:1109-1121 (1984).

26. Johnson HB. In: *Proc. VI Internat. Symp. Biol. Control Weeds* (ES DelFosse, Ed.), pp. 25-56. Vancouver, Canada, 1985.

27. Harris P. BioScience 38:542-548 (1988).

22

Chemical Control of Broom Snakeweed

Keith W. Duncan, Thomas D. Whitson, and Kirk C. McDaniel

Abstract

Efficacy of a herbicide in controlling target plants is influenced by the physiological, morphological, and phenological stage of the plant. Each of the various stages is in turn influenced by environmental conditions which affect plant growth. Picloram and metsulfuron have proven effective in controlling snakeweeds (*Gutierrezia* spp.) in New Mexico and Wyoming at any phenological stage if the rate of concentration is not limited. However, at lower rates of application both herbicides provide greater levels of control when applied in fall or spring compared to summer or winter. Picloram applied at 0.28 to 0.56 kg ae/ha or metsulfuron applied at 7.10 to 14.2 g ae/ha typically kills 90% or more of the snakeweed when sprayed postbloom in New Mexico or during the vegetative stage in Wyoming. Forage production is usually positive following snakeweed control. Herbage production increases of 100-800% have been recorded in New Mexico trials and 125-170% in Wyoming (1-3).

Introduction

Grassland communities exist in various stages of succession, depending upon historical grazing use, climate, and other factors. Reduced vigor of desirable, perennial forage species is one of the first signs of vegetation regression on rangelands (4). Proper grazing management is often successful in reversing or stopping the negative vegetative trends at this early stage. A change in use pattern or number of grazing animals may restore the vigor and stature of desirable species. Such a change may allow desirable species to compete with invading, undesirable plants. However, if vegetative regression progresses to the point where perennial, undesirable plants replace desirable species, then emphasis must often be placed on weed control methods rather than grazing management. Herbicide application, prescribed fire and various mechanical control methods are techniques which can be used to selectively remove undesirable plants.

Growth of desirable grasses and broadleaf plants is the primary purpose for herbicidal control of broom snakeweed (*Gutierrezia sarothrae*) and threadleaf snakeweed (*G. microcephala*). This approach to vegetation management is described by Scifres (4) as the process of expediting secondary succession on grasslands. When undesirable plants such as the snakeweeds increase to the point that forage production is retarded, the weeds must be controlled so as to shift productivity towards herbage more readily used by domestic and wild grazing animals. Snakeweed control can be viewed within the management framework as an effort to increase both plant and animal diversity and to improve efficient use of rangelands. Snakeweed control can be viewed in the ecological framework as an attempt to reverse a negative trend in succession, thereby developing a more diversified plant community and encouraging secondary succession.

Response to Herbicides

The activity of a herbicide on a plant is influenced by the physiological, morphological and phenological stage of development. Herbicide activity is further affected by the environmental constraints which directly influence plant growth (5). Assuming delivery of a biologically active herbicide to a target plant is not limited, then plant response to the chemical depends on the rate of concentration and stage of plant growth in the annual cycle. Picloram and metsulfuron will kill snakeweed at any particular phenological stage if there is no limit on rate of concentration. However, at lower rates of concentration, picloram and metsulfuron are more effective when applied in fall (post-flowering) or spring (early vegetative) compared to summer (late vegetative-prebloom) or winter (perennating bud) in New Mexico (Table 22.1) (1).

The importance of warm soil temperatures and relatively high plant water content which promote active vegetative growth was emphasized

TABLE 22.1 Snakeweed control one year following different dates of herbicide applications near Lovington, New Mexico, 1987.

| | | Snakeweed Mortality By Date of Application | | | | | | | |
Herbicide	Rate	Jan 14	Mar 03	Apr 16	May 26	Jul 15	Aug 19	Oct 01	Nov 10
	(kg/ha)					%			
Picloram	0.14	8	98	98	87	8	16	99	7
Picloram	0.28	29	99	99	97	19	19	99	93
Metsulfuron	0.021	99	63	83	35	38	12	65	98
Metsulfuron	0.042	57	92	99	80	73	42	98	97

in early studies utilizing phenoxy herbicides for snakeweed control (6-8). This period is usually between March and May under New Mexico conditions (1). However, snakeweed is usually sprayed commercially in New Mexico when the plant is in the full-bloom to post-bloom stage. This period is typically between October and December. Studies comparing the efficacy of picloram and metsulfuron have reported a higher percentage of snakeweed killed at comparable rates when applied in fall rather than in spring (9).

The efficacy of these foliar-absorbed herbicides has been attributed to an increase in total nonstructural carbohydrate storage in snakeweed (5). This process of carbohydrate storage occurs over a longer period and to a greater relative magnitude in fall compared to spring.

Most research attempts to control snakeweed with herbicides have the objective of selectively removing the snakeweed without harming associated grasses. Both picloram and metsulfuron have proven effective in removing snakeweed while not damaging the associated forage species. In 17 different experiments conducted at NMSU since 1983, snakeweed control averaged 96% with picloram applied at 0.28 kg/ha. Metsulfuron has also proven effective when applied at rates of 7.1 g ae/ha or higher. Picloram or metsulfuron applied at one-half the recommended rate has, in some cases, killed 100% of the snakeweed. However, in other trials the results have been extremely variable with few plants killed in some cases.

In Wyoming, Whitson and Freeburn (10) reported controlling greater than 95% of broom snakeweed with picloram applied at 0.28 to 0.56 kg/ha or metsulfuron applied at 11.34 to 28.63 g ae/ha (Table 22.2). Applications of metsulfuron appeared more efficacious when applied during the vegetative stage rather than during flowering. However, no

TABLE 22.2 Broom snakeweed controlled two years following herbicide treatments near McFadden and Wheatland, Wyoming.[1]

Herbicide	Rate kg/ha	% Control McFadden	Wheatland
Picloram	0.14	98	83
	0.28	100	96
	0.55	100	100
2.4-D(LVE)	2.2	96	0
Triclopyr	0.14	47	0
	0.55	29	0
Fluroxpyr	0.28	0	0
	0.55	75	0
	0.83	76	0
Dicamba + 2.4-D(A)[2]	0.28 + 0.82	60	0
Triclopyr + 2.4-D(LVE)	0.36 + 0.74	79	0
2,4-D[3]	2.2	94	0
Metsulfuron[4]	0.014	89	0
	0.028	99	87
	0.042	100	87
	0.055	100	91
	0.069	100	99
Check	--	0	0

[1] Treatments applied at McFadden on June 28, 1987, and at Wheatland on July 28, 1987. Evaluations (% control) calculated from total snakeweed plants/plot, counted July 10-11, 1989.
[2] Formulation Esteran 99 (isooctylester).
[3] Formulation Wedone 638 (butoxyethylester)
[4] Metsulfuron was applied with 0.25% v/v non-ionic (X-77) surfactant.

difference was detected between application times with the two picloram application rates. Trials evaluating postbloom applications of picloram and metsulfuron have resulted in little control of snakeweed. This difference in efficacy of post-bloom applications between New Mexico and Wyoming is probably related to differing phenological and physiological stages of snakeweed. In New Mexico, snakeweed maintains photosynthetically active tissue year around if soil moisture is not limiting. However, in Wyoming, snakeweed goes into winter dormancy following seed set. This dormant period possibly explains lack of control when spraying post bloom.

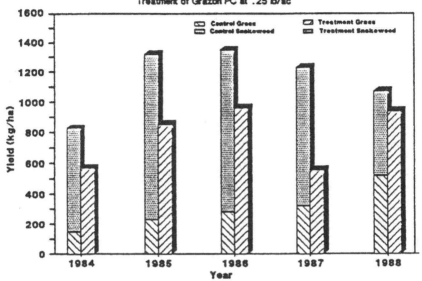

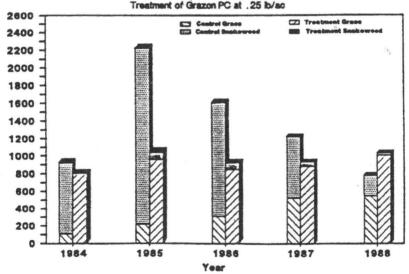

FIGURE 22.1 Grass production on sprayed and untreated rangeland near Vaughn and Roswell, NM. Picloram was applied at 0.28 kg/ha in Fall 1983.

TABLE 22.3 Production change in crested wheatgrass infested with broom snakeweed following herbicide applications near McFadden, Wyoming.

Herbicide[1]	Rate kg ai/ha	% Control[2]	kg/ha
Picloram	0.14	95	981
	0.14	92	864
	0.28	100	903
	0.28	99	864
	0.55	99	726
2,4-D(LVE)+ isooctylester	2.2	92	1099
Triclopyr	0.14	0	844
	0.55	16	766
Fluroxypyr	0.28	34	784
	0.55	58	706
	0.83	82	844
Dicamba+ 2,4-D(A)	0.28+ 0.82	75	688
Triclopyr+ 2.4-D(LVE)	0.36+ 0.74	34	961
butoxyethylester	2.2	93	1041
Metsulfuron[3]	0.014	92	1178
	0.028	99	1059
	0.042	100	961
	0.055	100	1177
	0.069	100	903
Check	---	0	688
LSD 0.05		33	314
CV		115	26

[1] Herbicides applied June 29, 1987.
[2] Evaluations made July 10, 1989.
[3] Metsulfuron was applied with 0.25 v/v non-ionic (X-77) surfactant.

Forage Response

Forage production is usually positive following snakeweed control with herbicides (5,9,11). Broom snakeweed and perennial grass yields have been recorded for 5 years in trials established in 1983 near Vaughn and Roswell, New Mexico (Figure 22.1) (1). Picloram applied at 0.28 kg ae/ha killed 100% of the snakeweed at both sites.

Sprayed plots remained free of snakeweed for five years at Vaughn, whereas untreated plots produced an average 862 kg/ha of snakeweed.

Grass yields averaged 784 kg/ha on sprayed plots and 302 kg/ha on unsprayed plots.

At Roswell, grass yields averaged 907 kg/ha in sprayed plots, while the production in unsprayed plots averaged 350 kg/ha. Some snakeweed plants reinvaded the sprayed plots but did not influence grass production. Snakeweed production on unsprayed plots was 2016 kg/ha in 1985 but subsequently declined to only 224 kg/ha in 1988. A corresponding increase in grass production was noted as snakeweed production decreased naturally in unsprayed plots near Roswell.

Production of crested wheatgrass was increased following control of broom snakeweed near McFadden, Wyoming (Table 22.3). However, production increases were statistically different only in those plots sprayed with 2.2 kg/ha of 2,4-D or with metsulfuron at 5.95, 11.34, 22.68, or 28.63 g ae/ha.

Treatment Longevity

The length of time before a reinvasion of snakeweed seedlings occurs following a successful herbicide treatment is unclear. Treatment longevity depends largely on initial treatment success and climatic conditions following control. Snakeweed propagates by seed and germination requirements have been described by Mayeaux (12). Adequate precipitation in fall, winter or spring is important for germination. However, the most critical period for seedling survival is undoubtedly during the usually hot, dry conditions of May, June and July in New Mexico. If soil moisture is lacking during this period when the seedling root system is not fully developed, plants will likely not survive.

Timing of herbicide applications based on the age of a snakeweed population is critical to the economic success of a treatment (2). This is illustrated in Table 22.4 from data collected near Vaughn, NM. Picloram was applied annually from 1979 to 1988, except in 1981. Snakeweed mortality was 83% or greater in each trial. Evaluations showed areas sprayed in 1979 and 1980 were reinfested by snakeweed in 1981. However, areas sprayed since 1982 remained snakeweed-free through 1988. These data suggest control of a snakeweed stand established only 1-2 years should be emphasized over older aged stands in a herbicide management program.

Future Direction

No method of brush and weed control, whether chemical, mechanical, biological or prescribed burning, is without its unique strong points

TABLE 22.4 Snakeweed control (%) one growing season after picloram application and when all years were evaluated in Fall 1988 near Vaughn, NM.

Year picloram was applied[1]	% Snakeweed control after one growing season	% Snakeweed control in Fall 1988
1979	98	25
1980	95	3
1981	_[2]	–
1982	83	83
1983	100	100
1984	99	99
1985	99	99
1986	100	99
1987	100	100
1988	100	100

[1] Picloram applied at 0.56 kg ae/ha in 1979, 1980, and 1982; applied at 0.28 kg ae/ha in 1983 to 1987.
[2] No treatment made in 1981.

and characteristic weaknesses (4). There are very few examples of a single-treatment approach which maximizes all long-term benefits. The use of herbicides as a panacea for snakeweed management is no exception. At present, however, aerial application of picloram is the primary tool selected by ranchers in New Mexico for removing snakeweed from heavily infested rangelands (Figure 22.2).

Efficiency of herbicides for snakeweed control can be improved. Research is needed to develop proper sequencing and application of herbicides in conjunction with other practices which might provide better long-term benefits than any single method alone. One such management system might include control of a dense infestation of snakeweed with a herbicide to be followed in subsequent years by prescribed burns to control seedlings. Such an approach would require flexibility in application timing and consideration of the grazing system upon which these practices are superimposed.

Additional research is also needed on herbicide fate and mode of action to improve application efficiency, thereby minimizing herbicide rate. For example, picloram applied at one-half the recommended rate can at times provide 100% mortality of snakeweed, but results are inconsistent. Snakeweed susceptibility to picloram, both phenologically and physiologically, has been only grossly defined.

BROOM SNAKEWEED SPRAYED

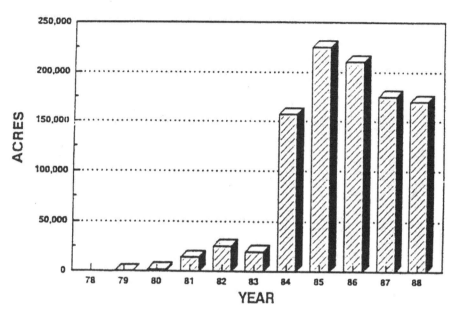

FIGURE 22.2 Acres of broom snakeweed sprayed commercially in New Mexico.

References

1. McDaniel KC. *Proc. for Broom Snakeweed Conf.*, Las Cruces, NM (1989).
2. Torell A, K Williams, and KC McDaniel. *Proc. for Broom Snakeweed Conf.*, Las Cruces, NM (1989).
3. Whitson TD, BR Shreve, and NR Adam. WSWS Res. Prog. Rep. Reno, NV, p. 57 (1990).
4. Scifres CJ. *Brush Management*. Texas A&M University Press, College Station, TX, 1980.
5. Sosebee RE. *Proc. of the Brush Manage. Symp.*, SRM Albuquerque, NM, pp. 27-43 (1983).
6. Allen TJ and JW Dollahite. Texas Agric. Exp. Sta. Prog. Rep. 2073:1-3 (1959).
7. Gesink RW, HP Alley, and GA Lee. J. Range Manage. 26:139-143 (1973).
8. Schmutz EM and DE Little. J. Range Manage. 23:354 (1970).
9. McDaniel KC and KW Duncan. Weed Sci. 35:837-841 (1987).
10. Whitson TD and JW Freeburn. WSWS Res. Prog. Rep. Reno, NV, p.55 (1990).
11. McDaniel KC. New Mexico State Univ. Exp. Sta. Bull. 706 (1984).
12. Mayeaux HS. *Proc. for Broom Snakeweed Conf.*, Las Cruces, NM (1989).

23

Toxicology of Snakeweeds, *Gutierrezia microcephala* and *G. sarothrae* [1]

G. Stanley Smith, Timothy T. Ross, Gonzalo I. Flores-Rodriguez, Bryan C. Oetting, and Thomas S. Edrington

Abstract

Snakeweeds infest vast areas of rangeland in North America. Although unpalatable, snakeweed foliage is grazed by livestock when range forage is scarce, and seasonally when early greening makes it attractive among dormant vegetation. Ingestion of snakeweeds by pregnant animals can cause abortion or untimely birth of weak offspring. In experiments at New Mexico State University, ovulation and embryonation were impaired in female rats fed snakeweed foliage as 12.5 or 25% of diet for 10 days before mating. Embryonic death was caused within four to six days after onset of feeding rats 10 or 15% snakeweed. Normal female rats, bred to males fed snakeweed, produced fewer litters and fewer offspring per litter. Their offspring had high infant mortality. Penned ewes fed 10% snakeweed mixed with poor quality grass hay had fewer mountings by vasectomized and intact rams. Snakeweed ingestion impaired fertility and reproduction of female rats and sheep before signs of toxicosis developed (elevated serum enzyme activities; hyperbilirubinemia). Snakeweeds contain saponins, alkaloids, terpenes, flavonols, and other substances that could cause toxicosis.

[1]Scientific paper 391 of the New Mexico Agric. Exp. Sta.

Introduction

Perennial snakeweeds in North America include threadleaf snakeweed (*Gutierrezia microcephala*) and broom snakeweed (*G. sarothrae*), also known as broomweed. These species were previously included in the genus *Xanthocephalum*. Ecology, biological control, and toxicology of snakeweeds were reviewed recently (1). Snakeweeds infest vast areas of rangeland in western USA, Canada, and northern Mexico, and economic losses related to diminished availability of forage for livestock amount to several millions of dollars annually (1).

Snakeweed foliage is unpalatable to horses, cattle, and sheep and usually comprises a small portion of the diet for these species. Pieper (2) reviewed literature about snakeweed content of herbivore diets and concluded that snakeweeds comprise less than 5% of diets grazed by cattle, sheep, goats, mule deer, and pronghorn antelope when suitable forage species are available. Blacktail jackrabbits consumed snakeweed amounting to 10% of total feed under some conditions.

Snakeweeds were not listed among the principal stock-poisoning plants of New Mexico in 1922 (3), of Wyoming in 1939 (4), or the western states of the USA (5,6), but they have long been recognized not only as noxious weeds but as plants responsible for poisoning of livestock on dry ranges in Texas, New Mexico, Arizona, California, Colorado, Idaho, and northern Mexico (7).

Snakeweeds Cause Livestock Poisoning
and Abortion

Ingestion of snakeweeds as a substantial portion of grazed forage caused widespread poisoning of cattle and sheep during the 1920s and 1930s in western Texas (8-10). Symptoms reported by various investigators (8-13) were summarized by Kingsbury (7):

> In acute cases, death occurs, but more commonly the major result...is abortion... Symptoms in severely poisoned sheep and cattle include listlessness, anorexia, rough coat, diarrhea or constipation, mucus in feces, vaginal discharge, and often hematuria. In cattle a nasal discharge accompanied with crusted and peeling muzzle has been noted, while sheep have displayed minor icterus. Lesions are those of gastroenteritis accompanied by degeneration change in liver and kidneys...[and] severe toxic nephritis with necrosis if the disease progresses sufficiently before the animal dies... The uterus is edematous, and hydrops may be observed in the fetal membranes. [pp. 407-408]

In the 1950s, Dollahite and coworkers (11-13) examined the abortifacient effects of ingested snakeweeds. Wide variability was found. Some cattle consumed more than 1,000 pounds (450 kg) fresh snakeweed foliage during 117 d (≈ 4 kg daily) before aborting, but others aborted in 7 d after consuming about 9 kg fresh foliage. In cattle, sheep, and goats, death was produced by feeding fresh foliage amounting to 10 to 20% of body weight over periods of 1/2 to 2 wk. Most commonly among cattle on range, snakeweed ingestion caused abortion at various stages of fetal development, or untimely birth of weak calves. Retained placenta was common. Lesser dosage caused vulvular swelling and premature enlargement of udders.

Norris and Valentine (14) included snakeweeds among the principal plants causing livestock poisoning in New Mexico, but questioned abortifacient effects:

> Broomweed is not readily eaten by livestock except when forage is scarce. Although it is frequently reported as producing abortion in cattle, feeding tests with *Gutierrezia microcephala* have not borne out the abortion-producing action; but sheep, cattle and goats were poisoned. Feeding 8 pounds... over...5 days was fatal to a sheep, and 24 pounds in 3 days was fatal to steers... Bloody urine is characteristic of severe cases, and the kidneys and liver are severely damaged. [pp. 20,21]

Toxicants in Snakeweeds

Dollahite and associates (15) extracted snakeweeds and injected a saponin fraction (150 g obtained from 22.7 kg = 0.66% of plant material) into pregnant rabbits, goats, and cows. Abortion and death resulted in most cases. They stated:

> The saponins studied are toxic when injected intravenously, and small amounts are abortifacient. The signs of anorexia, mucopurulent nasal discharge, diarrhea followed by constipation with feces containing large amounts of similar appearing mucus from the vagina are similar to the signs seen in cows with acute snakeweed poisoning.

Shaver et al. (16) further characterized the toxicants of "*Xanthocehalum* species" and reported that a saponin isolated from snakeweeds had abortifacient properties when administered orally to pregnant rabbits. This saponin preparation stimulated contractions in uteri isolated from rats and rabbits, and upon hydrolysis by weak acid, it yielded a triterpenoid sapogenin. Shaver et al. (16) noted that snakeweeds contain flavonoid material, which might be estrogenic, but their fractions were not estrogenic.

Flournoy et al. (17) injected pregnant rats (i.p.) with saponins isolated from *G. microcephala*, using dosages of 5, 15, and 25 mg/kg BW daily from day 9 of gestation through parturition. All rats given 5 mg/kg littered normally and all given 25 mg/kg had fetal resorptions and some died.

Because snakeweeds, ponderosa pine, Monterey cypress, and juniper all contain considerable quantities of essential (i.e., volatile and fragrant) oils, and all produce abortions under range conditions, Molyneux et al. (18) characterized the volatile constituents in *G. sarothrae*. They found both qualitative and quantitative differences in the mono- and sesquiterpene composition of broom snakeweed and ponderosa pine, and concluded it seems unlikely the similarity of abortifacient symptoms induced by these two plants can be attributed to their essential oil fractions.

Roitman and James (19) identified numerous highly oxygenated flavonol methyl esters from *G. microcephala*, some of which were present in amounts to warrant toxicological examination. The flavonols and flavonoids are usually considered in relation to polyphenolic substances (20), among which are several highly estrogenic (and anti-estrogenic) compounds.

Snakeweeds can accumulate selenium if grown on seleniferous soil (7). They contain relatively large amounts of saponins (some of which are triterpenoid), essential oils, mono- and sesquiterpenes, flavonols and flavonoids, and, undoubtedly, numerous other compounds that could contribute to their toxicity. Tannins (1.7%) and alkaloids ($\approx$ 1%) were reported in snakeweeds of New Mexico (21).

Recent Studies at New Mexico State University

The earlier literature characterized late stages of acute snakeweed toxicosis in livestock and laboratory animals in terms of hepatobiliary damage and renal failure. But initial and early phases, which might be reversed, remedied, or prevented by appropriate intervention, were not well characterized. Moreover, it had not been determined whether ingestion of snakeweeds at subtoxic levels before mating might adversely affect reproduction of livestock.

In 1986, New Mexico State University (NMSU) initiated studies with rats to further characterize initial and early aspects of snakeweed toxicosis, and to measure reproductive responses to graded dosages of dietary snakeweed. Experimentation through 1989 was summarized by Gonzalo Flores-Rodriguez (21). Some of the data have been shown in preliminary reports (22-27) which provide information about experimental procedures. For these NMSU experiments, foliage of threadleaf snakeweed (*G. microcephala*) was collected at bud stage from sandy soil at the NMSU College Ranch near Las Cruces, NM. The foliage was oven dried (50°C), ground to pass a 2 mm screen in a Wiley Mill, and stored at 4°C until mixed with freshly ground (2 mm) commercial rat feed and fed immediately thereafter.

Trial 1

The first experiment was conducted with male and female rats fed 0, 12.5, 25, or 50% snakeweed (SW), replacing ground rat feed, from weaning (≈ 50 g BW) through 49 d. Objectives were to establish the dose:response relation in terms of growth, morbidity, and mortality; to characterize serum clinical profiles in SW toxicosis; to determine effects of SW on reproduction; and to determine whether reproductive failure caused by SW could be reversed by removal of SW from the diet.

Weanling male and female rats provided 50% SW had very low feed intakes and died (10/12 males; 8/12 females) within 8 d. Rats fed 25% SW had no mortality, but growth was impaired severely and many were morbid at 49 d. At 12.5% of diet, SW caused no mortality or morbidity, but growth was impaired (BW ≈ 66% of controls at 49 d).

After consuming diets with 12.5 or 25% SW for 35 d, rats exhibited mild to moderate toxicosis with differences in blood serum electrolytes, metabolites, and enzyme activities (Chem-25 or Chem-30 clinical profiles) that reflected dose:response relationship to dietary SW. Hepatobiliary changes occurred early, and renal damage followed. By day 35, rats fed 25% SW had serum bilirubin elevated fourfold over controls and enzymes of biliary origin elevated slightly (alkaline phosphatase, ALK-P, twofold) or markedly (gamma-glutamyltranspeptidase, GGTP, 485-fold).

At the first mating, when males and females were paired within diet groups after 49 d, 5 of 12 females fed 0% SW became pregnant, but only 1 of 24 females fed 12.5 or 25% SW produced offspring. After 49 d, six rats that had consumed 12.5% SW and six rats that had consumed 25% SW were then fed only commercial rat feed for an additional 25 d to determine if the impaired fertility could be reversed. Ten of these twelve rats became pregnant and produced normal offspring, and (overall) eight of twelve controls produced litters. Of six rats continued on 12.5% SW throughout 70 d, only one became pregnant, and of six fed 25% SW throughout, none became pregnant. Thus, only 2 of 18 matings yielded pups from rats fed 12.5% SW, and 0 of 18 for rats fed 25% SW. Yet, 25 d after removal of SW from their diets, 10 of 12 females formerly fed SW became pregnant and subsequently produced litters. Numbers of pups per litter and size of pups at birth and weaning were not affected by prior consumption of SW.

This trial showed that SW is highly unpalatable to rats and that malnutrition contributes to toxicosis of animals forced to consume SW. Consumption of SW markedly diminished reproduction, but the impairment was reversible after removal of SW from the diet.

Trial 2

Objectives of this experiment were to determine how much SW can be ingested by normal, mature female rats, and for how long, before reproduction is impaired. SW was substituted into mixtures with ground rat feed as 0, 12.5, or 25% and fed to young adult female rats (32/diet) for 10, 20, 30, or 40 d before pairing females (eight/diet, at each time) with fertile males fed only commercial rat feed. Eight days after females were caged with males, serum progesterone concentrations averaged 92, 50, and 41 ng/ml (SE = 6.5, P < .05), for rats fed 0, 12.5, or 25% SW, respectively, reflecting diminished numbers of females impregnated. Percentages of females that later produced litters were 66, 47, and 28, respectively, for females fed 0, 12.5, or 25% SW before mating. Numbers of pups per litter averaged 11.5, 10.1, and 9.1, respectively, for dams fed 0, 12.5, or 25% SW before mating. Ingestion of SW for 10 d before mating impaired reproduction almost as much as ingestion for 20, 30, or 40 d. Dietary SW caused dose-related, progressive changes in serum electrolytes, metabolites, and enzyme activities (Chem-25 or Chem-30 profiles) that reflected mild to moderate hepatobiliary involvement. Serum changes at day 10 reflected mainly undernutrition in rats offered diets with 12.5 or 25% SW, reflecting poor feed intake related to unpalatability of diets containing SW. By day 20, hepatobiliary changes were evident in rats fed 25% SW [elevated unconjugated bilirubin; elevated enzymes (ALK-P, GGTP, and alanine aminotransferase, ALT)], and changes were progressive through days 30 and 40, not only for rats fed 25% SW but also for those fed 12.5% SW.

These results suggest that impairment of fertility in females forced to consume SW short-term relates mainly to undernutrition, whereas prolonged SW ingestion causes tissue damage that exacerbates infertility related to undernutrition.

Trial 3

Objectives of this trial were to determine whether ingestion of SW by adult male rats would impair male reproduction. Snakeweed foliage was fed mixed with commercial rat feed at 0, 12.5, or 25% of diet for 20 or 40 d, after which males were caged in pairs with virgin adult females fed only commercial rat feed. Male reproduction, determined as percentage of exposed females impregnated, was not impaired by ingestion of SW by males for 20 d. But males (six per diet) fed 12.5 or 25% SW for 40 d before pairing with females (five per male) failed to impregnate females successfully. Controls (0% SW) produced 10 litters/30 females exposed to males; males fed 12.5% SW for 40 d produced 3 litters/30 females exposed; and males fed 25% SW for 40 d produced 1 litter/30 females exposed. Mortality of pups through weaning was 24% for controls, 54% for progeny of males fed 12.5% SW, and 100% for progeny in the single litter from males fed 25% SW for 40 d. Such results suggest genotoxicity of SW for offspring of males; but no such suggestion derived from offspring of females fed SW in Trials 1 and 2.

In the continuation of this trial, blood serum testosterone concentrations at 0, 1, 2, and 4 hr after injection of 0.5 µg gonadotropin releasing hormone (GnRH) were not affected by SW ingestion as 12.5 or 25% of diets for 50 or 100 d. Blood collected from these male rats after 20 or 98 d of SW ingestion as 12.5 or 25% of diet showed only mild toxicity, whereby serum GGTP increased 16-fold at 20 d but subsided to control levels at 98 d. Ingestion of SW for 50 d did not affect sperm concentration or percentage of abnormal sperm, but SW ingestion for 102 d markedly increased the percentage of sperm that were visibly abnormal by microscopic examination.

These findings suggest that short term ingestion of SW would have little, if any, effect on reproduction by male livestock; however, long term ingestion would diminish semen quality and might cause genotoxicity.

Trial 4

To determine whether dietary snakeweed might cause embryonic or early fetal mortality in rats already pregnant when first exposed to SW, 40 mature female albino rats were paired individually with males for five days. Three days after breeding, 10 of these females were assigned randomly to each of four diets having 0, 5, 10, or 15% SW, replacing commercial rat feed. At days 4, 6, 8, 10, and 12 after onset of experimental diets, two rats from each group were bled, decapitated, and examined for numbers of live and dead embryos and fetuses. Live offspring averaged 8.0, 7.3, 6.6, and 6.3, and embryonic or fetal deaths averaged 0.7, 4.3, 4.6, and 7.2, respectively, for females fed 0, 5, 10, or 15% SW. The unmistakable dose:response relationship in this study (27) shows that ingestion of SW by pregnant rats caused embryonic and early fetal death. In these pregnant rats fed SW for 4 to 12 d before examination, serum concentrations of conjugated bilirubin, urea (BUN) and creatinine increased linearly ($P < .01$) with increasing dietary SW, without concomitant increases in serum enzymes. This suggests that renal involvement and mild cholestasis occur at very early stages in SW toxicosis, and that embryonic and fetal effects can be serious before drastic changes in blood profiles of the dam would signify overt toxicosis.

Trial 5

Objectives of this trial were to characterize serum profiles in fine-wool ewes fed SW, and to determine whether dietary SW would affect estrous cycle length, serum progesterone levels, or receptivity to rams. In a preliminary study, 15 mature ewes were fed chopped alfalfa hay in isonitrogenous mixtures (18% crude protein) that contained SW as 0, 12.5, or 25% of mixtures for 30 d. Intakes were equalized at 2.5% of BW daily for 30 d. Serum profiles showed no changes attributable to SW, except that triglycerides were lower (22, 14, and 11 mg/dl, respectively). Estrous cycle length tended to increase in relation to dietary level of SW (17.0, 17.4, and 17.8 d, respectively).

In the main trial, seven ewes were fed moderate quality blue grama hay (11% crude protein) and another seven ewes were fed the same hay (90%) in mixture with 10% SW. This 10% SW mixture contained 10% crude protein. These diets were fed through 102 days during the breeding season, and feed intakes by ewes fed 10% SW amounted to 1.5% of BW daily. Intakes by ewes fed only blue grama hay were restricted at 1.5% of BW daily to equalize intakes by ewes fed 10% SW. Estrus was synchronized in all ewes at day 0 and again at day 41.

A vasectomized ram detected estrus during days 1 to 30 in all seven control ewes, but only in four of seven ewes fed 10% SW. During days 40 to 90, an intact ram mounted all seven control ewes, but mounted only five of seven ewes fed 10% SW. Blood collected at days 26, 52, and 78 had serum profiles reflecting mild, progressive increases in an enzyme of hepatobiliary origin (ALT), and elevated levels of BUN and creatinine, suggesting renal involvement. Serum cholesterol levels were slightly lower in ewes fed 10% SW than in pair-fed controls.

These results indicate that well nourished females should tolerate SW ingestion for brief periods without impairment of reproduction so long as diet palatability is sufficient to sustain adequate nutrition.

Discussion

Early studies with livestock and laboratory animals showed clearly that ingestion of snakeweed foliage in substantial amounts by pregnant females will cause abortion and acute toxicosis, characterized by hepatic damage with cholestasis and renal damage. Saponins were implicated as the primary toxicant. The saponins extracted from threadleaf snakeweed caused fetal death (17) when administered i.p. to pregnant rats at dosages of 15 or 25 mg/kg BW.

In the studies at NMSU (21-27), ingestion of snakeweed foliage before mating seriously impaired ovulation and embryonation in females, even when ingested at low dosage and for short term. Moreover, ingestion of snakeweed foliage by newly pregnant females caused embryonic and fetal mortality, in a dose-dependent manner, even when ingested at low dosage for short term. These drastic effects on female reproduction occurred with onset of snakeweed ingestion either before or after breeding, and at stages when clinical signs of toxicosis in the dam were undetected or mild. Impaired fertility of females consuming SW occurred before tissue damage was evident and seemed dependent on undernutrition.

In male rats, reproduction was not impaired until SW ingestion was prolonged and preceded by clinical signs of toxicosis. Sperm quality was impaired and mortality of offspring from males fed snakeweeds became evident before histological evidence of testicular damage could be detected (unpublished data).

The snakeweed foliage that was fed in rat studies at NMSU contained saponins that were extracted into 50% aqueous ethanol and eluted from activated charcoal using pyridine:ethanol mixture (3:7, v/v), according to Van Atta et al. (28). Saponins accounted for 6.1 to 6.8% of plant dry matter by this analysis. Undoubtedly, these amounts are sufficiently large to account for toxicosis, if absorbed from the intestinal tract. But saponins in plant tissue vary widely in their effects on animals (29,30), and it remains questionable whether saponins in ingested snakeweeds act as the primary toxicant causing tissue damages. Some, but not all, clinical signs of early toxicosis seem compatible with effects of absorbed saponins. But few of the numerous rats examined post mortem in the NMSU studies had signs of gastrointestinal damage observable, either grossly or histologically. Moreover, few of the rats exhibited hemolysis or enlarged spleens, which would be expected in animals subjected chronically to saponins in their blood.

The saponins in snakeweeds are triterpenoidal glycosides, and if hydrolysed in the gut, the triterpenoid sapogenins would be readily absorbed. Because such sapogenins might resemble steroidal hormones, even small amounts in the blood could have profound effects on ovulation and embryonation, such as were observed and reported here.

Some types of dietary saponins (such as those in alfalfa) escape hydrolysis in the gut and bind cholesterol from the bile, thereby contributing to lowered levels of cholesterol in blood. Evidence of this was lacking in all except one of the rat studies at NMSU, because undernutrition of rats fed SW contributed to elevation of serum cholesterol levels, rather than lower levels. But in Trial 5 with ewes fed 10% SW and controls that were pair fed, serum cholesterol levels were lower in ewes fed 10% SW. This indicates that ingested SW affects blood serum cholesterol levels, and possibly steroidal hormone metabolism, in ways that are not dependent on level of nutrition.

Further research is required to identify the toxicant(s) in snakeweeds, to determine which are primary under varied circumstances, to elucidate modes of action, and to develop protocols and treatments that may prevent or alleviate effects of snakeweed when ingested by livestock grazing rangeland.

Summary

Although unpalatable to livestock, snakeweeds are ingested by grazing animals, and cause abortion and toxicosis when intake is excessive. In female rats and undernourished ewes fed snakeweed foliage, reproductive impairment occurred rapidly, whether ingested immediately before or immediately after breeding, and reproductive impairment was not dependent on overt signs of clinical toxicosis. Male rats tolerated dietary snakeweeds for longer time than females without impairment of breeding, but prolonged ingestion by males caused abnormal sperm and contributed to increased mortality of offspring from dams fed conventional feed. Saponins contribute to toxicity of snakeweeds, but terpenes, flavonoids, resins, and other compounds also may be important.

References

1. Huddleston EW and RD Pieper. *Snakeweeds: Problems and Perspectives*. New Mexico Agric. Exp. Sta. Bull 751 (1990).
2. Pieper RD. *Broom Snakeweed Content of Herbivore Diets*. pages 203-210, In: (1)(1990).
3. Johnson EP and WA Archer. *The Principal Stock-poisoning Plants of New Mexico*. Ext. Circ. 71, New Mexico Agric. Ext. Ser., New Mexico College of Agric. and Mechanic Arts (1922).
4. Beath OA, HF Eppson, CS Gilbert, and WB Bradley. *Poisonous Plants and Livestock Poisoning*. Wyoming Agric. Exp. Sta. Bull. 231 (1939).
5. USDA. *16 Plants Poisonous to Livestock in the Western States*. Farmers' Bull. 2106, U.S. Dept. Agric. (1958).

6. James LF, RF Keeler, AE Johnson, MC Williams, EH Cronin, and JD Olsen. *Plants Poisonous to Livestock in the Western States.* U.S. Dept. Agric., Agric. Bull. 415 (1980).

7. Kingsbury JM. *Poisonous Plants of the United States and Canada.* Prentice-Hall, Inc., Englewood Cliffs, NJ, 1964.

8. Schmidt H. *Poisonous Plant Investigations.* Texas Agric. Exp. Sta. Annu. Report 45:11 (1931).

9. Boughton IB. In: *Forty-sixth Ann. Rep.,* Texas Agric. Exp. Sta. p. 211 (1933).

10. Mathews FP. J. Amer. Vet. Med. Assoc. 88:55-61 (1936).

11. Dollahite JW and VW Anthony. Southwest Vet. 10:128 (1957).

12. Dollahite JW and VW Anthony. J. Amer. Vet. Med. Assoc. 130:525-530 (1957).

13. Dollahite JW and JT Allen. Texas Agric. Exp. Sta. Prog. Rep. 2105 (1959).

14. Norris JT and KA Valentine. New Mexico Agric. Exp. Sta. Bull. 390 (1954).

15. Dollahite JW, S Shaver, and BS Bennie. Amer. J. Vet. Res. 23:1261-1263 (1962).

16. Shaver TN, BJ Camp, and JW Dollahite. Ann. New York Acad. Sci. 111:737-742 (1964).

17. Flournoy R, W Fortson, and B Camp. Texas Reports on Biol. Med. 30:319-326 (1972).

18. Molyneux RJ, KL Stevens, and LF James. J. Agric. Food Chem. 28:1332-1333 (1980).

19. Roitman JH and LF James. Phytochem. 24:835-848 (1985).

20. Cheeke PR. *Toxicants of Plant Origin, Vol. III. Phenolics.* CRC Press, Inc., Boca Raton, FL, 1989.

21. Flores-Rodriguez GI. Ph.D. Dissertation, New Mexico State Univer-sity, Las Cruces (1990).

22. Smith GS, KC McDaniel, and GI Flores-Rodriguez. In: *Proc. Broom Snakeweed Conf. (Ft. Sumner, NM; March 1988),* pp. 11-14. Coop. Ext. Service Report, New Mexico State University, Las Cruces, 1988.

23. Flores-Rodriguez GI, GS Smith, and KC McDaniel. Proc. West. Sect. Amer. Soc. Anim. Sci. 40:217-221 (1989).

24. Flores-Rodriguez GI, GS Smith, DM Hallford, and TS Edrington. Proc. West. Sect. Amer. Soc. Anim. Sci. 41:95-101 (1990).

25. Edrington TS, GI Flores-Rodriguez, and GS Smith. Proc. West. Sec. Amer. Soc. Anim. Sci. 41:88-89 (1990).

26. Oetting BC, TT Ross, W Walraven, P Kloppenburg, GS Smith, and DM Hallford. Proc. West. Sect. Amer. Soc. Anim. Sci. 41:23-26 (1990).

27. Edrington TS, GS Smith, MD Samford, J Medrano, TT Ross, and JP Thilsted. Proc. West. Sect. Amer. Soc. Anim. Sci. 42:(In press).

28. Van Atta GR, J Guggolz, and CR Thompson. J. Agric. Food Chem. 9:77-79 (1961).

29. Oakenfull D. Food Chem. 7:19-40 (1981).

30. Oakenfull D and GS Sidhu. In: *Toxicants of Plant Origin, Vol. II, Glycosides* (PR Cheeke, Ed.), Chapter 4. CRC Press, Inc., Boca Raton, FL, 1989.

24

Weedy Thistles of the Western United States

Steven A. Dewey

Abstract

The majority of thistles found in the western United States are native species that pose little or no threat as weeds. However, thistle species introduced from Europe and Eurasia during the past two centuries have proven to be very aggressive and invasive, spreading rapidly throughout much of the West, crowding out and replacing desirable plant species in cropland, pastures, rangeland, and forests. Nineteen thistle species have been declared noxious according to state weed control laws or seed purity laws. Though Canada, bull, musk, and Scotch thistles are most common, all noxious species are known to occur in the West and will become widespread if not controlled. More intensive field monitoring, mapping, and early control programs are required to stop the spread of thistles on public and private lands. Programs to help more individuals recognize noxious thistles will be key to the success of all thistle control efforts.

Introduction

Most people would argue that all thistles are weeds. However, if one considers the criteria most often used to classify weeds, many thistle species don't fit the mold. For example, if you agree that "a weed is a plant out of place" or " a plant that interferes with land management objectives", then you need to realize that more than half of all our thistle species are native to the western U. S., occurring naturally in low numbers in their native habitat on rangelands, foothills, or mountains. Rarely, if ever, do they interfere with management objectives on those lands. On the other hand, there are a few other thistle species that constantly cause serious problems. These tend to be very aggressive, invasive thistles that spread rapidly and reduce or completely crowd out desirable plant species. A principle difference between these two groups of thistles is the fact that the latter is comprised almost entirely of foreign introduced species.

The thistle situation in Utah is probably representative of many Western states. Here, we have 30 recognized species or varieties of thistle (Table 24.1), 24 of which are native to the region. The majority of our native species are never a concern to anyone. For example, ranchers or public land managers occasionally will express curiosity or mild concern about meadow thistle or gray thistle, but rarely does a situation warrant control measures. On the other hand, four of the five introduced species have become very serious weeds, spreading rapidly over much of the state, and adversely impacting agriculture, wildlife, and recreation.

Four introduced species that always come to the forefront of thistle problems in Utah or most other Western states, are Canada thistle (*Cirsium arvense*), musk thistle (*Carduus nutans*), Scotch thistle (*Onopordum acanthium*), and bull thistle (*Cirsium vulgare*). Canada thistle and bull thistle were probably introduced into Utah by early settlers in the mid 1800s. These weeds have been around so long that they are almost considered "native". Statewide surveys as early as 1932 and 1940 found bull thistle established in all 29 Utah counties. By 1940 it covered an estimated 9590 ha, statewide. Canada thistle apparently has spread at a slower rate, reported to infest 620 ha in 19 counties in 1932; increasing to 2300 ha in 20 counties in 1940; and 43,800 ha in 23 counties in 1963. By 1981 Canada thistle was believed to exist in all 29 counties.

Bull thistle and Canada thistle are probably approaching the limits of their potential distribution in Utah, so rate of spread is decreasing. However, the more recently introduced thistles have far to go before reaching their limit, and continue to spread at alarming rates. Scotch thistle was first found in Utah in 1963. By 1981 it had invaded approximately 6070 ha in 17 counties. Eight years later the infestation had increased to over 22,540 ha in 22 counties. Musk thistle was first reported in Utah in 1963. By 1981 it had spread over approximately 10,927 ha in 22 counties. By 1989 it infested over 173,210 ha in 28 of Utah's 29 counties.

The story is similar on public lands on a regional scale. Noxious weed surveys conducted by the U.S. Forest Service showed only 46 ha of Scotch thistle and 162 ha of musk thistle on forest land in the Intermountain

TABLE 24.1 Thistles of Utah.

Common Name	Scientific Name	Origin
Arizona thistle	*Cirsium arizonicum*	native
	var. *arizonicum*	
	var. *nidulum*	
Canada thistle	*Cirsium arvense*	Eurasia
	var. *mite*	
	var. *horridum*	
Barneby thistle	*Cirsium barnebyi*	native
Cainville thistle	*Cirsium calcareum*	native
	var. *bipinnatum*	
	var. *calcareum*	
	var. *pulchellum*	
Fringed thistle	*Cirsium centaureae*	native
Fish Lake thistle	*Cirsium clavatum*	native
Eaton thistle	*Cirsium eatonii*	native
	var. *eatonii*	
	var. *harrisonii*	
	var. *murdockii*	
New Mexico thistle	*Cirsium neomexicanum*	native
	var. *neomexicanum*	
	var. *utahense*	
Owenby thistle	*Cirsium owenbyi*	native
Rothrock thistle	*Cirsium rothrockii*	native
Rydberg thistle	*Cirsium rydbergii*	native
Meadow (elk) thistle	*Cirsium scariosum*	native
	var. *scariosum*	
	var. *thorneae*	
Carmine thistle	*Cirsium scopulorum*	native
Snow thistle	*Cirsium subniveum*	native
Gray (wavyleaf) thistle	*Cirsium undulatum*	native
	var. *tracyi*	
	var. *undulatum*	
Virgin thistle	*Cirsium virginensis*	native
Bull thistle	*Cirsium vulgare*	Eurasia
Wheeler thistle	*Cirsium wheeleri*	native
Blessed thistle	*Cnicus benedictus*	Europe
Musk thistle	*Carduus nutans*	Eurasia
Scotch thistle	*Onopordum acanthium*	Europe

Region in 1969. By 1985 there had been nearly a three-fold increase in Scotch thistle (137 ha), and approximately a twelve-fold increase in the musk thistle infestation (1860 ha). Surveys of BLM lands in the Northwest indicate Canada thistle spreading at an annual rate of 10%, causing an average 42% reduction in range carrying capacity on infested lands. The same surveys indicate that musk thistle and Scotch thistle are spreading at annual rates of 15% and 16%, respectively, reducing carrying capacity by an average of 38%.

Of 19 thistles currently declared noxious in various states (Table 24.2), only one is a native species; all others have been introduced to North America. Partly because Canada thistle, musk thistle, Scotch thistle, and bull thistle are the most common, they have been declared noxious in more states. The other thistles in Table 24.2 are less widely distributed and are typically declared noxious in only one or two states. However, there is little doubt that most or all of these species have the capacity to spread to the same extent observed for musk, Scotch, bull, or Canada thistles. Careful field monitoring, extensive mapping, and intensive early control programs are essential if further invasion and spread of these weeds on public and private lands are to be prevented. Programs to help more individuals recognize noxious thistles will be key to the success of any thistle control effort.

Canada Thistle

Canada thistle (*Cirsium arvense*) is an erect creeping perennial, typically growing 0.5 to 1 m tall. Upper and lower leaf surfaces are usually smooth and glabrous, and margins are only moderately spiny. Flower heads are relatively small (1 to 1.5 cm wide), with small appressed involucre bracts. Flower heads are borne in clusters. Flower color varies from pink or lavender, to pure white. Male and female flowers are produced on separate plants. Stout spreading lateral rootstocks are probably its most distinguishing characteristic. At least three varieties have been recognized in the West. These are var. *mite* Wimmer & Grab., var. *horridum* Wimmer & Grab., and var. *arvense* (L.) Scop.

Canada thistle is common throughout most of the United States north of the 35th parallel. It has been declared noxious according to weed control laws or seed purity laws in at least 35 states. Typically, it is found in pastures, roadsides, cultivated fields, and other disturbed areas, but also is known to invade native plant communities on range and forest sites at elevations in excess of 2500 m.

TABLE 24.2 Noxious thistles of the United States.[1]

Common Name	Scientific Name	Noxious Designation
Artichoke thistle	*Cynara cardunculus*	CA
Bull thistle	*Cirsium vulgare*	IA, MN, PA, WA
Blessed thistle	*Cnicus benedictus*	GA, SC, TX
Canada thistle	*Cirsium arvense*	AZ, CA, CT, CO, DE, GA, HI, IA, ID, IL, KS, MA, MD, ME, MI, MN, MT, ND, NE, NH, NV, OH, OK, OR, PA, RI, SC, SD, TN, TX, UT, VA, WA, WI, WY
Carthamus	*Carthamus oxyacantha*	CA
Golden thistle	*Scolymus hispanicus*	CA
Illyrian thistle	*Onopordum illyricum*	CA
Italian thistle	*Carduus pycnocephala*	CA, OR
Milk thistle	*Silybum marianum*	OR
Musk thistle	*Carduus nutans*	CA, CO, ID, IA, IL, KS, MD, ND, NE, NV, OR, PA, UT, VA, WA, WY
Plumeless thistle	*Carduus acanthoides*	CA, MN, NE, WY
Scotch thistle	*Onopordum acanthium*	CA, ID, NV, OR, UT, WA, WY
Slenderflower thistle	*Carduus tenuiflorus*	CA, OR
Smooth distaff thistle	*Carthamus baeticus*	CA
Taurian thistle	*Onopordum tauricum*	CA
Wavyleaf thistle	*Cirsium undulatum*	CA
Whitestem distaff thistle	*Carthamus leucocaulo*	CA
Woolly distaff thistle	*Carthamus lanatus*	CA
Yellowspine thistle	*Cirsium ochrocentrum*	CA

[1] Declared noxious according to state weed control and/or seed purity laws.

Scotch Thistle

Scotch thistle (*Onopordum acanthium*) is a robust biennial, often growing to heights in excess of 2 m. Main stems may be up to 10 cm wide at the base. Upper and lower leaf surfaces are covered with a thick mat of cotton-like or wooly hairs, giving the foliage a characteristic gray-green color. Stems are lined with vertical rows of prominent, spiny, ribbon-like, leaf material, which extend all the way to the base of flower heads. Flower heads range from 2 to 5 cm wide, with long, stiff, needle-like involucre bracts. Flower color may vary from dark pink to lavender.

Scotch thistle is described by some authors as a "ruderal" weed, meaning it is often associated with waste places, rubble, or rubbish. It is typically found associated with rivers, streams, canals, or other waterways, but can also become abundant in dry pastures, fields, and rangeland. It is not commonly found at high elevations (above 2200 m). Scotch thistle occurs in most of the West and is considered noxious in at least seven states. Other common names for Scotch thistle include cotton thistle and winged thistle.

Musk Thistle

Musk thistle (*Carduus nutans*) is a biennial or winter annual growing one to two meters tall. Upper and lower leaf surfaces are usually smooth, but sometimes hairy. Leaves are dark green, with a broad, light-colored, midvein. Leaf margins are spiny, varying from deeply toothed to nearly entire. Flower heads are large (5 to 8 cm wide), often nodding, and borne singly at the ends of naked stalks. Involucre bracts are broad-based and wedge-shaped, usually flaring out from the flower head. Flower color may be pink to lavender, but is most often a bright rose-pink.

Musk thistle is typically found on disturbed sites, waste areas, along roads, in pastures, and on rangeland. It can become a severe problem in alfalfa and grass hay fields. It is also found spreading into sagebrush, pinyon-juniper, and mountain sagebrush communities up to elevations in excess of 2500 m. In some Midwestern publications *Carduus nutans* has been subdivided into *Carduus nutans*, *Carduus macrocephalus*, and *Carduus thoermeri*. Authors refer to these as the musk thistle group. Musk thistle has been reported in 42 states and is considered noxious in at least 16. Another common name for musk thistle is nodding thistle.

Bull Thistle

Bull thistle (*Cirsium vulgare*) is a bushy, taprooted biennial, normally growing 0.5 to 1 m tall. The upper surface of leaves is usually dark green, rough-textured, with few to no hairs. The lower surface of leaves is often densely covered with soft hairs, giving the underside a white-green cast. Leaf margins are deeply toothed, and toothed again (double dentate), with prominent stiff spines at the tips of each tooth. Flower heads average 2 to 5 cm wide, having long, stiff, needle-like involucre bracts. Flower color is usually a bright rose-purple.

Bull thistle is reported to occur in all 48 contiguous states. Though common, it is not generally considered as serious a weed as Canada, musk, or Scotch thistles. The fact that it is designated noxious in only four states seems to support this observation. Typical habitat includes meadows, pastures, roadsides, cultivated fields, and rangeland. Infestations have been reported at elevations as high as 2800 m.

Summary

All thistles are not bad. Learning to recognize and distinguish between the many thistle species allows public and private land managers to make sound weed control decisions. In most cases, programs to control native thistles are unnecessary, and a waste of time and resources. On the other hand, failure to detect, recognize, and immediately control new infestations of invasive foreign thistle species almost certainly will result in very large and costly weed infestations within a very short time.

Bibliography

Dennis LJ. *Gilkey's Weeds of the Pacific Northwest*. Oregon State Univ. Press, Corvallis, 1980.

Hitchcock CL and A Cronquist. *Flora of the Pacific Northwest*. University of Washington Press, Seattle, 1978.

Holmgren AH and BA Anderson. *Weeds of Utah*. Utah State Univ. Agric. Exp. Sta. Special Report 21, Logan, 1970.

Nelson EW and O Burnside. *Nebraska Weeds*. Nebraska Dept. of Agric., Lincoln, 1979.

Robbins WW, MK Bellue, and WS Ball. *Weeds of California*. California Dept. of Agric., Printing Division, Sacramento, 1951.

US Department of Agriculture, Forest Service. *Intermountain Region Final Environmental Impact Statement on Noxious Weed and Poisonous Plant Control Program*. Intermountain Regional Office, Ogden, UT, 1986.

US Department of Interior, Bureau of Land Management. *Northwest Area Noxious Weed Control Program Environmental Impact Statement*. Oregon State Office, Portland, 1985.

Welsh SL, ND Atwood, S Goodrich, and LC Higgins. *A Utah Flora*. Brigham Young Univ. Press, Provo, UT, 1987.

Whitson TD, SA Dewey, MA Ferrell, JO Evans, SD Miller, and RJ Shaw. *Weeds and Poisonous Plants of Wyoming and Utah*. (TD Whitson, Ed.). Cooperative Extension Service, Univ. of Wyoming, Laramie, 1987.

25

Biennial Thistle Control with Herbicides

K. George Beck

Abstract

Biennial thistles can be effectively controlled with picloram (0.13–0.5 lb ai/A), dicamba (0.5-2.0 lb ai/A), 2,4-D (1.5-2.0 lb ai/A), 2,4-D + dicamba (1.0 + 0.5 lb ai/A), dicamba + picloram (0.5-1.0 + 0.13-0.25 lb ai/A), and picloram + 2,4-D (0.13 + 0.25 lb ai/A). Metsulfuron (0.3 oz ai/A) and chlorsulfuron (0.75 oz ai/A) will control musk thistle and plumeless thistle is controlled by chlorsulfuron. Chlorsulfuron for plumeless thistle control and the auxinic herbicides for all biennial thistles should be applied when weeds are in the rosette growth stage. Metsulfuron or chlorsulfuron for musk thistle control should be applied when the weed is in the bolting to bud growth stages.

Introduction

Biennial thistles such as musk thistle (*Carduus nutans*), plumeless thistle (*Carduus acanthoides*), bull thistle (*Cirsium vulgare*), and Scotch thistle (*Onopordum acanthium*) are considered noxious in many states and usually are viewed as undesirable wherever they occur. Biennial thistles are not as difficult to manage as creeping perennial thistles such as Canada thistle (*Cirsium arvense*). Biennial thistles live for two growing seasons. These plants germinate in spring or fall, develop into a rosette, and overwinter as such. In their second growing season, plants resume vegetative growth in spring, then bolt (shoot elongation), flower, set seed, and die after completing their life cycle. Biennial thistles reproduce solely from seed; therefore, the key to their successful management is to prevent seed production.

Although this paper will discuss only chemical control of biennial thistles, it is important to recognize that herbicides are only part of good weed management. An integrated weed management system that combines two or more control methods (i.e., cultural, mechanical, chemical, and biological control) generally provides for the most effective reduction of weed populations.

Musk Thistle

Musk thistle is effectively controlled with several herbicides. The auxinic herbicides such as picloram, dicamba, or 2,4-D are commonly used and effective if applied in spring *before* musk thistle bolts. Musk thistle susceptibility to these herbicides decreases as bolting, flowering, and seed set occur. Herbicide application rate varies (Table 25.1) and is largely affected by stand density and environmental conditions. Herbicide rate should be increased as stand density increases or if environmental conditions are not conducive to active weed growth, e.g. drought or cold temperatures. Picloram is probably the most effective of these herbicides and may be applied in spring or fall to musk thistle rosettes. Dicamba also may be applied in spring or fall to rosettes, but is most often tank-mixed with 2,4-D at a ratio of 2:1 2,4-D to dicamba. 2,4-D is most effective in spring.

Fall is a good time to treat musk thistle with herbicides, particularly the auxinics, because the weed is in the rosette stage and will not bolt. Picloram is the most effective of the auxinics in fall because this herbicide is less affected by cold temperatures compared to dicamba and particularly 2,4-D. Fall applications of 2,4-D alone often fail due to cold temperatures.

Recent research conducted in Colorado and Nebraska indicates that the sulfonylurea herbicides chlorsulfuron and metsulfuron will effectively control musk thistle (1). Our results indicate that chlorsulfuron at 0.75 oz ai/A or metsulfuron at 0.3 oz ai/A will eliminate seed production if applied in the bolting to bud growth stages. Seed production still occurs if the auxinic materials are applied this late in musk thistle's life cycle. The sulfonylureas add a new dimension of extended time for effective musk thistle control in spring. Furthermore, chlorsulfuron and metsulfuron may be compatible with the musk thistle seedhead weevil (*Rhinocyllus conicus*) by a temporal separation of herbicide application and seedhead weevil presence on the weed; i.e., treat when lateral shoots are

TABLE 25.1 Herbicides and rates for musk or plumeless thistle control in pastures, rangeland, and non-crop areas.

Herbicide	Rate	Timing/comments
	(lb ai/A)	
Picloram	0.125-0.25	spring before bolting or in fall
Dicamba	0.5-2.0	spring before bolting or in fall if good growing conditions exist
2,4-D	1.5-2.0	spring before bolting
2,4-D + dicamba	1.0+0.5	spring before bolting or in fall if good growing conditions exist
metsulfuron	0.3 oz ai	*musk thistle*: spring from bolting to bud growth stage; *plumeless thistle*: spring before bolting; add non-ionic surfactant 0.25% v/v; rate is ounces ai/A
chlorsulfuron	0.75 oz ai	*musk thistle*: spring from bolting to bud growth stage; *plumeless thistle*: spring before bolting; add non-ionic surfactant 0.25% v/v; rate is ounces ai/A

in the bolting to early bud stage and early developing heads are in the bloom stage. This will allow the seedhead weevil to complete its life cycle in the early developing heads thereby reducing or eliminating seed formation; later seed development in lateral heads will be eliminated by chlorsulfuron or metsulfuron. This management system combines biological and chemical control, which is a rather new concept that has been previously viewed as incompatible. Future research will address the direct effects of chlorsulfuron and metsulfuron on the seed head weevil, and if these herbicides do not disrupt insect development or completion of its life cycle, the temporal separation may not be necessary.

Plumeless Thistle

Plumeless thistle is closely related to musk thistle and responds similarly to herbicide application. Picloram, dicamba, and 2,4-D are most often used, and like musk thistle, should be applied in spring before plumeless thistle bolts or in fall to rosettes. Rates of these herbicides vary (Table 25.1) with weed stand density or environmental conditions. Chlorsulfuron and metsulfuron also control plumeless thistle.

TABLE 25.2 Plumeless thistle control with various herbicides and rates on Colorado rangeland.[1]

Herbicide	Rate	Plumeless thistle control				
	(lb ai/A)	----------------% of non-sprayed check-------------				
		15 Sep 1988[2]		13 Sep 1989		24 Jul 1990[3]
picloram	0.13	100	100	100	100	91
picloram	0.25	100	100	90	90	96
picloram	0.5	100	100	100	100	93
dicamba	0.5	100	100	100	93	70
dicamba	1.0	100	100	100	100	85
dicamba + picloram	0.5 + 0.25	100	100	100	100	93
dicamba + picloram	1.0 + 0.125	100	100	65	60	100
2,4-D	1.0	68	39	94	94	12
dicamba + 2,4-D	0.5 + 1.0	100	100	100	96	53
chlorsulfuron[4]	0.023	54	28	15	17	18
chlorsulfuron	0.047	95	81	88	88	82
metsulfuron	0.009	71	50	34	15	23
metsulfuron	0.019	84	76	58	63	50
dicamba + chlorsulfuron	0.5 + 0.023	100	98	98	100	95
dicamba + chlorsulfuron	0.5 + 0.019	93	93	100	100	53
dicamba + chlorsulfuron	1.0 + 0.023	100	100	91	86	78
LSD (0.05)		19	16	27	24	34

[1] Treatments applied 15 Jun 1988.
[2] Second and first year plumeless thistle rosette control in first and second columns at each evaluation data, respectively.
[3] First and second year plumeless thistle rosettes grouped as one.
[4] X-77 added at 0.25% v/v to all chlorsulfuron and metsulfuron treatments.

A research project was conducted in Colorado to evaluate picloram, dicamba, 2,4-D, dicamba + picloram, dicamba + 2,4-D, chlorsulfuron, metsulfuron, and dicamba + chlorsulfuron efficacy on plumeless thistle (2). Very good to excellent control still was apparent two years after treatments were applied with picloram (0.13, 0.25, and 0.5 lb ai/A), dicamba (1.0 lb ai/A), dicamba + picloram (0.5 + 0.25 and 1.0 + 0.13 lb ai/A), chlorsulfuron (0.75 oz ai/A), and dicamba + chlorsulfuron (0.5 lb + 0.38 oz and 1.0 lb + 0.38 oz ai/A) (Table 25.2). Research has not been conducted to verify if late applications (bolting to bud growth stages) of metsulfuron or chlorsulfuron will eliminate plumeless thistle seed formation.

Bull Thistle

Bull thistle is less aggressive than other biennial thistles and apparently is easier to control. Dicamba + 2,4-D (0.25 + 0.5 lb ai/A) or 2,4-D (1.0 lb ai/A) alone are most often used and very effective. As with other biennial thistles, these compounds should be applied in spring before bolting or in fall to rosettes. If fall applications are to be made, dicamba + 2,4-D should be used to avoid poor control from cool temperatures.

Scotch Thistle

Onopordum acanthium is the predominate Scotch thistle species that occurs in the western United States. It has very pubescent leaves. A glabrous species, *Onopordum tauricum*, also occurs but much less frequently. Scotch thistle control with herbicides is similar to musk and plumeless thistles. Picloram, dicamba, 2,4-D, and dicamba + 2,4-D are most often used. Again, herbicide rate will vary (Table 25.3) depending upon stand density and environmental conditions. These herbicides should be applied in spring before Scotch thistle bolts or in fall to rosettes.

TABLE 25.3 Herbicides and rates for Scotch thistle control in pastures, rangeland, and non-crop areas.

Herbicide	Rate (lb ai/A)	Timing/comments
picloram	0.25-0.5	spring before bolting or in fall
dicamba	0.5-1.0	spring before bolting or in fall if good growing conditions exist
2,4-D	2.0	spring before bolting
2,4-D + dicamba	1.0+0.5	spring before bolting in fall if good growing conditions exist
2,4-D + picloram	0.25+0.125	spring before bolting or in fall

Research conducted in Idaho indicates that 2,4-D (2.0 lb ai/A), picloram (0.25 and 0.5 lb ai/A), dicamba (2.0 lb ai/A), dicamba + 2,4-D (0.5 + 1.5 lb ai/A), and picloram + 2,4-D (0.13 + 0.25 lb ai/A) reduced seed formation from 80 to 100% four months after treatments were applied (3). However, only picloram or picloram + 2,4-D provided adequate control one year after treatment.

Summary

Biennial thistle populations can be effectively reduced by good weed management, particularly if herbicides are a component of that system and they are applied at the correct rate and time. Often, seeding desirable plant species after biennial thistle have been controlled is necessary to provide plant competition with thistles that survive treatment. If little to no desirable plant species are left after thistles have been controlled, re-invasion of the area by biennial thistles is very likely to occur.

References

1. Beck KG, RG Wilson, and MA Henson. Weed Tech. 4:482-486 (1990).
2. Sebastian JR, KG Beck, and DE Hanson. Res. Prog. Rep. West. Soc. Weed Sci. p. 95 (1990).
3. Belles WS, DW Wattenbarger, and GA Lee. Res. Prog. Rep. West. Soc. Weed Sci. p. 53 (1980).

26

Canada Thistle Control

Lloyd C. Haderlie, Ray S. McAllister, Ray H. Hoefer, and Phil W. Leino

Abstract

Canada thistle (*Cirsium arvense*) control is easier once its biology and physiology are understood. This weed is a perennial which spreads by seed and underground roots. It is dioecious, so not all plants produce seed. Canada thistle occurs in the Northern Temperate Zone due to day length response and a high temperature limitation on growth. Flowering occurs when day length is 14 hr or more and temperatures are high enough. There is considerable variability in types of Canada thistle, and they can differ in their response to herbicides. Canada thistle seedlings can become perennial within three weeks after germination, producing buds on roots. Buds can develop at any location on the root and at any time if environmental conditions are favorable. Root bud growth was maximized when air temperatures were 15/5 C and root temperatures were 20 C under a 13-hr day length. As soil fertility increases, so does Canada thistle growth. Photosynthesis was as great in October as in June. Several herbicides are effective when applied to Canada thistle regrowth in the fall. However, on rangeland, soil moisture is often limiting in the fall and no regrowth occurs; hence, the best time for control is at or before flower bud stage, normally in June.

Biology and Physiology

Canada thistle is a Northern Temperate Zone plant because extended high temperatures inhibit growth and it has a long-day flowering requirement. Its growth is limited or stopped when temperatures exceed 30 C for extended periods of time. The requirement for flowering is at least a 14-hr photoperiod (1).

Variability is extensive in this plant and not all 'types' have been studied. There may well be exceptions to research results presented here.

Temperature

Three temperature regimes were tested for several growth characteristics, including regrowth (1). Plants were grown in growth chambers at 30 C day and 22 C night, 25 C day and 15 C night, or 15 C day and 5 C night temperatures. Higher temperatures caused earlier flowering and produced more shoots. More total growth occurred at the medium temperature. Low temperatures increased the proportion of root or root shoots.

Plants grew more when root temperature was maintained at 17 C and ambient or foliage temperature was changed than when root temperature was either lower or higher, regardless of air temperature, under a 13-hr photoperiod. Also, when root temperature was 20 C, ambient temperature was 15/5 C, and day length was 13 hr, root bud growth was greater (relative to other plant parts) than at any other environmental condition (2).

Nitrogen

Growth increased for Canada thistle as nitrogen was increased from a low (8 ppm) level in the soil (1). A decrease in root-to-shoot ratio occurred as nitrogen was increased.

Root Buds

Root bud development and growth make Canada thistle a perennial. Within three weeks of germination, root buds form on roots of Canada thistle seedlings (unpublished data of authors). Hence, seedlings have to be tilled or killed before three weeks in order to stop the plant from becoming perennial.

Weed science literature for many years has incorrectly referred to the perennial roots as rhizomes in this and other similar species. Rhizomes are underground stems with nodes and internodes. Buds on rhizomes develop at the nodes. Anatomically, the perennating, overwintering tissue in Canada thistle is a root and does not have nodes and internodes as do stems (3).

Root buds can develop along the root at any location or at any time of the year with favorable growing conditions (2). There appears to be no seasonal dormancy except that induced by limiting growth temperatures. Root buds do remain dormant during the winter and perhaps in response to 'apical' dominance exerted by other developing roots.

Root bud development or growth is highest in the fall when soil temperatures are warm and ambient temperatures are falling (1,2,4). Photoperiod was 13 hr when root bud growth was greatest.

Photosynthesis

Photosynthesis rate was found to be as high during late October in plants with new leaves (regrowth) as it was during June (5). Plant growth in the fall is not obvious because most occurs under the soil surface. Little foliar growth will result in the fall when warm soil temperatures, cool ambient temperatures, and short day lengths all signal the plant to send photosynthate to the roots for growth and development of buds. In an old plant that has not been cut, tilled, or otherwise disturbed during the summer, photosynthesis in the fall may be much different than for a plant producing new leaves.

Root carbohydrate levels in an undisturbed Canada thistle patch are lowest near the end of June or at flower bud stage (6,7). These levels can be further decreased by appropriately timed tillage and/or mowing. Lower root carbohydrate levels will weaken the plant and enhance control.

Control

Control of Canada thistle must be focused on preventing seed production and controlling or killing the root. In a rangeland ecosystem, the only reasonable way to control Canada thistle is to use biological control and chemical control. Insects, diseases, foraging animals, and rodents may weaken the plant somewhat. Chemicals are often costly and may not control the weed because of improper timing.

Chemical control will usually be maximized by applying the herbicide no later than the end of June when root carbohydrate levels are least and when soil moisture is not too low. Leaves are young enough to absorb the chemicals, as well. Herbicide application in the fall is ideal if plants have young leaves and there is adequate soil moisture for good growth. Translocation of herbicides to the roots is higher during the fall when good growing conditions exist than at any other time of year. Always consult the herbicide label for specific instructions.

References

1. Hoefer RH. PhD Thesis, Univ. of Nebraska, 1981.
2. McAllister RS. PhD Thesis, Univ. of Nebraska, 1982.
3. McAllister RS and LC Haderlie. In: *Proc. North Cent. Weed Contr. Conf. 36*, p. 157, 1981.
4. McAllister RS and LC Haderlie. Weed Sci. 33:148 (1985).
5. Leino PW. PhD Thesis, Univ. of Idaho, 1987.
6. Hodgson JM. USDA Tech. Bull. No. 1386 (1968).
7. McAllister RS and LC Haderlie. Weed Sci. 33:44 (1985).

27

Biological Control of Thistles

Norman E. Rees

Abstract

Musk (*Carduus nutans* group) and Canada (*Cirsium arvense*) thistles differ from each other in that they are in separate genera and have different growth habits. Musk thistle can act as an annual, a winter annual, or a biennial, and reproduces totally by seed, whereas Canada thistle is a perennial and reproduces mainly by vegetative regrowth and continues to emerge year after year from the same plant root material. *Rhinocyllus conicus* is a seed head attacking weevil which was introduced into North America for the biocontrol of musk thistle. Larvae of a second weevil, *Trichosirocalus horridus*, feed on the apical meristem of the musk thistle rosette and developing stems. Larvae of *Cheilosa corydon* mine directly into the tender, young musk thistle shoots, then mine up and down the stem, causing the stems to become easily broken and/or prematurely dried. The weevil *Ceutorhynchus litura* was released in Canada in 1965 for the biological control of Canada thistle. Larvae of *Urophora cardui* burrow into the stem where they induce the plant to create large galls, which act as a nutrient sink and stress the plant.

Introduction

Many of the thistle species which are currently problems in the United States are exotic and are believed to have been introduced into North America in a variety of ways, ranging from seed in the ballast of ships to being brought to North America by early immigrants in unclean seeds.

Although horses will occasionally nibble the flowers of young thistle plants and cattle will reluctantly feed on thistles during extreme drought conditions, livestock generally avoid areas where thistles grow. Therefore, by their presence, they occupy space and deny access to food plants rendering them unavailable for livestock and some forms of wildlife.

The introduced thistle species presently being studied include musk thistle (*Carduus nutans, C. macrocephalus, C. nutans leiophyllus*), Italian thistle (*Carduus pycnocephalus*), plumeless thistle (*Carduus acanthoides*), Canada thistle (*Cirsium arvense*), bull thistle (*Cirsium vulgare*), milk thistle (*Silybum marianum*), and Scotch thistle (*Onopordum acanthium*). Their environmental requirements, reproductive methods, and natural enemies differ somewhat. As examples of the general interactions of bioagents on thistles, and the interrelationships of these bioagents among themselves, we will examine two distinctly different thistle species, musk thistle and Canada thistle.

Musk Thistle

Musk thistle (also called nodding thistle in some countries) is the common name for several species or subspecies of plants belonging to the large-headed *Carduus nutans* group. Taxonomists do not agree on the level of nomenclature that each musk thistle type should occupy. Most classify *C. nutans* at the species level, while *C. macrocephalus* is sometimes presented at the subspecies level, *C. nutans* ssp. *macrocephalus*. *Carduus thoermeri* has been changed to *C. nutans* ssp. *leiophyllus* and as *C. nutans* ssp. *macrolipes* (2-6). To the novice taxonomist, leaves of *C. macrocephalus* are light green and very hairy; while at the other end, leaves of *C. nutans* ssp. *leiophyllus* are blue green and waxy-smooth.

Musk thistle can be found in the United States on all types of land except deserts, dense forests, high mountains, and newly cultivated lands (1). It is an introduced plant possessing long sharp spines that physically protect both itself and any other plants growing in close proximity to it from the grazing of livestock and other animals. A dense population of the plant discourages animals from occupying that portion of the field in which it grows. By its presence, it reduces the amount of available forage for livestock consumption.

The earliest collection on record for musk thistle in the United States came from Harrisburg, PA, in 1853 (6). Introductions on the eastern seaboard are reported to have been started by European ships dumping ballast soil infested with musk thistle seed. Since that time, this plant has spread to 12% or more of the counties of the United States, infesting over

730,000 ha of land in 40 states by 1981 (7,8). Trumble and Kok (9) established that one musk thistle plant per 1.48 m^2 will reduce forage production by 23%. Reproduction of musk thistle is totally by seeds, which are produced from June and July until death of the plant from either desiccation, freezing, or other lethal factors. Only 2% of these seeds are attached to the plumeless pappus and are borne on the wind; the rest fall to the ground close to the parent plant and/or are distributed by birds, small animals, and running water.

Musk thistle plants can act either as annuals, winter annuals or biennials. Seeds can germinate 6 to 8 weeks after they have fallen to the ground, or can remain dormant in the soil for years. In Montana, about 69% of the seeds of C. macrocephalus germinate during the first year, 20% germinate the second year, and the remainder at a later time or not at all.

Seedlings that germinate in the fall attain a rosette stage before winter. The numerous roots of the fall rosette arise from the crown of the plant and are located near the surface. In the spring, a tap root has formed and the diameter of the rosette may attain a size of 1/3 to 1 m before the plant bolts. Although each plant typically sends up one main stem, damage to the terminal meristem, either by frost or by mechanical means (such as insect or livestock feeding), will generally cause groups of stems to develop, usually in multiples of five. Plants attain heights of 0.3 to 2.3 m in Montana and sometimes exhibit considerable variation in height, form, and color between years, locations, and individual plants.

Flowers of musk thistle are reddish purple. Each terminal flower produces about 1,000 seeds; each lateral bud produces smaller flowers with fewer seeds, averaging about 840 seeds per head for secondaries and 640 for tertiaries. Flowering continues until plants are killed by adverse conditions; in Kansas, this often occurs in August to October, while in the Gallatin Valley of Montana, plants usually survive until sometime in November (1).

A discussion of the natural enemies of musk thistle follows.

Rhinocyllus conicus

Rhinocyllus conicus is a weevil which was introduced into North America for the biocontrol of musk thistle (10,11) and spends its larval, pupal and two weeks of its adult stage living within the developing host plant. It originated in Eurasia, and although the weevil utilizes species of four genera of thistles, it prefers the Carduus nutans group. In the Gallatin Valley, R. conicus utilizes C. macrocephalus and C. thoermeri equally well, with no significant differences in the number of weevils inhabitating each seed head or weevil survival in either thistle species (12).

Since *R. conicus* has been introduced from several European locations and established at various locations in the United States, scientists have been able to observe how strains vary. It appears that *R. conicus* collected in Europe from Italian thistle will also utilize scotch, slender flowered, and milk thistles,[1] whereas the same weevil species collected from musk thistle will not. There may also be a separate biotype specifically for Scotch thistle.[2]

Primary musk thistle buds receive the greatest number of eggs, followed in order by secondary buds, tertiary buds, and stems of musk thistle. Eggs are also deposited in buds of Canada, bull and wavy-leaved thistles to a lesser extent. Although a primary bud may receive in excess of 300 eggs, only the first 20 to 48 larvae that hatch and begin their cells will survive while the "locked out" larvae will perish. Nevertheless, unless the plant is moisture stressed or the seed head damaged, survival of the first larvae within the seed heads generally exceeds 98%.

Feeding by larvae in the seed head of musk thistle not only damages developing seeds which are contacted by the cell but also reduces the viability of remaining seeds. For example, uninfested seed heads produce seeds with a first year viability of 69%, while seed heads with four to five larvae have viability of undamaged seeds reduced to 45%, seed heads with more than nine larvae have viability of undamaged seeds reduced to less than 2%. Therefore, *R. conicus* functions characteristically like a gall-producing insect in that it causes the plant to furnish the feeding larvae with an enlarged area in which to live and the nutrients which would normally have been utilized by the developing seeds (13).

Larval development is synchronized with the development of the seed head in which it lives. This becomes evident by examination and comparison of developing larvae within primary, secondary, and tertiary seed heads at the same point of flower development. I have two theories to explain the synchronization. The first theory states that the primary buds begin to swell first and so receive the first eggs, which hatch first, the larvae develop first, and the adults emerge first. This theory is plausible since the primary buds are vulnerable to oviposition first and receive eggs over a longer period of time and do receive the greatest number of eggs. However, the second theory states that eggs are deposited somewhat randomly and hatch at all locations as stated before, (first deposited hatch first etc.), but development within the plant is retarded by plant hormones at all locations other than the primary flowers until after seed maturation and dispersal. Observations of egg deposition, dissections of all stages of flower development, and the fact

[1]Slender-flowered and milk thistle suggested by Eric Coombs.
[2]This information suggested by Jeff Littlefield.

that the period of primary, secondary, and tertiary development extends far beyond the period taken for egg deposition, all support the latter theory.

When egg densities on musk thistle buds are high, females deposit eggs on the stems of musk thistle. Larvae in the stems do not create cells as do larvae in the seed heads, but rather remain close to their oviposition site, usually just under the seed head. Larvae within the stem survive and mature at a slightly slower rate than the developing larvae in seed heads with uninfested stems. However, if feeding larvae sever the vascular bundles of the stem, the seed head above dies and all larvae within the head die from desiccation.

After both preferred and non-preferred oviposition sites have been utilized, gravid adults begin looking for sites where eggs are less dense. This occurs approximately 3 wk after oviposition begins in the core area. This pattern is obvious near recent release sites, where the outer boundaries or limits of the weevil infestation increase by 8 to 40 km each year. Emergence of weevils in the new infestation area is about 3 wk later than that in the core area. In the following year, however, emergence in this new infestation core area is synchronized with that of the old core, while emergence in the new infestation area (i.e., that area beyond last year's infestation) is again 3 wk later than that of the two cores. This demonstrates that the weevils utilize oviposition sites in the core area for 3 wk before migrating away from the center into new, unpopulated areas where they deposit their remaining eggs (1).

Experiments by T. J. Miller (14) in 1978 at Montana State University and Richard Lee (15) in 1980 at Utah State University have shown that mortality of R. conicus is not significantly increased when the herbicide 2,4-D is sprayed on bolting and bolted musk thistle plants. From only one year's data in 1987, I found that maturation time of the weevil was shortened by 4-1/2 d in plants sprayed with 2,4-D, and the resultant adults were active and appeared normal. I suspect that a plant hormone is involved since 2,4-D is a growth stimulant, and as stated earlier, larval development within the plant appears to be the result of plant and insect interaction.

Trichosirocalus horridus

Larvae of a second weevil, Trichosirocalus horridus, feed on the apical meristem of the musk thistle rosette and developing stems. There is only one generation per year and the life cycle and action on the plant are early and offset sufficiently so that T. horridus does not compete with R. conicus. The resultant increased number of smaller musk thistle seed heads produce fewer seeds and provide more niches, which is beneficial to the *Rhinocyllus* population. In this way, these two insects are complementary and not antagonistic to each other; both gain the greatest usage from the plant, and fewer seeds are produced.

To date, *T. horridus* has been reported to be established only in Virginia, Kansas, Missouri, and Wyoming, although it has also been released many times in Montana and other states (16).

Cheilosa corydon

In Italy, adult *Cheilosa corydon* flies emerge and begin oviposition in early to mid-March. Eggs are deposited on the young leaves and shoots, usually near the center of the plant. Newly hatched larvae mine directly into the tender, young shoots, then mine up and down the stem, causing the stems to become easily broken and/or prematurely dried. By summer, the third instar larvae are in the root of the senescent host, where they remain until fall rains begin.

Stresses to the infested plants include: 1) interruption of the water and nutrient transport system, 2) loss of infloresences, 3) lower seed production, and 4) invasion by soil microorganisms through lesions on the roots. When two or more larvae reach the root, the plant often dies (17).

This insect was approved for release in the fall of 1989. However, attempts at Bozeman to obtain large numbers of adults from imported pupae this last summer proved impossible. This condition also occurred in Maryland, where Phil Tipping was only able to make several very small releases. Since all Bozeman adults emerged in the refrigerator while being held as pupae for spring, it appears that the adults may naturally emerge and deposit eggs during cold, spring periods.

Canada Thistle

Canada thistle, *Cirsium arvense*, is an aggressive perennial weed found growing throughout the northern half of the United States in cultivated fields, pastures, rangelands, forests, and along roadsides, ditches, and river banks. It was reported in 1952 to infest more acreage than any other noxious weed in the states of Idaho, Montana, Oregon, and Washington (18). It is indigenous to north Africa, Europe, and Asia, and ranges from Scandinavia through Siberia into China and Japan, and southeast into Afghanistan (3). This plant can reproduce by seed, but mostly reproduces vegetatively by lateral roots (19, 20).

The Canada thistle plant consists of several well-separated shoots connected by a lateral root system. The above-ground portion of the shoot dies over the winter, but the underground part generally survives to send up regenerated shoots the following season. New shoots are also produced each year from the lateral root system.

A discussion of the natural enemies of Canada thistle follows.

Vanéssa cárdui

Besides aphids, leaf hoppers, and other native insects which frequent Canada thistle, the painted lady butterfly (*Vanéssa cárdui*) can be very effective, but only on an intermittent basis. It resides in the southernmost states, such as Arizona and New Mexico, and is commonly kept in low numbers by a virus. However, every eight to eleven years the population of the painted lady increases dramatically and large numbers migrate as far north as southern Canada. Stands of Canada thistle containing large numbers of larvae one year are often missing the following year. There are three generations of the butterfly per year. Commonly, the virus is detectable during the second generation, and numbers of the moth are generally very low or sharply declining during the third generation.[3]

Ceutorhynchus litura

The weevil *Ceutorhynchus litura* was released in Canada in 1965 for the biological control of Canada thistle, but did not become established until a fourth release in Ontario in 1967 (21). Within the United States, 18 releases consisting of 2461 adults were made in eight states (California, Colorado, Idaho, Maryland, Montana, New Jersey, South Dakota, and Washington) between 1971 and 1975.[4]

Life history, habits, and host plants of *C. litura* are described by a number of workers (21-23). *C. litura* has one generation per year. Overwintered adults begin to emerge from hibernation prior to the early rosette stage of Canada thistle and are present in the field in the Gallatin Valley for 4 to 6 wk in April and May. Females feed mainly on leaf tissue of the host plant and deposit their eggs in the feeding cavities. Emerging larvae mine the leaf tissue, migrate toward the midrib, and then down the inside of the stem to the root collar. Larval feeding and resultant black granular frass cause the inner portion of an infested stem to blacken, an appearance which is in sharp contrast to the white interior of noninfested stems. Mature third instar larvae burrow out of the plant ust below the ground level and pupate in the soil, emerging as adults in late June through July.

---

[3]Laboratory noted.

[4]Unpublished data, USDA, ARS, Biological Control of Weeds Research Unit, Albany, CA.

Studies at Bozeman established that: 1) *C. litura* spread up to 9 km in 10 years; 2) *C. litura* increased to infest over 80% of Canada thistle stems in a ten year period; 3) the infestation level of Canada thistle stems is not influenced by the presence or absence of surrounding vegetation; 4) many secondary organisms (mites, spiders, springtails, nematodes, and fungi do occur in *C. litura* mines; 5) the underground parts of shoots attacked by *C. litura* generally do not survive the winter; 6) underground parts of some unattacked plants also die in winter if they are connected by lateral roots to attacked plants; and 7) the roots of plants with at least one shoot attacked by *C. litura* produce less than two shoots the following year compared to over nine shoots for unattacked plants. It is concluded that *C. litura* is an effective agent as it does reduce the overwintering survival of Canada thistle. Thistle stands are often able to maintain themselves by recruitment from unattacked plants, but if the effects of the weevil were supplemented by another biocontrol agent, recruitment in most stands should be depressed well below the replacement level (24).

Urophora cardui

Adults of the gall-producing fly *U. cardui* have unique black banded wings. Females deposit eggs at various locations on Canada thistle plants at the apical meristem of developing stems. Larvae burrow into the stem where they induce the plant to create large galls, which act as a nutrient sink and stress the plant (25, 26). Galls near the terminal meristem can keep the plants from flowering.

Pupae overwinter in the galls and await periods of moisture and prolonged rainfall the next season which causes the softer gall tissue to disintegrate. The flies are then able to migrate to freedom. There is only one generation per year. However, because moisture levels and cell depth within the galls vary, numerous peaks of emergence may be seen throughout the season.

Rhinocyllus conicus

As stated earlier, *R. conicus* attacks the seed heads of Canada thistle. However, since only a small amount of new plants are derived from seed production, this agent produces but limited impact on the Canada thistle population.

Cassida rubiginosa

This beetle was evidently introduced accidentally into North America; therefore, no formal host specificity testing has been conducted on it.

The larvae are distinctive in that they carry their last shed skin and some fecal material suspended by their abdomen over their head and thorax, apparently to discourage any would-be predators. It is a defoliator and attacks both Canada and musk thistles. Although it is present in the eastern and central parts of the United States and can legally be moved around, it seems advisable to test this beetle on native *Cirsium* species and related crop species such as safflower to determine its host range before making releases in the west; host range testing is in progress in the Insect Quarantine Laboratory of Montana State University, Bozeman, MT.

Summary

Musk and Canada thistles differ from each other in that they are in separate genera and have different growth habits. Musk thistle reproduces totally by seed and dies at the end of the season in which the seeds are produced, whereas Canada thistle reproduces mainly by vegetative regrowth and continues to emerge year after year from the same plant root material. Therefore, a single species of bioagent that attacks the seed head of musk thistle, a plant which reproduces totally by seed, has a greater effect on the plant than a bioagent species that attacks the seed head of Canada thistle, a plant which supplements seed production with vegetative regrowth. Differences in when and where agents attack a plant allow them to be complementary and not antagonistic, because they are not competing with each other.

Seed heads of both musk and Canada thistles are attacked by *R. conicus*, while *T. horridus* larvae feed on the growing points and *C. corydon* burrow within the stems and roots of musk thistle. *Ceutorhynchus litura* burrow in the stems and roots, and *U. cardui* galls Canada thistle.

References

1. Rees NE. USDA-ARS Agric. Handbook No. 579 (1982).
2. McGregor RL. In: *Great Plains Flora Assoc., Flora of the Great Plains* (TM Barkley, Ed.), pp. 895-897. Univ. Press of Kansas, Lawrence, 1986.
3. Moore RJ and C Frankton. Canad. Dept. Agric. Monogr. 10:54-61 (1974).
4. Kazmi SMI. Teil II, Mitterillungen aus der Botanischen Stratssammlung. Munchen 5:279-550 (1964).
5. McCarty MK. In: *Biological Control of Thistles in the Genus Carduus in the United States: A Progress Report*, pp.7-10. USDA-SEA, 1978.
6. Stuckey RL and JL Forsyth. Ohio Sci. 71:1-15 (1971).
7. Dunn PH. Weed Sci. 24:518-524 (1976).

8. Batra SWT, JR Coulson, PH Dunn, and PE Boldt. USDA-SEA, Tech Bull. 1616:100 (1981).
9. Trumble JT and LT Kok. Weed Res. 22:345-359 (1982).
10. Hodgson JM and NR Rees. Weed Sci. 24:59-62 (1976).
11. Surles WW, LT Kok, and RL Pienkowski. Weed Sci. 22:1-3 (1974).
12. Rees NE. Weed Sci. 34:241-242 (1986).
13. Rees NE. In: *Biological Control of Thistles in the Genus Carduus in the United States: A Progress Report*, pp.31-38. USDA-SEA, 1978.
14. Miller TJ. MS Thesis. Montana State Univ., Bozeman, 1978.
15. Lee R. MS Thesis. Utah State Univ., Logan, 1980.
16. Kok LT and JT Trumble. Environ. Entomol. 8:221-223 (1979).
17. Rizza A, G Campobasso, PH Dunn, and M Stazi. Entomol. Soc. Amer. Ann. 81:225-232 (1988).
18. Hodgson JM. Weeds 12:167-171 (1963).
19. Gill HS and RK Bhatia. J. Res. Punjab Agric. Univ. 21:127-130 (1984).
20. Wilson RC. Weed Sci. 27:146-151 (1979).
21. Peschken DP and RS Beecher. Canad. Entomol. 105:1489-1494 (1973).
22. Peschken DP and ATS Wilkinson. Entomol. 113:777-785 (1981).
23. Zwölfer H and P Harris. Canad. J. Zool. 44:23-38 (1966).
24. Rees NE. Weed Sci. 38:198-200 (1990).
25. Peschken DP and P Harris. Canad. Entomol. 107:1101-1110 (1975).
26. Shorthouse JD and RG Lalonde. Canad. Entomol. 120:639-646 (1988).

28

Identification, Introduction, Distribution, Ecology, and Economics of *Centaurea* Species

Ben F. Roché, Jr., and Cindy Talbott Roché

Abstract

Numerous Old World *Centaurea* species (knapweeds and starthistles) have naturalized in western North America and are recognized by various states and provinces as noxious weeds. The fifteen species discussed in this paper include spotted knapweed (*Centaurea maculosa*), diffuse knapweed (*C. diffusa*), squarrose knapweed (*C. virgata* ssp. *squarrosa*), Russian knapweed (*Acroptilon repens*), yellow starthistle (*C. solstitialis*), bighead knapweed (*C. macrocephala*), short-fringed knapweed (*C. nigrescens*), brown knapweed (*C. jacea*), black knapweed (*C. nigra*), meadow knapweed (*C. jacea* X *C. nigra*), featherhead knapweed (*C. trichocephala*), purple starthistle (*C. calcitrapa*), Iberian starthistle (*C. iberica*), Malta starthistle (*C. melitensis*), and Sicilian starthistle (*C. sulphurea*). Within this genus, the species evolved diverse adaptations to different conditions of environment and disturbance in Europe and Asia. These adaptations foster their success in the new environments of North America, but it is apparent that each species must be managed differently because of differences in ecologic amplitude, response to disturbance, and methods of dispersal and establishment. This paper summarizes the introduction, distribution, and ecology of knapweeds and starthistles in western North America and includes a key for identification of the species.

Introduction

Centaurea is a large, complex genus with 1,350 validly published names of species and another 232 recognized hybrids (1). Many, but not all, are known in the English language as knapweeds and starthistles. Much confusion still exits in the taxonomy of this genus and changes occur as interpretations of the relationships between species are refined. For example, Russian knapweed (*Acroptilon repens*) and common crupina (*Crupina vulgaris*) were at one time named *Centaurea*.

The genus *Centaurea* is believed to have originated in the eastern end of the Mediterranean region (2). The oldest and most primitive *Centaurea* species are found near the Mediterranean coast and have short complete bract appendages (3). As the glaciers in Europe melted and watered the plains (about 10,000 years ago), the knapweeds began to migrate, evolving over time to adapt to different environmental conditions (3). Human development in the same general area during the Neolithic Revolution (4,000 to 12,000 years ago) and the Bronze Age (4,000 to 6,000 years ago) provided these species with additional selection pressures as mankind developed early cropping systems and closely supervised herds (4, 5). The various *Centaurea* species show distinctly different adaptations both in morphology and in preference for habitats.

The *Centaurea* species introduced to North America are detrimental to range resource values because they have low palatability to livestock and wildlife; they are able to invade relatively undisturbed native plant communities (6-8); they are poor protectors of soil and water resources (9); and they pose a wildfire hazard (10).

Spotted Knapweed

Spotted knapweed (*Centaurea maculosa*), a native of Europe (11), was collected on Vancouver Island, British Columbia, in 1905 and San Juan Island, Washington, in 1923 (12). By the 1930s it was common in Montana and northern Idaho. It is suspected to have been introduced along with diffuse knapweed in alfalfa seed, either from Asia Minor-Turkmenistan or with hybrid alfalfa seed from Germany (13).

Spotted knapweed is a short lived perennial. Boggs and Story (14) found 5- to 9-yr old plants by counting xylem root rings of mature plants. They reported high mortality among rosettes and that rosettes did not always flower during their second growing season.

Approximately 2.85 million ha in the western United States are infested with spotted knapweed. The amount reported by state includes 1,000 ha in Colorado; 928,000 ha in Idaho; 1,911,000 ha in Montana; 1,200 ha in Oregon; 1,000 ha in South Dakota; 200 ha in Utah; 11,800 ha in

Washington; and 40 ha in Wyoming (15). It has also been reported in Nevada (16) and California.

Spotted knapweed has been as influential in the building of the vegetation management philosophy in Montana as diffuse knapweed has in Washington or yellow starthistle in Idaho. It is obviously better adapted to the summer moisture pattern of Montana than to the more droughty shrub steppe of Washington or the bluebunch wheatgrass communities of the major river breaks in north central Idaho. Spotted knapweed was introduced in both Montana and Washington in the 1920s (12) but by the early 1980s had spread along roadsides and railroads to occupy an estimated 800,000 ha in Montana (17). The differences in rate of spread with presumably similar opportunities indicate differences in adaptability to habitat. The largest infestations in Montana grow in the forest steppe (with ponderosa pine and Douglas-fir) and the adjacent natural grasslands previously dominated by bluebunch wheatgrass, rough and Idaho fescue, and needle-and-thread (18). In Washington, spotted knapweed is primarily found in openings in forest habitats of ponderosa pine/bunchgrass or Douglas-fir/shrub, especially on gravelly soils derived from glacial till or outwash (19).

In the pastures of northeastern Washington, spraying the spotted knapweed with a phenoxy herbicide followed by annual applications of nitrogen fertilizer to invigorate the residual perennial grasses tipped the ecological balance on these forest soils to favor the forage species (20). As with other forage management programs, a change in grazing practices is required to effect long term control of the knapweed.

Diffuse Knapweed

Diffuse knapweed (*Centaurea diffusa*) is native to Eurasia, being common in Romania, Yugoslavia, northern Italy, the eastern shore of the Mediterranean, Turkey, Greece, Bulgaria, Asia Minor, Syria, and the U.S.S.R., especially in the Ukraine and the Crimea (21). The earliest record of diffuse knapweed in western North America is from an alfalfa field at Bingen, Washington, in 1907 (22). It may have been introduced with Turkestan alfalfa from the Caspian Sea region (23). Subsequent collections in the Northwest indicate spreading along roadsides and irrigation ditches (12).

Approximately 1.26 million ha in the western United States are infested with diffuse knapweed. The amount reported by state includes 12,000 ha in Colorado; 587,000 ha in Idaho; 4,200 ha in Montana; 485,000 ha in Oregon; 400 ha in South Dakota; 10 ha in Utah; 173,000 ha in Washington; and 2,000 ha in Wyoming (15). It has also been reported in Nevada (16) and California.

Diffuse knapweed is normally a biennial, but may behave as an annual or a short-lived perennial (24). In replicated spaced plantings under garden conditions, 10% of 400 plants flowered the first year, and only three plants died following flowering. Twenty-two percent of another 100 plants were still growing in the fourth year of mowing to 5 cm height each month of the growing season of April through October (25).

Diffuse knapweed is ideally suited to spread by vehicles and by tumbling in the wind. It evolved to spread by the wind blowing the ball-shaped plants in the same manner as tumble mustard (*Sisymbrium altissimum*). The seeds, held in urn-shaped heads which do not open widely, are lost gradually, giving the plant the advantage of far distant distribution. This technique adapts extremely well to modern vehicles and roads. Plants are also carried in rivers and irrigation systems.

Diffuse knapweed may have the widest ecological amplitude of the *Centaurea* species introduced in the Pacific Northwest, occurring in habitats ranging from scab sage on shallow soils to bitterbrush/needle-and-thread on deep loamy sands to openings in mid to upper elevation forests on silt loam soils. However, its zone of maximum competitiveness appears to be in the shrub steppe, with superior invasiveness in the bitterbrush/bunchgrass communities (19).

Squarrose Knapweed

Squarrose knapweed (*Centaurea virgata* ssp. *squarrosa*) is native to Bulgaria, Lebanon, northern Iraq, Iran, Turkey, Afghanistan and Turkestan (26). Squarrose knapweed was found in Big Valley, Lassen County, California, in 1950 (27) and at Tintic Junction, Juab County, Utah, in 1954 (22). It was found near Long Creek, Grant County, Oregon, in 1988 (28).

The squarrose knapweed infestation in Utah was estimated at 40,500 ha in five counties, including Juab, Tooele, Millard, Utah, and Sanpete (S. Dewey, personal communication). In northern California, infestations in Lassen, Siskiyou, Shasta, and Modoc counties were estimated at 642 ha and about 250 ha in central Oregon (28).

Squarrose knapweed is a perennial with multiple stems from a stout taproot. The rosette and mature leaves are similar to those of diffuse knapweed. The flowers are pink-purple. The slender urn-shaped flower head is similar to diffuse knapweed in size and shape and also remains closed at seed maturity. Unlike diffuse knapweed, the terminal spine of the involucral bracts curves backwards toward the base of the bract and the entire head is deciduous (29), creating the knapweed version of a cocklebur. The spiny heads fasten to wool, fur, or clothing like Velcro®, so that passing animals carry a little packet of seeds away with them.

Much of the historic movement of squarrose knapweed has been by sheep, but it appears that vehicles are also moving it as new infestations are found along roadsides.

The morphology of squarrose knapweed indicates that it may be a tough perennial relative of diffuse knapweed. It appears to develop slowly, spending years as a rosette, then producing seed for many years despite the rigors of a cold, dry climate that make seedling establishment precarious. In Utah, squarrose knapweed grows in basin big sagebrush (*Artemisia tridentata* var. *tridentata*) and adjacent juniper-sagebrush (*Juniperus osteosperma-Pinus monophylla*/*Artemisia tridentata* var. *tridentata*) communities (30). It is also present on the upper edges of the saltbush-greasewood (*Atriplex confertifolia-Sarcobatus vermiculatus*) shrublands found downslope from the basin big sagebrush. In northeastern California, squarrose knapweed grows in degraded juniper-shrub savanna with scattered western juniper and ponderosa pine and a chaparral-type understory of *Ceanothus cuneatus*, manzanita, Oregon white oak, and rabbitbrush on dry rocky terrain (personal communication, Ed Hale, Siskiyou County Ag Commissioner). In Oregon, squarrose knapweed grows in a juniper-Idaho fescue (*Juniperus occidentalis*/*Festuca idahoensis*) community, evidence that, in the absence of a better adapted competitor, it does well in a more hospitable environment.

Russian Knapweed

The native distribution of Russian knapweed (*Acroptilon repens*) includes Mongolia, western Turkestan, Iran, Turkish Armenia, and Asia Minor (31). Russian knapweed, along with several other species with a haploid chromosome number of 13, have been segregated out of the genus *Centaurea* (31). Other differences include the sub-basal rather than lateral attachment scar on the achene and the rust *Puccinia acroptili*, which attacks Russian knapweed, is morphologically unlike other *Puccinia* species on *Centaurea* species (32). Russian knapweed was introduced in the U.S. as a contaminant in alfalfa seed. Weed content of imported alfalfa seed varied, but figures in a report by Rogers (33) indicate that for seed released from the Denver Customhouse in 1928, 500 Russian knapweed seeds were present per pound of alfalfa seed, or one weed seed to each 450 alfalfa seeds. Once imported, it was spread widely by the sale of domestically produced alfalfa seed or hay containing weed seeds (12).

Until recently, Russian knapweed was the most widespread of the knapweeds, occurring as a problem weed in every western state (34). Although not the most abundant of the knapweeds, it is the most persistent. Current infestations are estimated at 552,000 ha, including

20,200 ha in Colorado; 360,000 ha in Idaho; 19,400 ha in Montana; 100 ha in North Dakota; 6,000 ha in Oregon; 1,200 ha in South Dakota; 60,700 ha in Utah; 3,300 ha in Washington; and 81,000 ha in Wyoming (15).

Russian knapweed has demonstrated an adaptability to poorly drained and saline/alkaline affected soils. It is often found in areas with some source of supplemental water, e.g. river or stream banks, irrigation, irrigation waste water above or below ground, along the dry bed of seasonal or underground streams, on toe slopes, and on deep fine-textured soils. In Washington, the primary indicator species for sites susceptible to Russian knapweed invasion appears to be basin wildrye (*Elymus cinereus*) because it occurred in 75% of the habitat types invaded by Russian knapweed (35). The most frequent habitat types were big sagebrush/bluebunch wheatgrass with basin wildrye, greasewood/saltgrass, and basin wildrye/saltgrass. Culver (36) correlated basin wildrye with site conditions related to either greater effective rooting depth or greater effective moisture than adjacent areas. Russian knapweed has not been found in forest habitats nor west of the Cascade Mountains in Oregon (personal communication, Dave Humphrey, Oregon Department of Agriculture) or Washington (37).

Russian knapweed is a long-lived perennial that forms dense clones by lateral root extension. Dense patches may exceed 100 shoots/m^2 (38). Roots grow to a depth of 2 to 2.5 m in the first year and 5 to 7 m in the second year (32). Russian knapweed spread is hastened by cultivation that breaks up and moves root fragments. It is weakened by dense shade (39), and crops that produce dense shade under irrigation (e.g. alfalfa hay) are used to suppress Russian knapweed.

Yellow Starthistle

Yellow starthistle (*Centaurea solstitialis*) is native to dry open habitats in southern Europe and has naturalized in parts of central Europe (40). In the U.S., it was collected from ballast grounds near western seaports including Oakland, California, in 1869 (22) and Seattle, Washington, in 1898 (12). It was introduced in southeastern Washington as early as 1900 as a contaminant in alfalfa seed (41). Alfalfa and clover seed continue to be important in the spread of yellow starthistle, even though in some states it is a restricted noxious weed in certified seed.

An estimated 3.7 million ha of rangeland in the western U.S. are infested by yellow starthistle, including 3.2 million ha in California (42); 457,000 ha in Idaho; 4,000 ha in Oregon (15); and 54,000 ha in Washington (37). Yellow starthistle also occurs in Colorado, Montana, Utah (15), and Nevada (16). Maddox and others (34) also reported yellow starthistle in Arizona, New Mexico, and Wyoming.

Yellow starthistle produces two types of achenes. The plumed achenes are produced in the center of the head and are extruded upon maturity. These achenes germinate during the first favorable conditions. The plumeless achenes are produced by the outer ring of flowers and remain in the head until the involucral bracts deteriorate during the fall or winter. These achenes require a higher temperature for germination and produce a seedling with a significantly greater rate of radicle elongation (41).

Yellow starthistle, while capable of establishment and short-term persistence in any of Washington's major plant communities below subalpine, demonstrates its maximum potential in the bluebunch wheatgrass-Idaho fescue (*Agropyron spicatum-Festuca idahoensis*) zone (43) of the steppe region of the Columbia Basin Province (41). In this community, on the southern slopes of the foothills of the Blue Mountains, it appears to reach its northern limits. Occasional small populations are found farther north, but the farther north the more they are restricted to steep south-facing slopes (19). About 49° North Latitude is the current known northern limit (37).

This weed is adventive and competitive, but it is dependent upon light on the soil surface for winter rosette and taproot development (Table 28.1) and a supply of soil moisture into early to mid summer (41). Its movement in the foothills of the Blue Mountains in southeastern Washington and adjacent central Idaho (Salmon and Clearwater drainages) has been fostered by the annual removal of the bunchgrasses, usually spring and fall, and hence the weakening of the dominant native species as competitors.

The limitation imposed by aspect seems related to yellow starthistle's dependency on winter sunlight. Taproot development between October and April provides it the ability to compete successfully for soil moisture. Because yellow starthistle flowers from June to August, soil moisture during summer becomes crucial for seed production (41); hence the more stable populations are restricted to deeper soils.

TABLE 28.1 Shade and starthistle vigor.

| | Light Reduction | | | | | | | |
	0%	30%	47%	55%	63%	73%	80%	94%
flowers/plant	6.0	2.0	2.0	0.5	0.6	0.2	0.2	0
weight/plant	40g	13g	11g	8g	7g	4g	2g	0.2g

As a winter annual, yellow starthistle is a vigorous competitor for spring moisture and continues its vigorous growth beyond the maturity of bluebunch wheatgrass until environmental conditions are no longer favorable, e.g. insufficient soil moisture or frost (41). Deeper soil compensates for lack of summer precipitation in supplying soil moisture.

Bighead Knapweed

Bighead knapweed (*Centaurea macrocephala*) is a perennial that reproduces by seed. Native to subalpine meadows in the mountains of Caucasia (26), it has been in cultivation for a long time, although it has not been as popular as cornflower (*C. cyanus*) or mountain bluet (*C. montana*). It is sold in flower seed catalogs as "lemon fluff" or "globe centaury". In the early 1980s, it was found to have escaped from abandoned gardens in two counties in Washington: Whitman and Pend Oreille. Since then, it has been reported as weedy in Okanogan County, Washington, and in Quesnel, British Columbia (44). Under favorable conditions it can grow to 1.5 m tall with many stalks, each terminating in a solitary head up to 7 cm in diameter. Under naturalized conditions in competition with other plants, bighead knapweed is generally less than 1 m tall with one or two stalks, each bearing a flower head about 4 cm in diameter. Leaves are lance-shaped, lower ones on petioles and upper leaves reduced in size up the flower stalk. The stem is swollen below the head and covered with soft cobwebby hairs. Flowers are yellow; peak flowering in northeastern Washington is in August. Bracts are golden brown, papery, with a toothed margin. Achenes are large (6-7.5 mm long) with a golden brown pappus 7-8 mm long.

The two largest populations in Washington were both growing in relatively favorable environments with annual precipitation in excess of 50 cm. In Whitman County, bighead knapweed was invading a Kentucky bluegrass sod in Palouse silt loam. In Pend Oreille County, it grew along the roadside and in a powerline clearing in what was a stand of Western hemlock/queencup beadlily (*Tsuga heterophylla/Clintonia uniflora*). It grew in competition with a stand of perennial grass, primarily hard fescue (*Festuca ovina* var. *duriuscula*), orchardgrass (*Dactylis glomerata*), and redtop (*Agrostis alba*). In the latter case, it was grazed by cattle during the late summer (45).

Short-fringed Knapweed

Short-fringed knapweed (*Centaurea nigrescens*) is native to south central and eastern Europe, extending southwards to southern Italy and northern Bulgaria (40). It is currently known only from Odell and Hood River in

Hood River County, Oregon, and Trout Lake in Klickitat County, Washington. The total population is less than 2 ha. It has persisted at Odell since 1919 (12,22). It grows in a pasture and along an irrigation ditch near Trout Lake.

Short-fringed knapweed bracts have a fringed black triangular tip similar to spotted knapweed bracts. This knapweed is also a perennial with multiple flower stalks from a woody crown. Flowers are rose-purple. The shape of the head is slightly more square than the urn-shaped head of spotted knapweed. Short-fringed knapweed leaves are entire and lance-shaped, occasionally lobed or lyrate.

Short-fringed knapweed seeds are carried in water of creeks or irrigation canals. They may also be moved by vehicles. This species has also been transplanted as an attractive ornamental.

Brown, Black, and Meadow Knapweed

Brown knapweed (C. jacea) and black knapweed (C. nigra) produce fully fertile hybrids (meadow knapweed) that appear to be weedier than either of the parents. All are perennials that reproduce by seed and to some extent by short lateral shoots from the parent plants, especially black knapweed (46). Brown knapweed bracts are brown, papery and round tipped. Black knapweed bracts have long fringes from a black or dark brown triangular center. Meadow knapweed is highly variable, exhibiting the full range of characteristics of the parents. Leaves are entire or shallowly lobed. The upright flower stems, mostly 5-10 dm tall, branch from near the middle. Flower heads are solitary at the tips of the branches, about the size of a nickel, much rounder than spotted knapweed heads. At flowering time, the bracts of meadow knapweed reflect a metallic golden sheen. Flowers are rose-purple, occasionally white.

Introduced from Europe in the early 1900s, by 1960 meadow knapweed was well established in the valley between the Coast and Cascade Ranges from British Columbia to northern California (12). By the early 1980s, major populations were established on the east slope of the Cascades in Kittitas and Klickitat counties. Meadow knapweed has spread eastward to extreme eastern Washington, northern Idaho, and western Montana as far as Bozeman. The total infestation in British Columbia, Idaho, Montana, Oregon, and Washington is approximately 1,900 ha with the largest infestations in Washington and Oregon (47). Meadow knapweed seeds are carried in rivers, irrigation water systems, and by vehicles.

Meadow knapweed is probably no better adapted to meadows than brown, black, or bighead knapweed. However, in the past 100 years, it has demonstrated a superior ability to move and occupy a yet undefined niche in the more mesic grasslands at lower and middle elevations. Daubenmire (48) defines meadow as dense grassland, usually rich in forbs, with the grasses having relatively broad and soft blades, and occurring in a relatively moist habitat. Our observations as well as the reports of cooperators in a recent survey (47) do not restrict meadow knapweed to grasslands, natural or created; it consistently grew in relatively moist habitats. This suggests that meadow knapweed, like others of its genus, is an aggressive invader of those unoccupied niches which meet its requirements.

The foliage of meadow knapweed is more abundant and much less harsh than the more drought tolerant species, and it was originally introduced as a potential winter forage species in western Oregon (22).

Featherhead Knapweed

Featherhead knapweed (*C. trichocephala*) is native to Romania and part of the Soviet Union (40). It is a perennial that spreads by seeds and by lateral roots. Leaves are linear and entire or toothed. Flowers are pink-purple with those in the center of the head almost white. Bracts are recurved with a delicate comb-like fringe on both sides. The only population in North America was found near Yakima, Washington, in the late 1970s in a poor condition bromegrass pasture on a ponderosa pine-white oak/bitterbrush habitat type (*Pinus ponderosa-Quercus garryana/Purshia tridentata*). It appeared to be self-sterile because no seeds were found and the clump appeared to be a clone. An eradication program was initiated and no plants have been found for four years at the site.

Purple and Iberian Starthistle

Purple starthistle (*Centaurea calcitrapa*) is native to the Mediterranean region, southern Europe, and northern Africa. A closely related species, Iberian starthistle (*Centaurea iberica*), is native to Asia Minor in the region between the Caspian and Black Seas (49). Both species were first introduced to North America in California. In the past two years, three new introductions of purple starthistle were found in Washington: in a dairy pasture in Island County and grass seedings in Asotin and Adams counties. Iberian starthistle was first reported in 1923 in Santa Barbara County, California, and subsequently spread to Alameda, Amador, Lake, Mariposa, Napa, San Diego, Santa Clara, Sonoma, and Yolo counties (49). It has been reported in Oregon and Washington but is no longer present. Purple starthistle is by far the more abundant of the two species and is particularly weedy in pastures and rangeland in Solano and Napa counties, California.

Purple starthistle grows to 3-6 dm tall, densely and rigidly branched. It is usually an annual or biennial. The stems and leaves are covered with cobwebby hairs frequently becoming almost smooth in age. The leaves are divided into narrowly-linear segments, with the uppermost ones narrow and undivided. Flowers are deep rose-purple. The bracts are tipped with a stout rigid straw-colored spine 10-25 mm long, with one to three prickles at the base. The achene is about 3 mm long, without bristles. The achene is straw colored, heavily mottled throughout with very dark brown (50). Iberian starthistle achenes have a plume of bristles.

Purple and Iberian starthistle spread by seeds. Purple starthistle's entrance to Washington is believed to have been via contaminated rice hulls or grass seed.

Malta Starthistle

Malta starthistle (*Centaurea melitensis*), also known as Tocalote, Napa starthistle, or Maltese centaury, is native to dry places and disturbed ground in southern Europe (40). It is weedy in California and Oregon. It was collected occasionally in Washington from roadsides, waste areas, beaches, and grainfields between 1880 and 1939 (12). It was also collected from Georgeson Island and Samuel Island, British Columbia, in 1980 (R. T. Ogilvie, Provincial Museum, personal communication).

Malta starthistle is an annual (or biennial), up to 80 cm tall, with numerous small yellow flowers (8-12 mm in diameter) and spiny bracts. The bracts can be differentiated from yellow starthistle bracts by the branches near the tip as well as near the base of the spine; also, the spine is flattened on top as viewed in cross section, compared to the round spine of yellow starthistle. Rosette leaves are lobed; the narrow linear stem leaves attach to the stem as wings. The entire plant is grayish green.

Malta starthistle spreads by seeds that may contaminate crop seed (51) or be carried along roadsides by vehicles.

Sicilian Starthistle

Sicilian starthistle (*Centaurea sulphurea*) is native to disturbed or rocky ground in southern Spain (40). It is known in the U.S. only from near Folsom, California (Doug Barbe, California Department of Food and Agriculture, personal communication), where it has persisted since 1923 (52).

Sicilian starthistle is an annual (50). Rosette leaves are entire; leaves of mature plants are winged down the stems. Flower heads are large (about

25 mm) with long (20 mm) dark brown to black spines at the tips of the bracts. Flowers are lemon yellow. The achene is 4-5 mm long with a bristly black plume.

Economics

The knapweeds and starthistles are less palatable than native or introduced forage species under traditional livestock management regimes. Thus, they have been considered to be entirely an economic liability. Losses in forage production due to diffuse and spotted knapweed in British Columbia were estimated at $320,000 annually, which equated to a theoretical loss in beef output value of over $1.5 million (53). Lacey and others (54) estimated that spotted knapweed caused a $4.5 million forage loss annually to the Montana range livestock industry. Losses from spotted, diffuse, and Russian knapweed and yellow starthistle were estimated at $2.9 million in Washington (36).

Methods of utilizing the *Centaurea* species and managing livestock to control the weeds are the subject of new investigations. In California, the effects of intensive grazing by cattle or sheep on yellow starthistle are being evaluated (55). Grazing management studies in Montana use a short duration grazing program designed to reduce the impact of spotted knapweed (56). A future evaluation of that project should significantly influence the "cattle management as a weed control method" philosophy.

Key to the *Centaurea* and *Acroptilon* Species
Naturalized in the Western United States

Unless stated otherwise, descriptions and measurements of bracts refer to the middle rows on the head.

1a. Bracts that surround the flower head are spine-tipped (≥5 mm)... 2

    2a. Stem leaves form wings down stems, flowers yellow... 3

        3a. Spine at tip of bract round in cross-section; yellow or straw-colored, with 2 pairs of lateral spinules at base of terminal spine; annual
            yellow starthistle (*C. solstitialis*)

3b. Spine flattened or grooved on top, brown or purple tinged, especially near the base . . .4

4a. Spine 5-10 mm long, with lateral branches near the top and the base; heads up to 10 mm in diameter
Malta starthistle (*C. melitensis*)

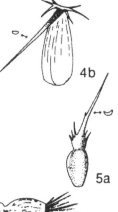

4b. Spine 15-20 mm long, dark purple, with 3-4 lateral branches on each side of the base of the main spine, heads larger, 25 mm in height and 20 mm in diameter; seeds 4-5 mm with dark brown pappus; plant annual
Sicilian starthistle (*C. sulphurea*)

2b. Stems not winged . . . 5

5a. Spine at tip of bract 10-25 mm long, often brown or purple tinged near base, obviously grooved on top; annual or biennial . . . 6

6a. Seed with pappus
Iberian starthistle (*C. iberica*)

6b. Seed without pappus
purple starthistle (*C. calcitrapa*)

5b. Spine at tip of bract up to 3 mm long, spreading or curved backward, longer than spines on sides of bract . . . 7

7a. Spine at tip of bract strongly curved backwards; flowers rose-purple; entire head falls off at maturity; plant tap-rooted perennial with many persistent flower stalks . . .
squarrose knapweed
(*C. virgata* ssp. *squarrosa*)

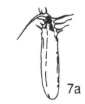

7b. Spine at tip of bract spreading; flowers white, cream, pale lavender or purple; plant normally biennial (sometimes annual or perennial), tap-rooted; bushy stalks break off at base to tumble with seed heads intact . . .
diffuse knapweed (*C. diffusa*)

7b

1b. Bracts not spine-tipped . . . 8

8a. Bract appendage is a comb-like fringe or raggedly torn, but not with a wide papery translucent margin . . . 9

9a. Appendages of adjacent bracts are not obviously overlapping . . . 10

10a. Edges of bracts awl-shaped or appear raggedly torn, not drawn out and needlelike; center part of bract has vertical veins, margins often brown . . . 10

11a. Leaf bases not winged, leaves mostly entire or lower ones lobed; flowers blue, purple, white or rose; upright annual
bachelor's button
(*C. cyanus*)

11a

11b. Leaf base winged, leaves entire or toothed; flower heads with outside ring blue and inner flowers violet; perennial with creeping rootstalks
mountain bluet
(*C. montana*)

11b

10b. Bract appendage has needle or comb-like fringe . . . 12

12a. Comb-like appendages on bracts equally spaced on each side, tip of bract recurved back over the base; flowers pink or purple; perennial
featherhead knapweed
(*C. trichocephala*)

12a

12b. Bracts with brown triangular tip . .13

13a. Leaves, except those on the uppermost stem, divided into narrow segments, lower leaves stalked; seeds normally plumed; perennial
spotted knapweed
(*C. maculosa*)

13a

13b. Leaves entire or lobed on margins but not narrowly divided, stalkless; seed plumeless; perennial
short-fringed knapweed
(*C. nigrescens*)

13b

9b. Fringes of bracts overlap those of adjacent bracts . . . 14

14a. Stem swollen beneath solitary head; head 25-75 mm in diameter; bracts have thin and somewhat translucent fringed margin; lower bracts have a weak central spine; perennial
bighead knapweed
(*C. macrocephala*)

14a

14b. Stems not obviously swollen beneath flower heads, heads multiple, less than 25 mm in diameter . . . 15

15a. Fringe of bract black, two or more times as long as the width of the center of the bract; outer ring of flowers in head are neither sterile nor enlarged; perennial
black knapweed
(*C. nigra*)

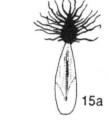
15a

15b. Fringe of bract brown to tan, highly variable in length and number; perennial
meadow knapweed
(*C. nigra* X *C. jacea*)

15b

8b. Bracts without a comb-like fringe, having a papery, translucent margin that may be split or somewhat raggedly torn . . . 16

16a. Bracts capped by a broad, thin, translucent tip, rounded or sharp-pointed, green straw-colored; perennial, spreading by dark brown lateral roots
Russian knapweed
(*Acroptilon repens*)

16a

16b. Bracts, especially central and lower ones, have widened papery margins that commonly appear more or less raggedly torn, color shades from dark brown center to a pale translucent fringe; perennial
brown knapweed
(*C. jacea*)

16b

References

1. Wagenitz G. Taxon 32:107-109 (1983).
2. Small J. New Phytologist 18:1-35 (1919).
3. Prodan I. Buletinul Academiei de Cluj Institulul de arte Grafice. *Ardealul* Strada, Memorandului No. 22, 1930.
4. Le Houérou HN. In: *Temperate Deserts and Semi-deserts, Ecosystems of the World No. 5* (NE West, Ed.), pp. 479-521. Elsevier Scientific Publishing Co., Amsterdam, 1983.
5. Lowdermilk WD. USDA Agric. Info. Bull. No. 99 (1953).
6. Lacey JR, P Husby, and G Handl. Rangelands 12:30-32 (1990).
7. Myers JH and DE Berube. Canad. J. Plant Sci. 63:981-987 (1983).
8. Tyser RW and CH Key. Northwest Sci. 62:151-160 (1988).
9. Lacey JR, CB Marlow, and JR Lane. Weed Tech. 3:627-631 (1989).
10. Xanthopoulos G. In: *Proc. Protecting People and Homes from Wildfire in the Interior West* (WC Fischer and SF Arno, Eds.), pp. 195-198. USDA Forest Serv. Gen. Tech. Rep. 251 (1988).
11. Harris P and R Cranston. Canad. J. Plant Sci. 59:375-382 (1979).
12. Roché BF Jr and CJ Talbott. Agric. Res. Center Res. Bull. XB0978, Washington State Univ., Pullman (1986).
13. Maddox DM. Rangelands 1(4):139-141 (1979).
14. Boggs KW and JM Story. Weed Sci. 35:194-198 (1987).
15. Lacey CA. In: *Proc. Knapweed Symposium* (PK Fay and JR Lacey, Eds.), pp. 1-6. EB45, Montana State Univ., Bozeman, 1989.
16. Hill K and I Hackett. Univ. Nevada-Reno Coop. Ext. Bull. 87-5 (1987).
17. French RA and JR Lacey. Coop. Ext. Serv. Circ. 307., Montana State Univ., Bozeman (1983).
18. Chicoine TK, PK Fay, and GA Nielsen. Weed Sci. 34:57-61 (1986).
19. Talbott CJ. MS Thesis, Washington State Univ., Pullman, 1987.
20. Roché BF Jr. In: *Proc. Wash. State Weed Conf.*, pp. 9-16. Washington State Weed Assoc., Yakima, 1990.
21. Popova AYA. Bot. Zh. (Moscow) 45:1207-1213 (1960).
22. Howell JT. Leafl. West Bot. 9:17-32 (1959).
23. Harris P and JH Myers. In: *Biological Control Programmes Against Insects and Weeds in Canada 1969-1980*, pp. 127-137. Commonwealth Agric. Bureaux, 1984.
24. Watson AK and AJ Renney. Canad. J. Plant Sci. 54:687-701 (1974).
25. Roché CT and BF Roché Jr. In: *Proc. Wash. State Weed Conf.*, pp. 19-23. Wash. State Weed Assoc., Yakima, 1990.
26. Wagenitz G. In: *Flora of Turkey and the East Aegean Islands, Vol. 5* (PH Davis, Ed.), pp. 465-485. Univ. Press, Edinburgh, 1975.
27. Bellue MK. Calif. Dept. Agric. Bull. 41:61-63 (1952).
28. Roché CT and BF Roché Jr. Northwest Sci. 63:246-252 (1989).
29. Tingey DC. Utah State Univ. Exp. Sta. Bull. No. 432 (1960).
30. Johnson KL. *Rangeland Resources of Utah.* Utah State Univ. Coop. Ext., Logan, 1989.

31. Moore RJ and C Frankton. Canada Dept. Agric. Monogr. No. 10, Information Canada, Ottawa (1974).
32. Watson AK. Canad. J. Plant Sci. 60:993-1004 (1980).
33. Rogers CF. Colorado Exp. Sta. Bull. 348 (1928).
34. Maddox DM, A Mayfield, and NH Poritz. Weed Sci. 33:315-327 (1985).
35. Roché CT. In: *Proc. PNW Range Shortcourse: Range Weeds Revisited* (BF Roché Jr and CT Roché, Eds.), pp. 23-28. Washington State Univ. Coop. Ext. MISC 0143, 1990.
36. Culver RN. MS Thesis, Oregon State Univ., Corvallis, 1964.
37. Roché CT and BF Roché Jr. Northwest Sci. 62:242-253 (1988).
38. Selleck GW. Proc. 7th British Weed Control Conf.:569-576 (1964).
39. Dall'Armellina AA and RL Zimdahl. Weed Sci. 36:779-783 (1989).
40. Dostál J. In: *Flora Europaea, Vol. 4* (TG Tutin, VH Heywood, NA Burges, DM Moore, DH Valentine, SM Walters, and DA Webb, Eds.), pp. 254-301. Cambridge Univ. Press, 1976.
41. Roché BF Jr. PhD Diss., Univ. of Idaho, Moscow, 1965.
42. Maddox DM, R Sobhian, DB Joley, and D Supkoff. Calif. Agric. 40:4-5 (1986).
43. Daubenmire R. Washington Agric. Exp. Sta. Tech. Bull. 62 (1970).
44. Cranston R. British Columbia Weed News. 88:2 (1988).
45. Roché BF Jr, GL Piper, and CJ Talbott. Coop. Ext. Bull. 1393, Washington State Univ., Pullman (1986).
46. Marsden-Jones EM and WB Turrill. *British Knapweeds: a Study in Synthetic Taxonomy.* Ray Society, London, 1954.
47. Roché CT and BF Roché Jr. Northwest Sci. 65:53-61 (1991).
48. Daubenmire R. *Plant Communities.* Harper & Row, New York, 1968.
49. Keffer M. Calif. Dept. of Food and Agric. Detection Manual 6:21 (1982).
50. Robbins WW, MK Bellue, and WS Ball. *Weeds of California.* Printing Div. Calif. State Dept. Agric., Sacramento, 1951.
51. Stanton TR and EG Boerner. Amer. Soc. Agron. J. 28:329 (1936).
52. Bellue MK. Calif. Dept. of Agric. Bull. 44:55-58 (1955).
53. Cranston R. In: *Proc. Wash. State Weed Assoc. 35th Ann. Weed Conf.,* pp. 27-31. Wash. State Weed Assoc., Yakima, 1985.
54. Lacey CA, JR Lacey, TK Chicoine, PK Fay, and RA French. Coop. Ext. Serv. Circ. 311, Montana State Univ., Bozeman (1986).
55. Thomsen CD, WA Williams, MR George, WB McHenry, FL Bell, and RS Knight. Calif. Agric. 43:4-7 (1989).
56. Robbins J. In: *Range Weeds Revisited* (BF Roché Jr and CT Roché, Eds.), pp. 39-41. Washington State Univ. Coop. Ext. MISC 0143 (1990).

29

Biological Control of
Centaurea Spp.

*Sara S. Rosenthal, Gaetano Campobasso, Luca Fornasari,
Rouhollah Sobhian, and C. E. Turner*

Abstract

For over 30 years foreign exploration and trap gardens have been used in Europe to find biological control agents for yellow starthistle (*Centaurea solstitialis*) and five knapweeds (*Centaurea* spp.) in North America. Two seedhead flies and two seedhead weevils have been imported from Mediterranean Europe and released against the annual, yellow starthistle. A gall wasp, another seedhead fly, two flowerhead weevils, and two pathogens are being studied as potential biological controls for this weed. Two seedhead flies and a seedhead moth have also been released against diffuse knapweed (*Centaurea diffusa*) and spotted knapweed (*Centaurea maculosa*) since the 1970s. However, research emphasis is now on two root boring beetles and three root boring moths that attack these biennial and perennial weeds in Eurasia. Some of these insects also attack squarrose knapweed (*Centaurea virgata* ssp. *squarrosa*), a weed that has recently increased in northwestern states. A rust fungus already attacks diffuse knapweed in North America. The most promising biological control for Russian knapweed (*Centaurea repens*), a perennial that grows from an extensive root system, is a stem and leafgalling nematode that is currently being used in the USSR as a bioherbicide. A gall mite, a gall wasp, and a rust fungus may be imported in the future to use against this weed.

Introduction

The main weedy knapweeds and starthistles targeted for biological control are yellow starthistle, diffuse knapweed, spotted knapweed, and Russian knapweed. However, some of the insects approved for release against these more serious weeds also attack the closely related purple starthistle (*C. calcitrapa*) and squarrose knapweed.

As these weeds originated in Eurasia, overseas research to find suitable biological control agents has been conducted by the Agricultural Research Service of the United States Department of Agriculture (USDA-ARS) at laboratories in Rome, Italy, beginning in 1959 and in Thessaloniki, Greece, since 1981. The International Institute of Biological Control (IIBC) has studied insects for knapweed control at their laboratory in Delemont, Switzerland, since the 1950s. Starting in 1984, research has also been conducted in Yugoslavia on yellow starthistle as part of a cooperative project with the USDA. From the papers cited herein, it is evident that many scientists have worked in Eurasia on these plants over the years. There are also many other federal, state, and university workers who have imported, established, and monitored knapweed biological control agents in the USA since the late 1960s.

For a discussion of biological control, it is convenient to divide the *Centaurea* spp. into three groups. There are different bioagents that appear appropriate for suppression of (a) the annual yellow starthistle; (b) the usually biennial or perennial diffuse knapweed and purple starthistle plus the perennial spotted and squarrose knapweeds; and (c) Russian knapweed, a perennial that sprouts from an extensive root system.

Yellow Starthistle

Yellow starthistle is mainly a problem in California where it infests over 3 million hectares (1) and in the northwestern states of Oregon, Washington, and Idaho (2). Not only is it competitive with good forage plants, but it is toxic to horses. As yellow starthistle is a winter annual, emphasis has been on finding flower and seedhead insects for its control. The larvae of about 20 insects feed in flowers of yellow starthistle in Greece (3). Of the 42 insects found attacking this plant in southern Europe, 18 are monophagous and six appeared especially appropriate for introduction into the USA as biological controls (4). Clement and Mimmocchi (5) concentrated on the flowerhead gall fly (*Urophora sirunaseva*), the seedhead weevil (*Bangasternus orientalis*), the seedhead fly [*Chaetorellia australis* (= *C. hexachaeta australis*)], the flowerhead feeding weevil [*Eustenopus villosus* (= *E. hirtus*)], a second seedhead fly (*Terellia uncinata*), and a second flowerhead weevil (*Larinus curtus*).

The first insect imported into the USA against yellow starthistle is the flowerhead gall fly *Urophora jaculata* that was released during the 1960s and 1970s as *U. sirunaseva*, but did not establish (6). In the research plots at Rome, Italy, the fly did not reproduce on yellow starthistle from California, Idaho, or Washington. During a survey of Greece, Sobhian and Zwoelfer (3) found six potential biological controls for this weed, including the real *U. sirunaseva*. This flowerhead gall fly was sent from Thermi, Greece, by Sobhian and released by USDA and state researchers in California and Idaho during 1984. Others were collected in Turkey that year and sent to the USA for release in Idaho (Rosenthal, unpublished). More were released in these states plus Oregon and Washington in 1985 (7). Adult flies deposit their eggs on the immature florets (3,5). Upon hatching, the larvae tunnel down to the receptacle and induce unilocular galls in which they develop. While this gall fly was also thought not to have survived anywhere, it was found by Charles Turner and Eric Coombs in 1989 to have established at a site in southern Oregon (7). By 1989, 5% of the seedheads in this Oregon site were infested and there was only limited availability of the insect for redistribution (7,8). In the meantime, several other organisms have been studied in Italy, Greece, and Yugoslavia for control of yellow starthistle.

The next insect to be released was the seedhead weevil, *Bangasternus orientalis*. While the adults feed a little on foliar material, the real damage is done by the larvae. They hatch from eggs laid on leaves near the flowerheads and burrow through the leaf and stem to the nearby head where they feed within the receptacle, complete development, and pupate (9). This weevil was first released by Don Maddox and state researchers in several western states during 1985 and it readily established in California by 1986 (9). It has also established in Idaho, Oregon, and Washington, and is slowly but surely increasing in all four states (10). Although it was found to be attacked by the native parasitic wasp *Microdontomerus anthonomi* in California, the impact of this parasite appears inconsequential thus far (10).

The seedhead fly *Chaetorellia australis* (= *C. hexachaeta australis*) was first released in the USA in 1988 (7). Eggs of this fly are deposited on the involucral bracts; the larvae burrow into the flowerheads and feed on the ovules and developing seeds (11). In Greece there are three generations/ year and the insect overwinters as a late instar larva in the seedhead (12). The larvae are able to destroy most of the seeds in a head (12). So far, there is no evidence that this fly has established in the USA.

The hibernating adults of the bud and flower-head feeding weevil *Eustenopus villosus* (=*E. hirtus*) emerge in the spring and the females oviposit in holes chewed in flowerhead buds (3). The larvae may destroy most of the achenes in a flowerhead. The host specificity was studied by Clement and Mimmocchi (5), by Clement et al. (13), and by Fornasari et al. (unpublished), who showed that adults feed on a wide range of thistles and knapweeds including a low level of feeding on safflower. This weevil was imported from Europe and released for the first time in the USA in 1990 (Fornasari et al., unpublished).

Terellia uncinata is a newly described species of seedhead fly (12) that is found in Italy, Greece, and Turkey. As it has only been reared from yellow starthistle and two other species in the same subgenus (*Solstitiaria*), it is being considered for study as another yellow starthistle biological control agent. The flowerhead weevil *Larinus curtus* is another potential biological control for the weed. Studies on the host specificity of this univoltine weevil from Greece have also been completed. Overwintering adults emerge during June and oviposit in open flowers. Larvae feed on the achenes. This weevil will probably be released to supplement the impact of the previous insects (Fornasari, unpublished).

The gall wasp *Isocolus jacea* (= *Anlax jacea*) forms galls in yellow starthistle heads and is expected to have a narrow host range (3). It is one of the most promising insects for use in the USA that was attracted to the yellow starthistle grown in experimental weed gardens in Jugoslavia. In 1987-88 it attacked 60% of the heads of this weed grown at Beograd, Zagreb, and Novisad (15). It was also found to be heavily parasitized in these studies.

The pathogen of most interest as a biological control of yellow starthistle is the rust fungus *Puccinia jaceae* that attacks the leaves and stems, reducing above-ground growth and, perhaps, root growth as well (16). An unidentified pathogen causes witches broom deformation of yellow starthistle growing in Greece (Sobhian, unpublished).

Diffuse, Spotted, and Squarrose Knapweeds
Plus Purple Starthistle

The various weedy knapweed species differ in the extensiveness of their geographic distributions. Diffuse and spotted knapweeds are widespread pests in both Canada and the USA. Squarrose knapweed is only found in the northwestern USA and is mainly a problem in Utah. Purple starthistle is only serious in California.

The IIBC initiated European research on diffuse and spotted knapweed in 1961 to find biological control agents for their control in Canada. While the USDA-ARS imported some of these insects for release in the

USA during the 1970s, US overseas research emphasis on these weeds did not begin until the 1980s.

As in the case of yellow starthistle, the first biological control agents developed for use against diffuse and spotted knapweeds were seedhead flies. The banded seedhead fly (*U. affinis*) was released in British Columbia during 1970 (17) and in Montana and Oregon beginning in 1973 (18,19). From 1974 to 1980 it was released repeatedly in these two states plus Washington, Idaho, and California (18). It readily established in Montana (19) and all the other western states (18). The UV knapweed seedhead fly (*U. quadrifasciata*) was also released in British Columbia in 1972 (17). By 1981 it had dispersed on its own as far as Montana (20). It is now found throughout the Pacific Northwest (21).

The banded seedhead fly oviposits into immature knapweed flower-heads and the larvae develop within hard galls that form from receptacle tissues. There is a partial second generation each year. The UV seedhead fly oviposits in more mature flowerheads, forms a thinner gall, and is bivoltine (22). As well as directly reducing seed production, the seed-head flies' galls, especially the hard galls of the banded seedhead fly, may devitalize the rest of the plant by acting as metabolic sinks for nutrients from other plant parts (17). The activities of these flies can cause up to 95% reduction in knapweed seed production (17). However, 1300 to 1600 seeds/m^2 remained undamaged by the flies in Canadian studies (17).

The spotted knapweed seedhead moth, *Metzneria paucipunctella*, was released on spotted knapweed in British Columbia in 1973 and in the USA in 1980. By 1988 it had spread to diffuse knapweed as well. The mature larva overwinters in the seedhead, pupation occurs in May, and the adult emerges at the end of the month and oviposits in June (23). In Europe, the univoltine larvae feed mainly on the achenes and may destroy up to 95% of the seed in a head. However, Harris and Myers (23) feel that a much lower proportion of seed was destroyed in Canada than in Europe. The larvae are aggressive and, especially in smaller heads, will kill one another or other seedhead larvae, including those of seedhead flies, if they encounter them (24). McEvoy (25) found that the gall flies were not vulnerable to attack by the seedhead moth in central Oregon and that the combined attack of the three species reduced seed production in attacked heads below the level caused by the two fly species alone.

After it was found to have a restricted host range in Greece (26), the biology of the seedhead weevil, *Bangasternus fausti*, was studied for control of diffuse knapweed. The seedhead weevil is univoltine and overwinters as an adult in the heads or in the litter (Sobhian et al., unpublished). Overwintering adults become active in May. They mate and oviposit on flowerhead bracts or terminal leaflets from then until mid-August with the peak oviposition period occurring in late June and July. Larvae mine through leaflets, petioles, and stems to the flowerbuds where they complete development in about 32 days. By that time, almost all of the florets and seeds have been consumed. Any seeds remaining in infested heads do not germinate. Adults of the overwintering generation begin emerging from the heads in August. Experimental releases of the seedhead weevil were made on diffuse and spotted knapweed in Oregon and Montana, respectively, in 1989 to determine if it would interfere with the seedhead flies (Rosenthal, Cuda, Lang, and Sobhian, unpublished).

While no special effort is being expended for the control of purple starthistle, the seedhead weevil attacks this weed in Turkey (27) and may help control it in North America. Other insects that destroy developing seeds and may be released in North America in the future include the weevils *Larinus minutus* and *L. obtusus* and the fly *Chaetorellia acrolophi* (28,29).

The leaf galling mite, *Aceria centaureae*, can cause severe damage to the rosettes and shoots of diffuse knapweed (30). As such mites tend to be host specific as well as destructive, it is now being studied in Italy, Greece, and, in quarantine, at Bozeman, Montana, for release in the USA. A bud galling mite, *Aceria* sp., causes witches-broom growth of the plants and is more common in Greece than *A. centaureae*. This species appears to reduce the growth and seed production of diffuse knapweed. In addition, it can be fatal to plants in the rosette stage. Studies by Castagnoli and Sobhian are ongoing in Europe.

The other insects being released against diffuse and spotted knapweeds are the root boring beetles and moths. Due to their direct damage and gall formation plus the pathogens that may enter through their tunnels, they are probably the most effective bioagents to use against these biennial or perennial weeds.

The diffuse knapweed root beetle, *Sphenoptera jugoslavica*, was the first such insect to be studied in Europe by the IIBC for use against this weed in Canada. The beetle oviposits between the rosette leaves and the hatching larva bores into the root where it mines and forms a gall within the central vascular tissue (23,31), leaving a cylinder of cortex undamaged. Attacked plants may not be killed but may remain as rosettes or short, bolted plants for another year and may be attacked again. This root beetle was released in British Columbia in 1976 and by 1981 was infesting from 25 to 50% of the plants at White Lake (23). In the USA, importations began in 1980 when beetles came from Greece through the USDA-ARS Biological Control of Weeds Laboratories in Rome, Italy, and Albany, California, and were distributed to California,

Idaho, Oregon, and Washington for release. Since then, with Canadian cooperation, more have been collected in British Columbia and released in Idaho (32), Oregon, Washington (33), and Montana during 1983 (34). The beetle is now established in all of these states except for California, where its status is unknown.

The root boring weevil, *Cyphocleonus achates*, was released against spotted knapweed in Canada during 1987 and in Montana during 1988. Like the root boring beetle in diffuse knapweed, this weevil has one generation/year, lays its eggs on the crown of the plant in the fall, and the hatching larvae mine down to eat and gall the central vascular tissue (31). It is established in Montana (21).

Several root boring moths have been released against diffuse and spotted knapweed in the USA and Canada. The first one was the yellow-winged knapweed root moth, *Agapeta zoegana*, that feeds as a larva mainly in the cortex of spotted knapweed rosettes but may also attack diffuse knapweed (35). As is common with the root boring moths, it has one generation/year and most of its life is spent as a larva within the root. Adult emergence, mating, and oviposition occur in the summer. The yellow-winged moth was introduced from Europe and released in Canada during 1982 and in Montana beginning in 1984. Subsequent US releases were made in Oregon and Washington in 1987 and in Idaho during 1989. While it survived the 1987-88 winter in Washington, it is now known to be established only in Canada and Montana.

The brown-winged root moth, *Pelochrista medullana*, also develops as a larva in the root cortex of diffuse and spotted knapweeds (31). It was released in British Columbia in 1982 and in Montana during 1984 but has not become established in North America.

The grey-winged root moth, *Pterolonche inspersa*, attacks the central vascular tissue of the roots of both diffuse and spotted knapweeds (31). It is particularly damaging to its hosts as infested roots become swollen and spongy and are also weakened by entering decay organisms (36). So far, releases in Canada, Idaho, Oregon, Washington, and Montana apparently have failed, according to sampling in summer 1990 after the severe 1989-90 winter.

Of the insects that have already been released against diffuse and spotted knapweeds in North America, the two seedhead flies, the seed-head weevil, and the grey-winged root moth have been found attacking squarrose knapweed in Turkey (Rosenthal, unpublished). Redistribution of these insects to US squarrose knapweed infestations may help contain or reduce populations of this pest with relatively little additional effort.

A rust fungus has been found infesting diffuse knapweed at 15 different locations in North America (37). It can infect cotyledons of safflower seedlings, but the true leaves of safflower are resistant to it (38).

Diffuse and spotted knapweeds are also attacked by *Sclerotinia* [A. K. Watson as cited by Ford (39)].

Russian Knapweed

Russian knapweed is a native of Central Asia that is now found on every continent. Since its introduction into North America in the early 1900s as a contaminant of Turkestan alfalfa seed (40), it has become widespread in the USA and Canada, particularly in the arid west (3). It is a perennial plant that spreads locally mainly by growth of its extensive budding root system and is a pest of grazing land, grain and other crops, waste places, and irrigation ditches (41). As its vast root system is relatively free of herbivores or pathogens, the most promising organisms considered for its biological control in the USSR, Canada, and the USA have been gall forming nematodes, mites, and insects that attack the stems, leaves, and flowers (42). Pathogens also look promising.

The first organism studied for its biological control in these countries was the gall forming nematode, *Subanguina picridis* (43-45). This nematode overwinters as second instar larvae in the galls in the upper soil layers. In early spring, these infective stage larvae are activated by the moisture and leave the deteriorating galls. They penetrate immature leaves and stems of the new shoots, and galls eventually form at the infected sites. The nematodes multiply within these galls until August, when the mature galls contain primarily second instar larvae. These larvae are released as the galls disintegrate on the soil surface and become infective after at least a month in soil moisture (46,47). *S. picridis* is considered so valuable as a biocontrol for its host in the USSR that it is mass reared in a special knapweed-infested field east of Samarkand, collected by soaking the galls in water, and sprayed as an augmentative biocontrol over knapweed-infested fields where control is needed (48). Releases have been made in Canada with limited success, but weed damage might have been increased with more plant competition in the knapweed-infested fields (49). The gall forming nematode was also released in northern Washington by Piper in 1984, but it did not establish (Rosenthal and Piper, unpublished). However, additional quarantine testing at warm temperatures on crops grown in warmer areas of the USA and some native *Centaurea* species was required before more releases could be made. Such an experiment was conducted in quarantine (45) and showed no widening of the host range. Therefore, more experimental releases were allowed during 1990 in the northwestern USA to determine the nematode's impact on artichoke and two native North American *Centaurea* spp. growing in the field.

Several other potential biological control agents for this weed are being studied in the USSR. The gall mite, *Aceria acroptiloni*, forms galls in the flowerheads; the gall wasp, *Aulacida acroptilonica*, forms stem galls; and the rust fungus, *Puccinia acroptili*, infects the foliage. This latter rust fungus already attacks Russian knapweed in North America (42). Also, the stem rot, *Sclerotinia sclerotiorum*, attacks this weed in North America (37).

Conclusion

In spite of the many scientist-years of research that have been conducted on biological control of knapweeds, these pests have not been completely overcome. However, biological control is expected to be a slow process and knapweeds are difficult targets. What little evaluation there has been of established biological agents indicates that it is necessary to do more than simply lower seed production to manage these weeds. A wide variety of natural enemies will be necessary to reduce weed populations to a tolerable density. Researchers have thus gone beyond traditional insect biological controls to study mites, nematodes, and pathogens. The integration of biological controls with other range management practices is the next important step toward knapweed suppression.

References

1. Maddox DM and A Mayfield. Calif. Agric. 39(11):4-5 (1985).
2. Maddox DM, A Mayfield, and NH Poritz. Weed Sci. 33:315-27 (1985).
3. Sobhian R and H Zwöelfer. Z. Ang. Entomol. 99:301-21 (1985).
4. Clement SL. Environ. Entomol. 19(6): In Press (1990).
5. Clement SL and T Mimmocchi. Proc. Entomol. Soc. Wash. 90:47-51 (1988).
6. White IM and SL Clement. Proc. Entomol. Soc. Wash. 89: 571-580 (1987).
7. Turner CE, JB Johnson, and JP McCaffrey. In: *W-84 Review Document* (LA Andres and Goeden RD, Eds.). Univ. Calif. Press., Berkeley, CA, 1991.
8. Coombs E, GL Piper, and JP McCaffrey. In: *Pacific Northwest Weed Control Handbook* (LC Burrill, RD William, R Parker, SW Howard, C Eberlein, and RH Callihan, Eds.), pp. 3-5. Oregon State Univ. Ext. Serv., Corvallis, 1990.
9. Maddox DM, R Sobhian, DB Joley, A Mayfield, and D Supkoff. Calif. Agric. 40(11):4-5 (1986).
10. Turner CE, EE Grissell, JP Cuda, and K Casanave. Pan-Pacific Entomologist 66(2):In Press (1990).
11. Maddox DM, A Mayfield, and CE Turner. Proc. Entomol. Soc. Wash. 92:426-430 (1990).

12. Sobhian R and IS Pittara. J. Appl. Entomol. 106:444-450 (1988).
13. Clement SL, R Mimmocchi, R Sobhian, and PH Dunn. Proc. Entomol. Soc. Wash. 90:501-507 (1988).
14. White IM. Entomol. Monthly Magazine 125:53-61 (1989).
15. Manojlovic B, M Maceljski, J Igrc, V Zlof, R Sekulic, B Talosi, and T Keresi. Zastita bilja 40:251-71 (1989).
16. Bruckart WL. Plant Disease 73:155-60 (1989).
17. Harris P. Z. Ang. Ent. 89:504-14 (1980).
18. Maddox DM. Weed Sci. 30:76-82 (1982).
19. Story JM and NL Anderson. Environ. Entomol. 7:445-448 (1978).
20. Story JM. Canad. Entomol. 117:1061-2 (1985).
21. Story JM. In: *Proc. Knapweed Symp.* (PK Fay and JR Lacey, Eds.), pp. 37-42. Montana State Univ., Bozeman, 1989.
22. Berube DE. Z. Ang. Entomol. 90:299-306 (1980).
23. Harris P and JH Myers. In: *Biological Control Programmes Against Insects and Weeds in Canada 1969-1980* (JS Kelleher and MA Hulme, Eds.), pp. 127-37, 1984.
24. Englert WD. In: *Proc. Second Int. Symp. Biol. Contr. Weeds* (PH Dunn, Ed.), pp. 161-5, 1973.
25. McEvoy PB. In: *Proc. Knapweed Symp. 1984* (JR Lacey and PK Fay, Eds.), pp. 57-65, 1984.
26. Maddox DM and R Sobhian. Environ. Entomol. 16:645-48 (1987).
27. Ter-minasyan ME. *Tribe Lixini* (Transl. from Russian). Amerind Publ. Co., New Delhi, 1978.
28. White IM and K Marquardt. Bull. Ent. Res. 79:453-487 (1989).
29. Clement SL. Environ. Entomol. 19(6):1882-1888 (1990).
30. Schroeder D. In: *Proc. VI Intern. Symp. Biological Control of Weeds* (ES Delfosse, Ed.), pp. 103-119. Canad. Govt. Pub. Centre, Ottawa, 1985.
31. Mueller H. Oecologia 78:41-52 (1989).
32. Harmon BL and JP McCaffrey. Univ. Idaho Agric. Expt. Sta. Bull. No. 707 (1989).
33. Piper GL. In: *Proc. Knapweed Symp.* (PK Fay and JR Lacey, Eds.), pp. 175-79. Montana State Univ., Bozeman, 1989.
34. Story JM. In: *Proc. VI Int. Symp. Biol. Contr. Weeds* (ES Delfosse, Ed.), pp. 837-842. Canad. Govt. Pub. Centre, Ottawa, 1985.
35. Mueller H, D Schroeder, A Gassmann. Canad. Entomol. 120:109-124 (1988).
36. Dunn PH, SS Rosenthal, G Campobasso, and SM Tait. Entomophaga 34:435-446 (1989).
37. Mortensen K and MM Molloy. Canad. Plant Disease Survey 69:143-145 (1989).
38. Watson AK and I Alkourey. In: *Proc. VI Int. Symp. Biol. Contr. Weeds* (ES Delfosse, Ed.), pp. 301-305. Canad. Govt. Pub. Centre, Ottawa, 1985.
39. Ford EJ. In: *Proc. of the 1989 Knapweed Symp.* (PK Fay and JR Lacey, Eds.), pp. 182-189. Montana State Univ., Bozeman, 1989.
40. Groh H. Sci. Agric. 21:36-43 (1940).
41. Reed CF and R Hughes. USDA-ARS Agric. Handbook No. 366, Washington, DC, 1970.

42. Watson AK. Canad. J. Plant Sci. 60:993-1004 (1980).
43. Kirjanova ES and TS Ivanova. Ushchel's Kondara (Akademii Nauk Tdzhikskoi SSR) 2:200-217 (Translation Bureau, Canada Department of Secretary of State) (1969).
44. Watson AK. In: *Proc. IV Int. Symp. Biol. Contr. Weeds*, Gainesville, FL (TE Freeman, Ed.), pp. 221-223, 1977.
45. Rosenthal SS. In: *Proc. Knapweed Symposium* (PK Fay and JR Lacey, Eds.), pp. 190-196. Montana State Univ., Bozeman, 1989.
46. Watson AK. J. Nematology 18:112-120 (1986).
47. Watson AK. J. Nematology 18:149-154 (1986).
48. Kovalev OV, LG Danilov, and TS Ivanova. Opisainie Izobreteniia Kavtorskomu Svidetel'stvu Byulleten 38 (Translation No. 619708, Translation Bureau, Canada Dept. of Secretary of State) (1973).
49. Watson AK and P Harris. In: *Biological Control Programmes Against Insects and Weeds in Canada 1969-1980* (JS Kelleher and MA Hulme, Eds.), pp. 105-110, 1984.

30

Chemical Control of
Centaurea maculosa
in Montana

*P. K. Fay, E. S. Davis, C. A. Lacey,
and T. K. Chicoine*

Abstract

The longevity of spotted knapweed (*Centaurea maculosa*) control following herbicide application was measured. 2,4-D provided excellent control for just one season, however, dicamba was more effective. The most effective herbicide was picloram, which provided four years of complete control. The longevity of control and rate of reinfestation depended heavily upon site conditions. The amount of plant competition and soil characteristics such as organic matter content and soil texture were important determinants of the length of spotted knapweed control with picloram.

Introduction

Spotted knapweed is a noxious weed infesting approximately 1.9 million ha of range and grazeable woodland in Montana (1). A total of 13.4 million ha may be vulnerable to spotted knapweed invasion in the state (2). The annual forage loss due to displacement of native forbs and grasses by spotted knapweed in Montana is estimated at $4.5 million (3).

The perennial growth habit, profuse seed production, and aggressiveness of spotted knapweed result in rapid establishment and spread. Initial infestations occur in disturbed areas such as roadsides, trails, construction sites, overgrazed land, and waterways (4). Once established, spotted knapweed often achieves near-monoculture stands.

Chemical Control

Herbicides are an effective means of controlling spotted knapweed. 2,4-D applied from the bolt to early flowering stage of growth provides adequate control of established plants and excellent control of seedlings (5). Annual applications of 2,4-D at rates of 1.1 to 2.2 kg a.i./ha halted seed production of spotted knapweed (6,7). Reapplications are necessary to control regrowth of older established plants since translocation of 2,4-D to the root system of spotted knapweed is minimal (8). The ester formulations are more effective than amines (9).

Dicamba has soil residual properties that make it more effective than 2,4-D, giving 2 to 3 yr of spotted knapweed control depending on the level of plant competition following treatment (10). Application rates of 2.2 kg/ha provided excellent control when applied in the rosette stage (5).

Picloram is the most effective herbicide for long-term control of spotted knapweed (7,8,11). Picloram applied at a rate of 0.28 kg/ha has provided up to 5 yr of complete control on certain sites (6,10). This extended period of control is due to soil persistence of picloram and plant competition after treatment which discourages reinfestation (11,12).

Several factors influence the degradation and movement of picloram in soil. The decomposition rate increases in the presence of plant roots, high soil organic matter levels, elevated moisture and soil temperatures, and acidic soils (13). Picloram is susceptible to leaching because of its relatively high water solubility and low adsorption to soil colloids (14). Resistance to leaching is correlated with adsorption. Leaching and sorption characteristics are influenced by soil properties and seasonal precipitation. Picloram has a pK_a of 3.6, therefore the proportion of ionized to non-ionized picloram decreases with decreasing soil pH. Maximum soil adsorption occurs in low pH soils with high organic matter and hydrated iron and aluminum oxides. Adsorption is minimal in neutral or alkaline sandy loam soils low in organic matter. Spotted knapweed thrives on neutral to high pH, well-drained sandy or gravelly soils in Montana, which may account for the shorter duration of control achieved with picloram at these sites.

Picloram is susceptible to rapid photodecomposition on plant and soil surfaces (15,16). Rice (16) applied picloram at 0.28 kg/ha and measured 68 and 42% loss from vegetation and soil surfaces, respectively after 7 d without precipitation during June in Montana. Although 90% of the picloram applied may intercept the vegetative canopy, only 1 to 2% of the foliar applied picloram is actually absorbed into the plant (17). Therefore, picloram is vulnerable to photodecomposition until the first rainfall occurs and moves it into the soil.

Spotted knapweed seedlings begin to reestablish when picloram concentrations in soil drop below 0.012 ppm (18). Picloram provides long-term control of spotted knapweed when site conditions and timing of precipitation following application are optimum (19). Precipitation is necessary for incorporating most of the picloram into the soil where it is available for root uptake and protected from decomposition by sunlight.

Clopyralid is a pyridine herbicide closely related to picloram. Clopyralid is more selective than picloram and has a shorter soil residue period. Rates of 0.28 kg/ha provided 100% control of spotted knapweed 1 yr following application at two sites in Montana without altering the native forb density or diversity (20). Seedling reestablishment occurs by the second year after application so reapplication is necessary to prevent seed production. The narrow weed control spectrum of clopyralid makes it a valuable herbicide for spotted knapweed control in environmentally sensitive areas where the broad spectrum herbicide picloram can't be used.

Spotted knapweed displaces native vegetation and often results in near monoculture infestations. In these situations, chemical treatment with picloram or clopyralid at a rate of 0.28 kg/ha resulted in greater plant community diversity than exists in untreated, knapweed-dominated sites (20,21). Reseeding infested areas following herbicide treatment increases forage production and augments control of spotted knapweed by suppressing seedling reestablishment (22,23).

Longevity of Control

Results of a study are reported here to illustrate the longevity of control of spotted knapweed using 2,4-D, dicamba, and quantities of picloram at and below the present recommendation of 0.28 kg/hectare.

Materials and Methods

Site Descriptions

Field trials were established at two sites in June 1982 (12). Site one was located on the Blackfoot-Clearwater Game Refuge, Ovando, MT. The study was established on gently rolling foothill grassland community dominated by timothy (*Phleum pratense*), smooth bromegrass (*Bromus inermis*), and Kentucky bluegrass (*Poa pratensis*). The dominant forb was spotted knapweed. The site had been cultivated in 1954 and seeded to timothy. There was no grazing use by livestock or wildlife for the duration of the trial. The elevation of the study site is 1837 m and average annual precipitation during the 7-yr study period was 32.6 cm. The soils are classified as loamy-skeletal, mixed, Frigid Typic Haplustols. The soil at Ovando had a pH of 7.3, and 5.7% organic matter. Soil analysis results were 76 ppm N, 26 ppm P (as P_2O_5), and 294 ppm K. This site represents a "best case" scenario since the soil was deep and fertile, and was in excellent range condition.

Site two was located 33 km south of Harlowton, MT on a floodplain grassland vegetation community dominated by western wheatgrass (*Agropyron smithii*), Kentucky bluegrass, needle-and-thread (*Stipa comata*), and blue grama (*Bouteloua gracilis*). Spotted knapweed was the dominant forb. The elevation at the site is 2041 m and average annual precipitation measured at Harlowton was 34.4 cm during the 7-yr study period. The soil is classified as a loamy-skeletal, mixed, Frigid Ustic Torrifluvent. The soil at Harlowton has a pH of 7.1, and 1.4% organic matter. Soil analysis results were 2.1 ppm N, 6.5 ppm P (as P_2O_5), and 326 ppm K. This site represents a "worst case" scenario since the soil is extremely rocky, well drained, and infertile. Range condition due to excessive livestock grazing was poor.

Herbicide Application

Herbicide treatments were applied by Chicoine (12) in June 1982, with a CO_2-pressurized backpack sprayer delivering 137 L/ha spray solution. Plots (3.4 X 12.2 m) were arranged in a split block design with 3 replications. 2,4-D amine and dicamba were applied at 2.2 and 1.1 kg/ha, respectively. Picloram was applied at 0.07, 0.11, 0.14, 0.22, 0.25, and 0.28 kg/ha as main plot treatments, and 2,4-D amine was applied at 2.2 kg/ha to half of each plot as the subplot treatment. Individual plots were spatially separated from each other by a 2.1 m buffer zone treated with a tank mix of picloram and 2,4-D amine at 0.37 and 2.2 kg/ha, respectively, to prevent contamination from spotted knapweed seed production in neighboring plots. The perimeter of the experimental area was further separated by a buffer zone 3.4 m wide which was sprayed with the same mixture of picloram and 2,4-D amine. The buffer zones were retreated in June 1986 with the same treatment described above.

Spotted knapweed plant counts were taken in July from 1983 through 1989. Four 1 m^2 frames were placed at random within each plot and mature spotted knapweed plants were counted.

Biomass production was measured at peak standing crop 2 mo, 1 yr 2 mo, 2 yr 2 mo, and 7 yr following herbicide application. Two 0.5 m^2

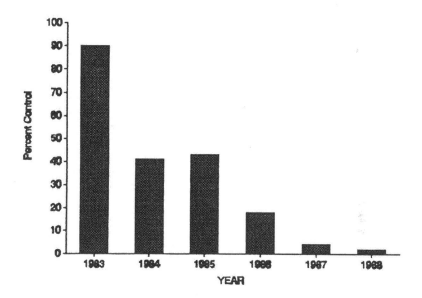

FIGURE 30.1 The percent control of spotted knapweed in mid-July for 6 yr as a result of a single application of 2.2 kg/ha 2,4-D amine in June 1983.

frames were placed at random locations within each plot. The standing crop was clipped at ground level and the current year's growth separated into grasses or spotted knapweed, the only forb present in significant frequency. Vegetation was oven-dried at 50 C for 48 hr to determine herbage dry weight production.

Results and Discussion

Many spotted knapweed plants are not completely killed by a single application of 2,4-D. The herbicide provides excellent top growth control but regrowth from the roots of some of the established spotted knapweed plants occurs after treatment. Annual applications of 2,4-D are necessary to provide satisfactory control of spotted knapweed (Figure 30.1).

While dicamba was initially more effective than 2,4-D, the degree of control continuously dropped from year to year (Figure 30.2). Dicamba needs to be applied every 2 to 3 yr to prevent seed production.

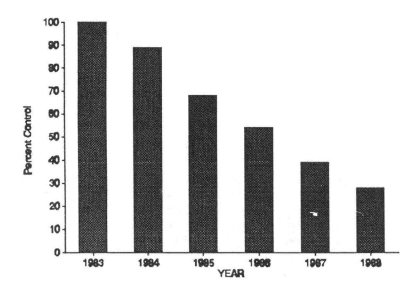

FIGURE 30.2 The percent control of spotted knapweed in mid-July for 6 yr as a result of a single application of 1.1 kg/ha dicamba in June 1983.

All rates of picloram reduced spotted knapweed density each year for 7 yr after application at Harlowton and Ovando, respectively (Tables 30.1 and 30.2). Complete control of mature spotted knapweed plants 3 yr following treatment was obtained with rates ranging from 0.14 to 0.28 kg/ha at Ovando (Table 30.1). Rates of 0.11 to 0.28 kg/ha provided 100% control of mature spotted knapweed 2 yr after herbicide application at Harlowton; however, only the 0.28 kg/ha rate provided complete control for 4 yr (Table 30.2). The addition of 2,4-D amine as a subplot treatment did not significantly improve control over picloram alone (8).

Picloram soil residues dropped below the level needed to control spotted knapweed seedlings by 3 yr after application of 0.28 kg/ha picloram at Ovando and Harlowton (8). Reinvasion by spotted knapweed seedlings started 2 to 4 yr after application, depending upon picloram rate in treated plots at Harlowton. By July 1988, the density of mature spotted knapweed plants within picloram-treated plots was not significantly different than in untreated control plots (Table 30.2).

Reinvasion also occurred at the Ovando site. Low densities of mature spotted knapweed plants were present in plots treated with picloram doses lower than 0.25 kg/ha 4 yr after treatment. Picloram ranging from 0.11 to 0.28 kg/ha resulted in less than one mature spotted knapweed plant per m^2 in 1989, 7 yr following application, compared to 47 plants per m^2 for the untreated control (Table 30.1).

Picloram at 0.14 to 0.28 kg/ha provided 100% control of spotted knapweed 2 yr after application, decreasing spotted knapweed herbage from 1990 and 2316 kg/ha to zero kg/ha at Harlowton and Ovando, respectively (unpublished results). Grass production increased 700% at Ovando and 200% at Harlowton 2 yr after treatment in response to elimination of spotted knapweed (Table 30.3 and 30.4).

Grass herbage production within all picloram-treated plots was not significantly different than in the untreated plots 7 yr following application at Harlowton (Table 30.4). Heavy grazing pressure by cattle during the spring of 1983, 1984, and 1985 provided enough soil disturbance to permit significant germination of dormant spotted knapweed seed to occur. With the picloram soil residues depleted by 1983, and grass competitiveness reduced, spotted knapweed quickly reinfested the treated plots.

While the recolonizing spotted knapweed plants at Harlowton were in low density, they were much larger than individual plants within the control plots due to reduced interspecific competition. The large plants produced a heavy seed crop, leading to increased spotted knapweed density. Powell et al., (24) measured the biomass, population density, and seed production of recolonizing diffuse knapweed (*Centaurea diffusa*) plants 3 yr following treatment with picloram. They found that while the density of knapweed was less than one tenth of the control population, individual plants were much larger. Plant biomass and seed production per m^2 was greater in sprayed plots than from plants in unsprayed areas. These results suggest that recolonization occurs rapidly once individual knapweed plants become established following dissipation of picloram residues on sites devoid of plant competition. The suspected allelopathic influence of spotted knapweed may also contribute to its ability to invade and displace native vegetation (25-27).

Perennial grass herbage production increased 700% at Ovando by 2 yr after herbicide application of picloram at 0.14 to 0.28 kg/ha (Table 30.3). Spotted knapweed herbage production averaged 178 kg/ha in plots treated with picloram at rates of 0.14 to 0.28 kg/ha compared to 1113 kg/ha for the untreated control (unpublished results).

Grass production at Ovando 7 yr after treatment was 2.5 times lower than it was 2 yr following treatment. This reduction was not due to reinvasion from spotted knapweed as at Harlowton. While the grass herbage production in picloram-treated plots was 11 times greater than in the untreated plot 7 yr after treatment, the reduction in grass

TABLE 30.1 Number of mature spotted knapweed plants per m^2 in July 1983 through 1989 following herbicide application in June 1982 at Ovando.

Picloram (kg/ha)	Year						
	1983	1984	1985	1986	1987	1988	1989
0.28	0.0a[1]	0.0a	0.0a	0.0a	0.6a	0.3a	0.2a
0.25	0.0a	0.0a	0.0a	0.0a	0.4a	0.2a	0.1a
0.22	0.0a	0.0a	0.0a	0.1a	0.5a	0.5a	0.6a
0.14	0.0a	0.0a	0.0a	0.2a	1.7a	0.8a	0.9a
0.11	3.2a	4.6a	2.2a	0.2a	0.2a	1.3a	0.5a
0.07	41.2b	21.1b	4.1b	1.4b	4.7b	2.0a	1.1a
0.00	65.8c	58.4c	57.2c	54.2c	41.7c	52.0b	46.5b

[1] Means in a column followed by the same letter are not significantly different at the 5% level using the LSD test.

TABLE 30.2 Number of mature spotted knapweed plants per m^2 in July, 1983 through 1989 following herbicide application in June, 1982 at Harlowton.

Picloram (kg/ha)	Year						
	1983	1984	1985	1986	1987	1988	1989
0.28	0.0a[1]	0.0a	0.0a	0.0a	8.8a	21.0a	15.2a
0.25	0.0a	0.0a	0.2a	0.0a	12.3a	15.8a	15.3a
0.22	0.0a	0.0a	1.1a	0.1a	6.5a	17.5a	9.4a
0.14	0.0a	0.0a	1.2a	0.2a	8.4a	25.4a	16.4a
0.11	0.0a	0.0a	1.2a	0.2a	10.8a	22.8a	11.8a
0.07	2.1a	2.1a	1.8a	1.4b	10.2a	12.3a	15.2a
0.00	126.7b	84.1b	75.7b	54.2c	61.9b	30.8a	15.3a

[1] Means in a column followed by the same letter are not significantly different at the 5% level using the LSD test.

TABLE 30.3 Perennial grass production at four times after herbicide application at Ovando.

Picloram Rate	Time After Herbicide Application			
	2 mo	1 yr 2 mo	2 yr 2 mo	7 yr
(kg/ha)	Perennial Grass Herbage Production (kg/ha)			
0.28	560.5a[1]	1630.8ab	2612.8b	803.3ab
0.25	307.1a	1474.0bc	2686.4b	1200.0ab
0.22	682.3a	1655.4ab	2739.2b	1330.0a
0.14	440.5a	1433.8b	2896.8a	1050.0ab
0.11	316.5a	1275.0b	2517.6b	910.0ab
0.07	350.2a	1021.2bc	1636.4c	766.7b
0.00	360.7a	332.0d	418.4d	93.3c

[1] Means in a column followed by the same letter are not significantly different at the 5% level using the LSD test.

TABLE 30.4 Perennial grass production at four times after herbicide application at Harlowton.

Picloram Rate	Time After Herbicide Application			
	2 mo	1 yr 2 mo	2 yr 2 mo	7 yr
(kg/ha)	Perennial Grass Herbage Production (kg/ha)			
0.28	898.5a[1]	1700.8a	807.2a	196.7a
0.25	727.9ab	1747.4a	795.2a	230.0a
0.22	516.4ab	1287.6a	841.6a	121.0a
0.14	693.1ab	1466.2a	702.0a	130.0a
0.11	698.9ab	1472.2a	640.4a	280.0a
0.07	704.4ab	1437.4a	659.6a	286.7a
0.00	429.5b	321.2b	256.7b	63.3a

[1] Means in a column followed by the same letter are not significantly different at the 5% level using the LSD test.

production in treated plots from 1984 to 1989 is substantial (Table 30.3). The overall decrease in grass production observed in all treatments at Ovando may be due to a sodbound condition of the smooth bromegrass-dominated site following years of heavy grass production without utilization as evidenced by the heavy thatch of dead grass in treated plots. In addition to reducing grass productivity, the heavy thatch may inhibit spotted knapweed seed germination by filtering out red light (28).

Conclusion

Long-term control of spotted knapweed with reduced rates of picloram appears to be more dependent on site conditions and management practices than on the rate of picloram applied. Picloram applied at 0.14 to 0.28 kg/ha provided the same level of spotted knapweed control for 7 yr at Ovando, a highly productive, low disturbance site. The same concentrations of picloram provided similar control for 3 yr at Harlowton, followed by a uniform rate of reinfestation throughout all treated plots.

The Ovando site, located at the Blackfoot Clearwater Game Range, contained a stand of improved pasture grass species. The deep soil was high in organic matter and represents a best-case scenario for long-term picloram performance. Picloram has a relatively strong affinity for soil organic matter which restricts leaching. This maintains the herbicide near the soil surface and results in long term knapweed seedling control. The grass stand, in response to application of picloram, provided intense competition for germinating knapweed seedlings. The result was 100% control for the first 3 yr and 85% control 7 years after application (Table 30.1).

The knapweed site at Harlowton represents a worse case scenario. The site was heavily grazed and the soil is a very shallow mixture of gravel and rocks with very little top soil. While complete control was maintained for 3 yr, control dropped below 50% by 1988 (Table 30.2). These two sites are excellent examples of the effect of soil type upon longevity of spotted knapweed control with picloram. While herbicide retreatment was needed at both sites by either year 3 or 4 to prevent seed production, picloram is an extremely effective herbicide for spotted knapweed control. A single application of picloram killed all the knapweed plants present and provided 3 yr of complete control of seedling establishment at two locations. This treatment is quite economical and will provide a "knapweed-free" period to permit a land manager to establish desirable plant species to inhibit reestablishment of spotted knapweed from seed.

References

1. Lacey CA. *Proc. Montana Weed Control Assoc.* Great Falls, MT, pp. 10-20 (1987).
2. Chicoine TK, PK Fay, and GA Nielsen. Weed Sci. 34:57-61 (1985).
3. French R and JR Lacey. Coop. Ext. Ser., Montana State Univ. Bull. 307 (1983).
4. Watson AK and AJ Renney. Can. J. Plant Sci. 54:687-701 (1974).
5. Lacey CA, JR Lacey, TK Chicoine, PK Fay, and RA French. Circ. Mont. State Univ. Coop. Ext. Serv., Bozeman, MT (1986).
6. Belles WS, DW Wattenbarger, and GA Lee. West. Soc. Weed Sci. Res. Prog. Rep., pp. 55-56 (1980).
7. Wattenbarger DW, WS Belles, and GA Lee. West. Soc. Res. Prog. Rept., pp. 63-64 (1980).
8. Lacey CA. M.S. Thesis, Montana State Univ. (1975)
9. Belles WS, DW Wattenbarger, and GA Lee. Proc. West. Soc. Weed Sci. 31:17-18 (1978).
10. Fay PK, ES Davis, TB Chicoine, and CA Lacey. Proc. Knapweed Symp. Montana State Univ., Bozeman, pp. 43-46 (1989).
11. Renney AJ and EC Hughes. Down to Earth 24:6-8 (1969).
12. Chicoine TK. M. Sc. Thesis, Montana State Univ. (1984)
13. Goring CA and JW Haymaker. Down to Earth 27:12-15 (1971).
14. Bovey RW and CJ Scifres. Tex. Agr. Exp. Sta. B-1111. 24p (1971).
15. Hall RC, CS Giam, and MG Merkle. Weed Res. 8:292-297 (1968).
16. Rice PM. *Proc. Knapweed Symp.* Montana State Univ. Dept. Plant and Soil Sci. and Coop. Ext. Serv. Bozeman. pp. 88-94 (1989).
17. Bovey RW, FS Davis, and MG Merkle. Weeds 5:245-249 (1967).
18. Watson VJ, PM Rice, and EC Monnig. J. Environ. Qual. 18: (1989).
19. Cranston R. *Proc. Knapweed Symposium.* Plant and Soil Sci. Dept. and Coop. Ext. Serv. Montana State Univ. Bull. 1315, p. 4-7 (1984).
20. Lacey CA, MB McKone, and D Bedunah. Proc. West. Soc. Weed. Sci. 42:280-284 (1989).
21. Bedunah D and J Carpenter. *Proc. Knapweed Symp.* Montana State Univ. Dept. Plant and Soil Sci. and Coop. Ext. Serv., Bozeman, pp. 205-212 (1989).
22. Fagerlie D. *Proc. Knapweed Symp.* Montana State Univ. Dept. Plant and Soil Sci. and Coop. Ext. Serv., Bozeman, pp. 220-221 (1989).
23. Hubbard WA. J. Range Manage. 28:406-407 (1975).
24. Powell RD, C Risley, and JH Myers. *Proc. Knapweed Symp.* Montana State Univ. Dept. Plant and Soil Sci. and Coop. Ext. Serv., Montana State Univ., Bozeman, pp. 58-66 (1989).
25. Harris P and R Cranston. Can. J. Plant Sci. 59:375-382 (1979).
26. Kelsey RG and LJ Locken. J. Chem. Ecol. 13:19-33 (1987).
27. Muir AD and W Majak. Can. J. Plant Sci. 63:989-996 (1983).
28. Nolan DG and MK Upadhyaya. Can. J. Plant Sci. 68:775-783 (1988).

31

Neurotoxicity of the Knapweeds (*Centaurea* Spp.) in Horses

Kip E. Panter

Abstract

Yellow starthistle (*Centaurea solstitialis*) and Russian knapweed (*C. repens*) induce a similar neurological disease in horses called equine nigropallidal encephalomalacia (ENE) ("chewing disease"). The disease appears in the horse suddenly after prolonged ingestion of the knapweeds for more than 30 to 60 days, in which cumulative plant quantities of 60 to 200% of the animal's body weight are consumed. The disease is characterized by an acute inability to eat or drink, and the muscles of the lips, face, and tongue are hypertonic, resulting from degenerative lesions in the globus pallidus and substantia nigra of the brain. Younger horses are most often affected; however, toxicoses in older horses are also reported. The toxin or toxins are believed to be of the sesquiterpene lactone group although the actual compound is not known. Progress is being made in this area using 8-day chick embryo sensory neurons *in vitro* as a bioassay to test purified components of the plant. Prevention of toxicity is accomplished through identifying the plants and management practices that provide adequate quality and quantities of feed.

Introduction

Plant-induced ENE was first described in horses in 1954 in central and northern California where yellow starthistle is prevalent (1). Cases of

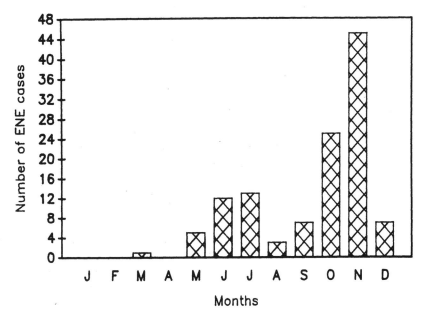

FIGURE 31.1 Seasonal pattern of 118 cases of yellow starthistle-induced ENE in horses in California adapted from Cordy (3).

yellow starthistle-induced ENE have been reported from southern California and southern Oregon (2). The disease has also been reported in the subhumid regions of Argentina and Australia (3).

The same disease induced by Russian knapweed (4) has been reported in western Colorado (5), eastern Utah (5), and eastern Washington (6).

In California, the disease follows a seasonal biphasic prevalence where a small peak appears in June-July and a larger peak in October-November (3) (Figure 31.1). This pattern is also reported in Argentina, where the first peak is thought to be due to regrowth of yellow starthistle plants behaving as biennials, while the second peak is due to those behaving as conventional annuals (3). In California, yellow starthistle plants behave as annuals, thus discounting the Argentina theory (3). The *Centaurea* species have a minimal moisture requirement and often may be the only green plant remaining in nonirrigated pastures in a dry season.

Yellow Starthistle Description

Yellow starthistle is an aggressive noxious weed introduced from the Mediterranean region, usually growing in uncultivated fields, roadsides, and ranges (2,7). It is abundant in many pastures of central and northern California and is a rapidly spreading noxious weed in many other parts of the United States. The plant begins growth in early spring (March or April) and continues throughout the summer and fall.

Yellow starthistle is a member of the sunflower family (Compositae). It is an annual weed, a prolific seed producer, and matures at about 30 to 76 cm tall in summer or early fall. The stems are rigid and branch out from the base. The leaves and stems are somewhat pointed, and the base of the leaf is attached to the stem and extends down the stem to form a wing-like structure. The flowers are about 25 mm in diameter, bright yellow, and are solitary at the end of the branching stems. The bracts of the head have rigid 6 to 25 mm spines radiating out, thus the name yellow starthistle (2,7).

Russian Knapweed Description

Russian knapweed is a perennial, forming dense colonies and inhibiting growth of other forage by its aggressive nature and allelopathic properties (8,9). It is a noxious weed of Eurasia, probably introduced into the U.S. in the late 1800s, and now widely established in the western U.S. (10).

Russian knapweed grows 45 to 91 cm tall. Stems are erect and openly branched; lower leaves are deeply lobed, 5 to 10 cm long; upper leaves are narrow to a broad base, entire or serrate. Flower heads are cone-shaped, 6 to 12 mm in diameter, solitary at the branch tips; flowers are pink to lavender and occur in June to September; seeds are 3 to 6 mm long, bearing numerous whitish bristles. The plant primarily spreads by black, deep-growing roots which may be over 2.4 m deep, thus resistant to dry conditions (10).

Toxicity

Yellow starthistle and Russian knapweed cause neurological disease in horses called equine nigropallidal encephalomalacia (ENE) or "chewing disease". The disease induced by these plants has only been observed in horses and resembles Parkinson's disease in humans. Feeding the plant to sheep (1), guinea pigs (3), rats (11), monkeys (11), dogs (3), chicks (3), or mice (3) has failed to induce the disease in these animal species. Cattle have grazed the knapweeds and knapweed-infested pastures with no evidence of toxicity.

The clinical syndrome in the horse is rather abrupt with the signs of nerve damage severe and often erratic the first few days. Violent,

FIGURE 31.2 Horse with ENE. Note the unchewed hay in the mouth and the dazed, depressed look of the horse. (Courtesy J. L. Shupe)

aggressive behavior may be a part of the early signs. In a few days, these early signs subside to a static level which persists without much change for the duration of the disease (1,3). Death usually results from starvation or dehydration unless intervened and the horse humanely euthanized. Affected horses have been kept alive for many days if nutritional requirements are met by gavaging through a stomach tube. Horses with small lesions in the brain may adapt to their neurological deficiency and survive for extended periods of time. The humane judgement in prolonging a horses life in these cases should be considered due to the permanence of the impairment.

The disease occurs suddenly after prolonged ingestion of the plant of at least 30 to 60 days and cumulative quantities of 60 to 200% of body weight consumed (3,5). Typically, the early signs of the disease will be reduced movement and spontaneous activity. Some horses may walk aimlessly with their muzzle to the ground, while others may stand for long periods in awkward-looking postures. Prehension and chewing are severely impaired; however, swallowing appears to be unaffected (3). Horses so affected appear to have a desire to eat and drink but are

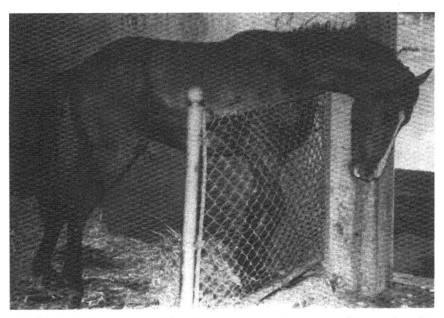

FIGURE 31.3 Horse with ENE. Note the unusual stance, pushing against the fence and profuse sweating on the neck and flank regions. (Courtesy J. L. Shupe)

unable to. Often an affected horse may be standing with hay or grass hanging from the mouth for many minutes to hours (Figure 31.2). Occasionally, some horses are able to push enough feed into the back of the mouth to swallow it. However, cases of macerated feed lodged in the back of the mouth and aspiration pneumonia have been reported (2). Even though drinking is impaired, some horses learn ways to accomplish getting water to the back of the mouth and swallow it. This may be accomplished in some horses by immersing the muzzle in the water up to the eyes or above, while some horses may chew at the water as if they are eating hay; others may lap at water like a dog. Affected horses are unable to purse their lips to drink normally (2).

Variations in the clinical syndrome have created differential diagnostic problems. Some horses may push against objects causing trauma to the head and shoulders (Figure 31.3). This behavior has been commonly associated with hepatic insufficiency or encephalomyelitis. Some horses have difficulty breathing, although no lesions were found in the lungs. Frothing at the mouth has been observed and may be confused with signs indicative of rabies (2).

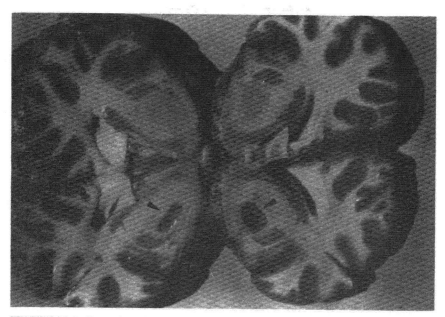

FIGURE 31.4 Brain from horse shown in Figures 31.2 and 31.3. Note the large symmetrical lesion in the globus pallidus (arrows). (Courtesy J.L. Shupe)

Data from rat studies suggest that increased muscular tone of the muscles innervated by the 5th, 7th, and 12th cranial nerves may account for the signs of poisoning. Mechanical destruction of the dopaminergic nigrostriatal pathway in the rats mimics the horse condition. Thus, the early signs are thought to result from massive release of the stored neurotransmitter, dopamine, from the nigrostriatal nerve endings; and the subsequent stabilized signs are thought to reflect the ensuing dopamine deficiency (3).

Young et al. (5) have calculated dosages and time periods for ingestion of these plants required to induce ENE. An intake rate of 1.8-2.5 kg/100 kg body weight per day (59-71% of the horse's weight) of fresh Russian knapweed ingested over a period of 28-35 (average 30) days or 2.3-2.6 kg/100 kg BW/day (86 to 200% of the horse's body weight) over a period of 26-81 (average 54) days of fresh yellow starthistle was required for the disease to occur (5). Thus, Russian knapweed is the most toxic. ENE has also been induced by feeding dried Russian knapweed (5).

Pathological lesions are typically found in any of four areas of the brain: the right and/or left sides of the globus pallidus and substantia nigra (Figure 31.4). No significant differences in the clinical signs of toxicity are apparent among horses with lesions in any of these four regions (3). This suggests a similar or single pathway at all levels and the severity of the signs apparently correlate with the size of the lesion rather than the specific site or number of sites. Occasionally, unilateral lesions in both nuclei will result in only one side of the horse becoming affected. Small atypical lesions in experimentally fed horses have been reported in other areas of the brain including the inferior colliculus, the dentate nucleus, the mesencephalic nucleus, and the tract of the fifth cranial nerve (3). In all cases, necrosis in the foci occurs suddenly and completely and lesions do not enlarge beyond their initial size even in horses able to ingest additional plants (3). Gray matter of the brain is predominantly affected; white matter is also involved. Fiber tracts passing through the foci undergo total necrosis in the same manner as gray matter (3).

Microscopically, the initial lesions are foci of pannecrosis and all elements are necrotic, including neurons, glia, and capillaries. Later changes are those associated with cleanup of necrotic tissue. The margin between living and necrotic tissue is usually sharply defined with narrow transition zones (3).

Conditions of Poisoning

Poisoning by knapweeds and yellow starthistle is generally caused by a lack of good quality forage in infested pastures. Thus, horses are forced to consume these types of plants or starve. However, a few cases of ENE have been reported in horses on good pasture where they apparently acquired a taste for yellow starthistle and ate it even though other forage was available (2).

The pathogenesis of the disease follows a similar pattern in all horses. It appears the plant is eaten without apparent harm until some critical threshold is exceeded, at which time the lesions and clinical signs appear abruptly. Accumulation of the toxin or a progressive nutrient depletion have been suggested as possible mechanisms (1). Cordy (3) suggests the disease may be due, instead, to the summative effects of small daily injuries to cell structures, and in effect, the lesions and clinical signs associated appear suddenly from a final dose in itself too small to be toxic (3).

Because only the horse appears to be susceptible to this toxicosis, it has been suggested that gastric or intestinal metabolism by a microorganism specific to the horse is a probable explanation. Thus, an innocuous plant substance may be converted to a toxin or antimetabolite (3). Similarly, why the specific site in the brain is affected is not understood. Morphologic and biochemical studies of these nuclei give no substantial clues to their selective liability to injury. Their function in dopamine production and release is probably a key factor.

Toxin

The toxic principle in these plants is not known, although progress has been made toward its elucidation by testing the neurotoxicity of individual sesquiterpene lactones in a bioassay utilizing a cell response of 8-day chick embryo sensory neurons *in vitro*. The most toxic component in this bioassay was repin (TD50 = 80nM) followed by subluteolide (TD50 = 300nM), janerin, and others (12). It is interesting to note that not only was repin three to four times more toxic than its closest rival, subluteolide, in the bioassay; but it was also determined that repin is the major lactone in *C. repens*, which is the most toxic of the two *Centaureas* discussed (12). Repin is toxic to neurons and glia when inoculated in nanomole doses into the intact brain of the rat (12).

Treatment and Control

There is no known treatment in the horse once the disease is manifest. The prognosis is poor, and in most cases, horses will die from starvation and/or dehydration. Therefore, euthanasia should be a consideration for humane reasons.

Successful chemical control of plants may be accomplished with picloram (Tordon) followed by grazing management practices to maximize grass production (13).

References

1. Cordy DR. J. Neuropath. and Exptl. Neurol. 13:330 (1954).
2. Fowler ME. J. Amer. Vet. Med. Assoc. 147:607 (1965).
3. Cordy DR. In: *Effects of Poisonous Plants on Livestock* (RF Keeler, KR Van Kampen, and LF James, Eds.), pp. 327-336. Academic Press, New York, 1978.
4. Larson KA and S Young. J. Amer. Vet. Med. Assoc. 156:626 (1970).
5. Young S, WW Brown, and B Klinger. J. Amer. Vet. Med. Assoc. 157:1602 (1970).
6. Farrell RK, RD Sande, and SS Lincoln. J. Amer. Vet. Med. Assoc. 158:1201 (1971).
7. Kingsbury JM. In: *Poisonous Plants of the United States and Canada*, pp. 396-397. Prentice-Hall, Englewood Cliffs, NJ, 1964.
8. Muir AD and W Majak. Can. J. Plant Sci. 63:989 (1983).

324

9. Stevens KL. This volume.
10. Whitson TD (Ed). In: *Weeds and Poisonous Plants of Wyoming and Utah,* pp. 166-169. University of Wyoming and Utah State University Cooperative Extension Services, 1987.
11. Mettler FA. J. Neuropath. and Exptl. Neuro. 22:164 (1963).
12. Stevens KL, RS Riopelle, and RY Wong. J. Natural Products 53:218 (1990).
13. Callihan RH, RL Sheley, and DC Thill. Univ. of Idaho, College of Agric. Coop. Exten. Service Curr. Info. Ser. No 634, Moscow.

32

Junipers of the Western U.S.: Classification, Distribution, Ecology, and Control

Neil E. West

Abstract

There are 13 species of *Juniperus* found west of the Mississippi River. The distribution of these small trees and shrubs was largely confined to steep, rocky fire-proof sites prior to European colonization. Since then, junipers have greatly expanded their distribution as fire frequencies and competition from herbaceous species were reduced. Climatic trends, allelochemic properties of litter, and animal dispersal may also be involved in these expansions and thickenings. Increases in trees were followed by diminishment of understories and accelerated soil erosion. Because juniper tissue is utilized by very few herbivores and detritivores, juniper dominance can persist for centuries unless man intervenes. Unfortunately, understory fine fuels are usually now too sparse to use prescribed burning. Soil seed reserves of desirable forages are also usually depleted. Thus, the most common treatments during the 1950s and 1960s were chaining or cabling followed by seeding of desirable forage plants. The longevity of these treatments depends on the thoroughness of the initial treatment and subsequent management. Enhanced environmental awareness in recent decades has voided further application has of these treatments on public lands. Increased costs for treatments and lowered demands for red meat have limited type conversions on private lands. Although herbicides are available to kill mature junipers, the amounts required make them too expensive for

TABLE 32.1 Common and scientific names of all species of junipers (*Juniperus* spp.) naturally occurring in the U. S. west of the Mississippi river. * = those resprouting after fire (1).

Common name	Scientific name	Western States Where They Occur
Ashe juniper	*J. ashei* Buchholz	TX, OK, MO, AK
California juniper	*J. california* Carr	CA, NV, AZ
common juniper	**J. communis* L.	WA, OR, CA, ID, NV, MT, WY, UT, AZ, CO, ND, SD, NM
alligator juniper	**J. deppeana* Steud.	AZ, NM, TX
drooping juniper	*J. flaccida* Schlecht	TX
creeping juniper	**J. horizontalis* Moench	MT, WY, ND, SD
one seed juniper	*J. monosperma* (Engelm.) Sarg.	AZ, NM, CO, TX, OK
western juniper	*J. occidentalis* Hook	WA, OR, CA, NV, ID
Utah juniper	*J. osteosperma* (Torr.) Little	CA, NV, ID, WY, MT, UT, CO, AZ, NM
redberry juniper	*J. pinchotii* Sudw.	TX, NM, OK
Rocky mountain juniper	*J. scopulorum* Sarg.	WA, OR, NV, ID, MT, WY, UT, AZ, CO, TX, ND, SD, NE
southern redcedar	*J. silicola* (Small) Bailey	TX, LA
eastern redcedar	*J. virginiana* L.	ND, SD, NE, KS, OK, TX, AK, MO, IA, LA

common use. Meanwhile, increased sediment production continues and water yields decline from juniper-dominated watersheds. Midsummer fire storms are beginning to convert dense stands to dominance by introduced annuals or warm season sod grasses. Recurring fires generally lead to site degradation. Proactive management, especially on the federal lands, is presently almost nil. Further reductions or even complete removal of livestock will have very little effect on this regional problem and wildfires may even be encouraged. Without more substantial financial return from the wood, there is little hope for treating mature juniper stands. We can, however, use prescribed burning where it is still possible, promptly reseed after wildfires, and retreat the earlier chainings and cablings that are going back to trees.

Classification and Distribution

The genus *Juniperus* is represented by 13 species west of the Mississippi River (Table 32.1). Their collective distribution involves all of the western states (Figure 32.1). Although all junipers are apparently native here, all are regarded as weeds, at least in some circumstances. This designation comes from the way in which junipers have expanded their distribution and dominance over the last several centuries.

Ecology

Most junipers were apparently found in pre-Columbian times only on steep and/or rocky sites with little herbaceous understory. Occasional ground fires in the surrounding vicinity apparently kept them confined to such sites. Only a few junipers can resprout following fire (Table 32.1). Junipers' generally slow growth rate probably kept them from shading out fine fuels before the next fire recurred. Fire regimes were

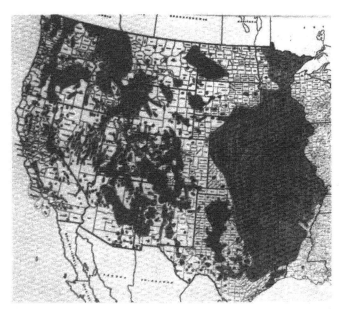

FIGURE 32.1 Collective distribution of all *Juniperus* spp. listed in Table 32.1. Compiled from maps of individual species distributions found in (1).

328

altered by European colonists through reducing numbers and areas influenced by Indians and grazing of livestock. These actions led to reduced fuels and herbaceous competition. Climatic trends around the turn of the 20th century may also have made it easier for trees to succeed (2).

Animal-scattered juniper seeds and allelochemic properties of juniper litter are also involved in their thickening in original pockets or belts of occupancy and invasion into adjacent areas. Although there was some locally important harvesting of these trees for fuel, charcoal, mine props, and fence posts, junipers generally increased much more rapidly than they were reduced. Thus, by the end of World War II, there were noticeably vast areas, once having only scattered or few trees, that had become woodlands of various density.

The long-lived tissues of junipers are very efficient in the capture of light, water, and nutrients. Accordingly, without the set-backs from fire, junipers can easily dominate in many semiarid environments. In addition to the shade and litter they cast, the root systems extend two to three times wider than the crown canopy. Thus, as the individual trees mature and the stand thickens, trees crowd out the understory even where there are openings in the canopy between trees. This results in a strongly

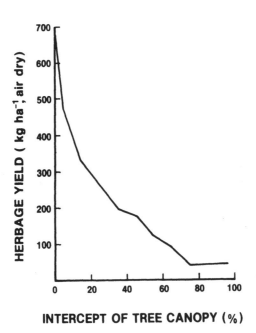

FIGURE 32.2 Relationship of herbage yield to intercept of tree canopy in a seral ensemble from 14 pinyon-juniper woodland locales in northern and central Arizona (3).

negatively exponential relationship of understory production to overstory cover over successional time (Figure 32.2). Even more pronounced reductions in understory with increasing tree cover is expected in the Great Basin and Pacific Northwest because both trees and understory grow during the same part of the year (spring).

As the trees come to dominate and understory cover declines, soils in the interspaces become less protected and erosion accelerates up to 400% (4). We now observe much of the juniper woodlands becoming a monoculture of trees on mounded microrelief, with rills and gullies forming in the interspaces. Although the vegetation seems to be barely changing when monitored in usual ways, I interpret the changes in microrelief to mean that the system is degrading to a new, lower level of potential because of decreasing overall soil depth, organic matter, and nutrient content. Because individual juniper trees can live for hundreds, if not thousands, of years, the successional trajectory is fixed for a long time unless other influences intervene (5).

Junipers are extraordinarily well protected from herbivory and decay by mechanical and chemical means. There is little likelihood, therefore, of those influences affecting much change. Recent reductions in livestock and increases in wood harvest have had no appreciable impact on recovery of understory vegetation, apparently because seed banks and sources have been lost (6).

Control

Land managers recognized that the just-described successional changes were negatively reducing livestock and big game forage and water production, but increasing sediment yields. Thus, in the 1950s and 1960s, agencies and ranchers began to use mechanical means to alter this vegetation on relatively gently sloped sites. Succession had usually gone too far to institute prescribed burning since the fine surface fuels lacked continuity. The end of World War II led to surplus large machinery and people who could operate it. National policy was to subsidize livestock operators to increase beef for then-growing consumer demands. The most popular approach was to connect a battleship anchor chain or cable between two large crawler tractors and pull down the trees. The second pass of the cable or chain may have been preceded by aerial seeding of grasses, or alternatively, the debris was ricked into piles and sometimes burned. Drills could then be used to implant the seeds of desired species. The short-term result usually was a 20- to 30-fold increase in grasses (5). The root sprouting junipers required root-plowing or hula-dozing. The longevity of such treatments depended on site characteristics, thoroughness of tree reductions, care in seeding, and subsequent management (7).

Most of the chaining and cabling operations were carried out before petroleum prices soared and environmental impact statements and archeological surveys were required. Early treatments were focused on livestock production. Large square blocks following legal boundaries were treated. Planted species usually didn't include plants desirable to wildlife. Archeological values were ignored. Enhanced environmental awareness during more recent decades forced re-examination of such approaches. Land management agencies began to choose only the best sites for treatment and leave less suitable sites. Contoured, curving treatment boundaries were employed to simultaneously create less visual impact and to afford escape cover for wildlife. Browse and forbs favored by wildlife were included in the seed mix.

Although chemical means of reducing junipers in both natural and formerly chained or cabled stands have been identified (8,9), very little of this technology is presently being applied. Large amounts of active ingredients are required, making such treatments too expensive to pay back through increased livestock usage alone. Some effective chemicals have not been licensed or have had their labels dropped because of concerns for human health. Furthermore, subsequent chaining and seeding is usually necessary to establish desirable forages.

Recent public outcry over a west-wide proposal produced for vegetation management by the Bureau of Land Management (BLM) indicates consid-erable public resistance to the kinds of type conversions done in the past. Many mistakenly believe that the present juniper woodlands are still pristine. They do not realize the extent or magnitude of change that has taken place. Furthermore, they do not recognize that accelerated soil erosion and reduced water yields are occurring.

Although unrestricted grazing undoubtedly had a role in creating these changes, removal of livestock will not automatically reverse these results. Instead, taking livestock off such lands will probably lead to more wildfire. Now fires can easily become hot crown fires. Mid-summer fire storms will lead to radical alteration of such lands. Unless we are ready to reseed immediately (10), the land can quickly convert to dominance by introduced annuals, at least in the Pacific Northwest and Great Basin (6, 11). The annuals will, in turn, lead to more frequent fires earlier in the growing season. When followed by winds and heavy rains associated with summer convectional storms, greatly accelerated soil erosion must be expected. Annuals recycle and store much less nutrients than in the original systems. The result will be a downward spiral of land degradation similar to what happened in the juniper woodlands of the Middle East several millennia ago. Global climatic change will only accelerate these trends.

Breaking the current "nonmanagement" posture by the federal land management agencies for juniper woodlands (12) will not be easy. Lack of current possibilities of some substantial financial return from tree harvest is preventing proactive management on federal lands. Current oversupply of red meat provides no incentive for federal subsidization of type conversions. Conventional economics wouldn't support the concerns for wildlife habitat and soil erosion alone. Federal subsidies for private land owners to reduce junipers are disappearing. In the meantime, retrogressional trajectories set in motion a century or more ago are leading to further tree dominance and soil erosion (4).

What can we do about the lack of proactive land management in juniper woodland? First of all, we need to tell environmentalists the truth about how most juniper-dominated areas came to be. We have considerable data on tree age class structure, old photos, relict areas, and records from early surveyors (13) that show how recent the expansion of trees has been. We have to acknowledge that unrestricted livestock grazing in earlier decades did play a role in causing these changes. We can, however, demonstrate that reductions or even complete removal of livestock will not lead to a quick reversal of tree dominance. In fact, removal of grazing may lead to larger, more severe summer fire storms.

We could also demonstrate the role of trees in reducing biotic diversity, accelerating soil erosion, and drying up stream flow. Again old data records, tree growth rings, old photographs, and public records demonstrate these changes (2).

A third thing we could do is lobby for support of research and development of equipment to harvest the trees and uses for the materials. Former support for such work has evaporated. Environmentalists will not tolerate simple toppling and burning of trees on federal lands, particularly now that they have become aware of global warming possibilities. We need to also further demonstrate that harvesting can be done in patterns sensitive to wildlife, watersheds, and aesthetics.

In the meantime, until more results on the above become available, we can focus our shorter term attention to retreatment of earlier chainings and cablings that are going back to tree dominance. These areas have already been altered and archeological, aesthetic, and wilderness values foregone. The public has already made a considerable investment in fences and water development. Hand application of pelleted tebuthiuron (N-[5-(1,1-dimethyl)-1,3,4-thiadiazol-2-yl]-N,N'-dimethylurea) can be used to kill the trees but will leave nearly all the browse and other forage plants (14,15).

Another action of relatively low cost, but high payoff, would be to promptly and carefully reseed areas covered by wildfire. There usually are not enough soil seed reserves present before the fire. Crown fires are so hot that what little seed remains is destroyed. Therefore, only very vagile weeds or resprouters typically come in. These weeds are usually annuals that present further fire hazards. Instead of letting Mother Nature choose the mix, we could promptly broadcast seeds of more desirable species onto the burned areas (10). I have seen superlative success where cattle trampling and/or subsequent chaining of the snags was done to cover the seed following aerial seeding. This requires planning to have an assortment of seed, equipment, and/or livestock on hand and acting expeditiously. Bureaucratic response time is now typically too slow to beat the weeds.

We should also encourage land managers to use the rapidly diminishing opportunities for prescribed burns that will stay as ground fires. This may cost a season or two of livestock grazing to build up fuels and rest afterwards, but yield at least several decades of better forage and protected soils. Wildlife and biotic diversity in general are favored by small burned patches. Prescriptions for successful prescribed burning of juniper woodlands have been available for some time (5,16-18).

Opportunities for public harvest of wood products should be better planned and controlled. Perhaps volunteer help could be engaged to police firewood harvests such that they open spots in the woodlands rather than simply high-grade some trees from extensive areas. Seed could even be broadcast before the harvest so that surface disturbances would cover some of them.

Summary

Junipers are often regarded as weeds by particular interests. Because of their growing abundance and dominance, junipers now control many important interactions in wildland ecosystems of the western U.S. Junipers have to be managed more sensitively on public lands because of misunderstanding by environmentalists. Unless we correct the mistaken assumptions of the public, our options for proactive management to prevent further degradation will be severely restricted. We could do this by marshalling the facts of undesirable change in juniper woodlands for our locales and begin small demonstrations of how the undesirable changes could be reversed. Once we rebuild confidence of the correctness of our understanding of juniper and its important role in ecosystem degradation, perhaps we will regain political and fiscal support to commence proactive management of the extensive juniper woodland rangelands now largely ignored at our long-term peril.

References

1. Little EL. USDA For. Serv. Misc. Publ. 1146 (1971).
2. West NE and NS Van Pelt. USDA For. Serv. Gen. Tech. Rep. INT-215 (1987).
3. Arnold JF, DA Jameson, and EH Reid. USDA Prod. Res. Rep. 84 (1964).
4. Carrera PE and TR Carroll. Earth Surface Processes 4:307 (1979).
5. West NE. In: *Developing Strategies for Rangeland Management* (Nat. Res. Council), pp. 1301-1332. Westview Press, Boulder, CO, 1984.
6. Koniak S and RL Everett. Amer. Midl. Nat. 108:295 (1982).
7. West NE. In: *Proc. Second Utah Shrub Conf.* (KL Johnson, Ed), pp. 21-33. Utah State Univ. Logan, UT, (1984).
8. Clary WP, S Goodrich, and BM Smith. J. Range Manage. 38:56.
9. Wilson J and T Schmidt. Rangelands 12:156 (1990).
10. Koniak S. USDA For. Serv. Res. Note INT-334 (1983).
11. Billings WD. In: *The Earth in Transition* (GM Woodwell, Ed.). Cambridge Univ. Press, New York, 1990.
12. Evans RA. USDA For. Serv. Gen. Tech. Rep. INT-249 (1988).
13. Sparks SW, NE West, and EB Allen. USDA For. Serv. Gen. Tech. Rep. INT-318 (1990).
14. Van Pelt NS and NE West. J. Envir. Qual. 18:281 (1989).
15. Van Pelt NS and NE West. J. Range Manage. 43:39 (1990).
16. Bruner AD and DA Klebenow. USDA For. Serv. Res. Pap. INT-219 (1979).
17. Ralphs M and FE Busby. J. Range Manage. 32:267 (1979).
18. Wright HA and AW Bailey. "Fire Ecology: United States and Canada". John Wiley and Sons, New York, NY, (1982).

33

Sagebrush: Classification, Distribution, Ecology, and Control

Thomas D. Whitson

Abstract

Artemisia is a large genus of the Asteraceae family with approximately 200 species found throughout North America, North Africa, and Eurasia. Sagebrush species are woody long-lived perennials living over 50 years on well-drained habitats. Big sagebrush (*Artemisia tridentata*) is the most abundant species of the genus with estimates of amounts of rangeland occupied exceeding 100 million hectares. Silver sagebrush (*Artemisia cana*), another highly competitive species, occupies over 13 million hectares of rangeland in North America. Other *Artemisia* species of less abundance that are also considered of low forage value and highly competitive in rangeland include common sagewort, fringed sagebrush, and sand sagebrush. Silver sagebrush is a resprouting species while other species reproduce from seed. Control of big sagebrush has been reported as a production tool since before 1914 when fire was used for its control. The use of 2,4-D was reported as an effective sagebrush control in 1952 and continues to be a practical control today. Pelleted tebuthiuron was first used selectively to control big sagebrush in rangeland in 1975 and has provided consistent control of big sagebrush in several experiments since that time.

Introduction

Until early in the nineteenth century, the American buffalo was the only large herbivore to exist in significant numbers on western rangelands (1). Because the buffalo was a relatively nonselective grazer, not preferring grass over other vegetation, the ranges remained in a state of equilibrium. That relative balance was maintained until settlers began moving into the western rangelands in the early 1800s. By 1839, populations of the American buffalo were exterminated west of the Rockies. The mining booms produced demands for meat and livestock products. Cattle and sheep herds were expanded in number (2). Soon after the coming of the railroads in the late 1800s, these lands were overrun with livestock (3). Sagebrush species in the plant community proved remarkably resistant to grazing. Grazing without control continued until the "dust bowl" days of the early 1930s which prompted the necessity for the preservation and management of public rangelands (4).

Sagebrush has a competitive advantage over other range plants because of the selective preference of livestock for grasses. Most species of sagebrush are unpalatable due to their essential oils which also inhibit the rumen microflora of both sheep and cattle. Big sagebrush also possesses allelopathic properties (1).

In the 1948 USDA Yearbook of Agriculture, sagebrush was listed as a noxious plant. The authors emphasized, "after deep rooted, undesirable plants (such as mesquite and big sagebrush) have become firmly established, moderation in grazing use alone is ineffective, and drastic measures must be used" (5). Fire has been commonly used as a control measure for big sagebrush (*Artemisia tridentata*). Other more recent measures of control have included 2,4-D and tebuthiuron (6,7).

Sagebrush Distribution and Ecology

Artemisia is a large genus with approximately 200 species found throughout North America, North Africa, and Eurasia (8). Big sagebrush inhabits 122 million ha in the western states, ranging from Nebraska to California and from the southern part of the United States to British Columbia. It grows on deep, well-drained soils with elevations ranging from 450 to 3050 m with annual precipitation ranging from 20 to 50 cm.

Silver sagebrush (*Artemisia cana*) is found on over 13 million ha in North America. The species occurs at 220 to 1830 m elevations from Canada through central Oregon, northwest Wyoming to Nebraska, Colorado, and New Mexico (9). Its habitat includes deep loamy to sandy soils (8). Eleven sagebrush species are found throughout Arizona, California, Colorado, New Mexico, Oregon, Utah, Washington, and Wyoming. Other *Artemisia* species considered competitive on western United States rangelands include common sagewort (*A. compestris*), sand sagebrush (*A. filifolia*), and fringed sagebrush (*A. frigida*) (10).

Sagebrush Phenology

Life processes of sagebrush occur at regular time periods each year. Seed germination occurs from February to April, depending on moisture availability and soil temperatures. Seed either results in sagebrush seedlings the first growing season after being shed or becomes nonviable. Extremely wet or dry soils prevent seedling survival (8). Seedlings may reach the flowering stage in as little as three years. In Washington, bud enlargement has been reported in early April with shoot emergence taking place in May (7). In Utah, a delay from Washington of approximately two weeks for these processes has been reported.

Spring leaves are shed by July with reproductive shoots reaching full size in late August. Floral buds found in late September are opened by mid-October. Seed development starts in late October and seed shed from mid-November into the winter months (1).

The sagebrush root system consists of three to six main branches subdividing according to soil types in which they are growing. Fan-shaped fibrous roots are located at the ends of primary roots (11). Lateral roots are concentrated at two soil depths, the first at 40 cm and the second at 80 cm, with tap roots extending to depths of over 27 m in deep soils. Root systems are reported to be much shallower in sandy soil (12). Soils in which sagebrush survive are: (a) coarse textured and well drained; (b) of depths greater than one meter; and (c) free from alkali salts (13).

Sagebrush is considered an evergreen. However, the leaves may function for only short periods and be shed at any time during the year depending on soil moisture. Leaves are variable in size and shape on the same plant and among plants of the same species. Leaf characteristics, while not definitive, are often used to distinguish between species. Leaves on flowering shoots are greatly elongated and entire, while leaves on vegetative shoots are in compact whorls (8,14). Hybridization is common between similar sagebrush species, which makes leaf shapes and types a difficult characteristic for distinguishing taxa of sagebrush (8).

Taxonomy of Sagebrush

All sagebrush species are members of the Asteraceae or sunflower family.

Silver sagebrush is a low erect shrub that is commonly 4 to 9 and up to 15 dm high (14). In addition to seeds, plants spread through stolons and root sprouts. It commonly resprouts after being burned or chemically treated (8,15). Leaves are a silver-conescent color and become slightly sticky with age. Leaf shapes are described as linear to linear-oblanceoloate, entire and with occasional leaves with one or two irregular teeth (9). Seed heads of silver sagebrush are erect, 4 to 5 mm high, and 3 to 4 mm wide. Seeds are enclosed by 8 to 15 bracts (14).

Sagebrush Control

Sagebrush control can best be planned if life and reproductive systems of the plant are understood. Control techniques are more effective when they are directed at one specific plant system. For example, when phenoxy herbicides are used, they must be applied as leaves start early spring growth.

Control of big sagebrush on heavily infested areas has been recognized as a production tool since before 1914 when the *Journal of Agriculture Research* reported the use of fire as a tool to control this species (16).

Railing and plowing have also been used extensively to help control this unpalatable, woody plant (16). Chemical control of big sagebrush was first introduced in Wyoming in 1952 when 438 ha were sprayed with 2,4-D [(2,4-dichlorophenoxy)acetic acid]. This practice was soon recognized as a management tool by the Bureau of Land Management and the U.S. Forest Service. Cost sharing was introduced in 1954 by the Agricultural Stabilization and Conservation Service to encourage control of this woody species. Control practices in Wyoming increased until 1967, when 70,114 ha were treated with 2,4-D. Treated acreages totaled 787,381 ha by 1976. Total acreages sprayed in Wyoming by federal agencies and individuals included 118,466 ha by the U.S. Forest Service, 543,281 ha by the ASCS and Great Plains Conservation Program, and 87,097 ha by private land holders (17).

Alley (18) reported on a study conducted from 1952 to 1955 on chemical control of big sagebrush and its effects on production of perennial grasses. Grass utilized by livestock increased proportionally from 166 kg/ha in to 0% control area (where sagebrush was not controlled) to 682 kg/ha in the 96-100% sagebrush control area. In other

TABLE 33.1 Big sagebrush control and perennial grass yields eleven years following tebuthiuron applications.

Herbicide[1]	Rate kg ai/ha	% Sagebrush[2] Control Applied		Perennial grass yield (kg DM/ac) Applied	
		May	Sept.	May	Sept.
Tebuthiuron	0.28	74	77	745	889
	0.55	90	91	908	890
	0.82	95	97	890	924
	1.10	98	99	851	788
Tebuthiuron	0.28	66	80	706	816
	0.55	87	89	822	980
	0.82	86	90	834	913
	1.10	96	94	820	806
Check	–	0	0	414	429

[1] Herbicides were applied May 29 and Sept. 16,1980.
[2] Evaluated July 25, 1990.

studies, he found even low production sites increased from 90 kg air-dried forage/ha to 270 kg/ha when 96-100% control was obtained.

Orpet and Fisser (19) monitored grass production for 15 consecutive years on five study sites where big sagebrush had been chemically controlled with 2,4-D butyl ester. During the study period, yearly perennial grasses increased 900% on the sprayed areas when compared to those untreated.

In a 1979 study conducted in Idaho, rangeland had a threefold production increase when big sagebrush was controlled with 2,4-D Low Volatile Ester (LVE) (20).

In 1973, tebuthiuron (N-[5-(1,1-dimethylethyl)-1,3,4-thiadiazol-2-tl]-N,N'-dimethylurea) was introduced by Elanco as a broad spectrum herbicide (12). In 1978, Bjerregaard et al. (21) found individual brush species to be rate specific to tebuthiuron.

Klauzer and Arnold (22) found sagebrush (*Artemisia* spp.) to be effectively controlled by applications of tebuthiuron of 1.1 kg/ha.

Experiments by Whitson and Alley (23) with 20% pelleted tebuthiuron near Ten Sleep, Wyoming, in 1979 showed that applications of 0.34, 0.73, and 1.0 kg ai/ha controlled 69, 96, and 99.5% of big sagebrush, respectively. The predominant perennial grass species, western wheatgrass, had a 397% greater live canopy within the area treated with tebuthiuron at 0.73 kg ai/ha than in the untreated control. Forage yields

TABLE 33.2 Control of big sagebrush with various herbicides and resulting forage production.

Herbicide[1]	Rate kg ai/ha	% Control[2] 1990	Air Dry Forage kg/ha 1990
Dicamba (Banvel)	1.1	19	587
	2.2	47	705
2,4-D ester	1.1	36	470
	2.2	95	685
2,4,5-T ester	1.1	85	783
	2.2	90	823
Tebuthiuron	0.14	39	587
	0.28	90	548
	0.55	91	529
	0.83	98	568
	1.1	98	587
Triclopyr	0.28	5	548
	0.55	87	607
	1.1	87	723
Triclopyr+2,4-D (Crossbow)	0.55+1.1	79	627
Check	--	0	372
(LSD 0.05)		32	
(CV)		27	

[1] Herbicide treatments applied June 10, 1982.
[2] Visual control evaluations 5/31/84 and 8/23/90. Production measurements 7/24/84 and 8/24/90 from six 0.25 m² diameter quadrats per treatment.

two years after the application increased more than twofold over the untreated control to 630 kg/ha (6).

Tebuthiuron formulations and application rates were compared eleven years following applications made in May and September 1980 near Bosler, Wyoming (Table 33.1). When tebuthiuron was compared at the same application rates, no differences were found between formulations. No differences in sagebrush control or production were found with equal rates applied in either May or September. Sagebrush control was significantly higher when tebuthiuron was applied at 0.55 kg ai/ha and above compared to treatments of 0.28 kg ai/ha or the untreated check. All treatments applied at rates greater than 0.28 kg ai/ha had perennial grass production increases significantly higher than the untreated check.

In 1982, a study was established to compare tebuthiuron to other herbicides for big sagebrush control and perennial grass production (Table 33.2). Tebuthiuron application rates greater than 0.28 kg ai/ha all provided greater than 90% control eight years following treatments. When sagebrush control was greater than 85%, forage yields were two times that of the untreated check.

Two years after treating with 2,4-D low volatile ester at 2.2 kg ai/ha, perennial grasses had significantly higher yields than in those areas treated with tebuthiuron at 0.825 kg ai/ha and above. Perennial cool season grasses such as western wheatgrass (*Agropyron smithii*) and green needlegrass (*Stipa viridula*) were thinned with the higher tebuthiuron application rates. The differences were not apparent after eight years.

Other sagebrush species such as fringed sagebrush (*Artemisia frigida*) can be controlled with picloram at 0.28 lb ai/ha applied in the vegetative stage. Sand sagebrush (*A. filifolia*) and silver sagebrush are best controlled with repeated annual applications of 2.2 kg ae/ha of 2,4-D (LVE). After three years, 85 to 95% control of these resprouting species can be attained. Common sagewort (*A. campestris*) control can be attained with the application of metsulfuron at 0.4 kg ai/ha or 2,4-D (LVE) at 2.2 kg ae/ha applied in the vegetative stage.

Conclusions

Big sagebrush is well established on rangelands of the western United States. Control can be obtained by using fire, mechanical, or chemical methods. After complete control is obtained, reinfestation takes place very slowly. Therefore, forage yields can often be twice or three times that of an untreated control for periods exceeding 20 years. Silver sagebrush cannot be controlled easily because of its resprouting characteristics. Therefore, without several repeated yearly applications of a herbicide, complete control cannot be obtained. Before establishing a control method, the species of sagebrush must be positively identified and a perennial grass inventory taken to determine which plant community species will replace the sagebrush after control is achieved.

References

1. Utah State University. *Sagebrush Ecosystem Symposium Proc.*, 1979.
2. Whitson TD. PhD Thesis, Univ. Wyo., Laramie, 1982.
3. Sampson AW. In: *Range Management Principles and Practices*, pp. 361-376. John Wiley and Sons, New York, 1952.

4. US Bureau of Land Management. *The Taylor Grazing Act*. US Govt. Printing Off., Washington, D.C., 1964.

5. Stefferud A (Ed.). In: *Grass--The Yearbook of Agriculture 1948*. US Govt. Printing Off., Washington, D.C., 1948.

6. Whitson TD, MA Ferrell, and HP Alley. Weed Technology 2:486 (1988).

7. Daubenmire R. Northwest Science 49:1975 (1975).

8. Beetle AA. Wyo. Agric. Exp. Sta. Bull. 368 (1960).

9. McArthur ED, AC Blaver, AP Plummer, and R Stevens. USDA Forest Ser. Res. Paper Int-220, 1979.

10. Whitson TD. Wyo. Agric. Ext. Serv. Bull. No. B-855-UW, pp. 146-155 (1987).

11. Sheets WB. PhD Thesis, Univ. Wyo., Laramie, 1968.

12. Perkings AT, GB Johnson, TC Gee, and DJ Stroud. *North Central Weed Cont. Conf.*, pp. 119-120, 1976.

13. Kearney TH, JL Briggs, HL Shantz, JW McLane, RR Piemeisel, and Bureau of Plant Industry. *Indicator Significance of Vegetation in Tooele Valley, Utah*, pp. 377-387, 1914.

14. Hall HM and FE Clements. In: *The Phylogenetic Method in Taxonomy*, pp. 135-153. Judd and Detweiler, Washington, D.C., 1921.

15. USDA. *Chemical Control of Range Weeds*. US Govt. Printing Off., Washington, D.C., 1966.

16. Pechanec JF, AP Plummer, JH Robertson, and AC Hull Jr. *Sagebrush Control on Rangelands*. US Govt. Printing Off., 1965.

17. Kearl GW and JW Freeburn. *Economics of Big Sagebrush Control for Mitigating Reductions of Federal Grazing Permits*, 1980.

18. Alley HP. MS Thesis, Univ. Wyo., Laramie, 1955.

19. Orpet JE and HG Fisser. *Proc. Soc. for Range Manage.* 29 (1979).

20. Wattenbarger DW, WS Belles, and GA Lee. *Proc. Western Soc. Weed Sci.*, 1979.

21. Bjerregaard RS, JA Keaton, KE McNeill, and LC Warner. *Proc. First Int. Rangeland Congress of the Soc. for Range Manage.*, pp. 654-656, 1978.

22. Klauzer JC and WE Arnold. *Proc. North Cent. Weed Cont. Conf.* 30:44 (1975).

23. Whitson TD and HP Alley. West. Soc. Weed Sci. 39 (1982).

34

Rabbitbrush: Classification, Distribution, Ecology, and Control

Christopher A. Call

Abstract

Rabbitbrush species (*Chrysothamnus* spp.) are widespread throughout western North America, occurring in most major plant communities from sea level to over 3,300 m elevation. The most current taxonomic revision of the genus *Chrysothamnus* recognizes 16 species and 41 subspecies. Rabbitbrush species are typically subdominant shrubs in most high-seral communities, but they can become seral dominants in communities that have been disturbed by fire, grazing, and range improvement practices. The increase in dominance is attributed to crown sprouting, prolific seed production, and ease of seedling establishment. Top removal practices (mowing, fire) generally have little effect on rabbitbrush plants due to their resprouting ability. Chemical control practices, mostly involving the simultaneous control of big sagebrush (*Artemisia tridentata*) on Great Basin rangelands, have been intensively researched and applied, but results have been variable. Rabbitbrush plants are susceptible to herbicides such as 2,4-D when available soil moisture and warm soil temperatures promote active growth, but these conditions can be quite transient or may not occur during certain growing seasons. Failure to recognize the appropriate time of herbicide application has resulted in the mortality of other shrubs and the release of rabbitbrush species.

Introduction

Rabbitbrush species (*Chrysothamnus* spp.) present a management paradox for range managers in western North America. Several species and subspecies can be considered noxious, yet the same taxa may have potential beneficial uses. Beneficial uses may include food and cover for livestock and wildlife, early successional species for revegetating disturbed lands, low-maintenance species for landscaping, and sources of natural rubber and resins and energy biomass (1).

From the standpoint of vegetation management, the most significant characteristics of many rabbitbrush species and subspecies are the ability of plants to become quickly established on disturbed areas and the resilience of plants to various control practices. One must better understand the taxonomy, distribution, and ecology of the rabbitbrush species to better manage them.

Classification

Rabbitbrush plants are much-branched subshrubs or shrubs up to 3 m high with ascending to erect stems bearing alternate, narrow, deciduous leaves, and flowers arranged in cymes, racemes, or panicles (2). Plants range from glabrous to densely tomentose, and are commonly resinous and aromatic. The common name rabbitbrush comes from the fact that rabbits use the shrubs for food as well as a favorite shelter.

The genus *Chrysothamnus* has undergone several taxonomic revisions since it was established by Nuttall in 1840. Hall and Clements (3) proposed natural sections that have provided the foundation for our current understanding of the genus. Most recently, Anderson (4) has recognized five sections containing 16 species and 41 subspecies (Table 34.1). The sections are distinguished by differences in floral, stem, and achene characteristics. Of the 16 species in the genus, three dominate in terms of diversity and distribution: low (green, Douglas) rabbitbrush (*C. viscidiflorus*); parry rabbitbrush (*C. parryi*); and rubber (gray) rabbitbrush (*C. nauseosus*) (5).

Considerable intraspecific variation in morphological characteristics, particularly nonfloral characteristics, makes subspecific taxonomic identification sometimes difficult in low and rubber rabbitbrush complexes. In recent years, laboratory analyses and plant-insect associations have been used to supplement morphological criteria in the classification of plants. Palatable and unpalatable subspecies of low and rubber rabbitbrush have been recognized by chromatographic patterns of phenolic compounds (6,7). Four common subspecies of rubber

TABLE 34.1 Sections with constituent species in *Chrysothamnus* as recognized by Anderson (4).

Section *Chrysothamnus*	Section *Gramini*
C. *albidus*	C. *eremobius*
C. *greenei*	C. *gramineus*
C. *humilis*	
C. *linifolius*	Section *Pulchelli*
C. *spathulatus*	C. *depressus*
C. *viscidiflorus* (5 ssp.)	C. *molestus*
	C. *pulchellus* (2 ssp.)
Section *Nauseosi*	C. *vaseyi*
C. *parryi* (12 ssp.)	
C. *nauseosus* (22 ssp.)	Section *Punctati*
	C. *paniculatus*
	C. *teretifolius*

rabbitbrush in Utah have been identified by different forms of tephritid fly (Diptera: Tephritidae)-induced galls on plant stems (8). White-stem rubber rabbitbrush (*C. nauseosus* ssp. *albicaulis*) has been differentiated from thin parry rabbitbrush (*C. parryi* ssp. *attenuatus*), and mountain low rabbitbrush (*C. viscidiflorus* ssp. *lanceolatus*) has been differentiated from spreading rabbitbrush (*C. linifolius*), by electrophoresis of isoenzymes (9). For a more detailed treatment of rabbitbrush taxonomy, see Hall and Clements (3), McArthur et al. (2), Anderson (4), and McArthur and Meyer (5). The remaining sections of this paper will focus on the low and rubber rabbitbrush complexes.

Distribution

Rabbitbrush species and subspecies are widely distributed in western North America, from British Columbia to Saskatchewan and North Dakota to Texas and Coahuila and Baja California (100° W - 124° W longitude, 27° N - 50° N latitude) (4, 5). They occur in deserts, plains, foothills, valleys, and mountains, from sea level to 3,300 m in elevation (2). Individual species and subspecies distributions have been mapped and described by Anderson (4).

Some subspecies of rubber rabbitbrush (*albicaulis, consimilis, graveolens, hololeucus,* and *nauseosus*) and low rabbitbrush (*lanceolatus* and *viscidiflorus*) are generalists with fairly wide ecological amplitudes, while other subspecies of rubber rabbitbrush (*arenarius, bigelovii, nanus, texensis,* and *psilocarpus*) and low rabbitbrush (*planifolius* and *axillaris*) are restricted to more narrow habitats (4,5). Widely distributed subspecies of rubber rabbitbrush may occur in different areas (allopatric), may be contiguous (parapatric), or may occur together in the same area (sympatric) (10). Each subspecies is apparently evolving on its own course with minimal interaction with other subspecies because of separation in space or because of breeding system incompatibility (5,10).

Ecology

In general, rabbitbrush species and subspecies are subdominants in several high-seral plant communities, including northern desert shrub, salt desert shrub, pinyon-juniper (*Pinus-Juniperus*), mountain brush, subalpine forest, grassland, southern desert shrub, and riparian areas (11). Some of the many subspecies of rubber rabbitbrush may dominate the vegetation in specific habitats. For example, salt rabbitbrush (*C. nauseosus* ssp. *consimilis*) is a dominant shrub species on many flood plains and desert flats with a shallow water table in salt desert shrub communities (12). Rabbitbrush species and subspecies can become seral dominants in many plant communities that have been disturbed by fire, livestock grazing, and/or shrub manipulation as a range improvement practice (13-16). The increase in dominance of low and rubber rabbitbrush in disturbed big sagebrush (*Artemisia tridentata*) communities is attributed to sprouting from basal stem and crown buds, and subsequent seed production and seedling establishment. The reproductive response of rabbitbrush following a disturbance depends upon environmental conditions and the impact of the disturbance on associated woody and herbaceous vegetation (17). Vigorous stands of young, low rabbitbrush plants have the potential to produce up to 20 million seeds/ha annually (15).

Seeds normally ripen over a 2- to 4-wk period during late fall or early winter, depending upon site conditions (18). Seeds are generally considered to be nondormant and probably do not form a persistent seedbank under natural conditions. Under dry storage conditions (temperature range from -29.9 to 38.3 C), germination of white-stem rubber rabbitbrush declined from 80 to 14 % from the second to the fifth year of storage (19). Ripe seeds have a well-developed pappus which enhances wind dispersal, seed placement in the seedbed, and seed anchorage when primary roots penetrate into the soil (20).

Seed germination requirements have been characterized for several subspecies of rubber rabbitbrush under controlled environment conditions. Sabo et al. (21) reported no effect of light or alternating temperature for subspecies *bigelovii*, while subspecies *consimilis* required light at high temperatures. Khan et al. (22) reported that seeds of subspecies *viridulus* germinated at both constant and alternating temperatures in light or dark, with maximum germination occurring between 25 and 30 C. In the same study, seed germination was progressively inhibited by increases in substrate concentrations of sodium chloride and polyethylene glycol. Romo and Eddleman (23) reported similar results for subspecies *graveolens* and *nauseosus*. Total germination and germination rate for both subspecies were highest at 20 and 30 C and declined as temperatures decreased to 10 C and substrate osmotic potentials (induced by polyethylene glycol) decreased to -1.5 MPa. Results from all of these germination studies were based on a single collection of each subspecies. Recently, Meyer and co-workers (24) reported that interpopulation differences in the germination rate of rubber rabbitbrush subspecies at low temperature (3 C) could be correlated with winter/spring frost and drought risks to seedlings at the site of seed origin.

Few shrub species establish as easy as rabbitbrush following a disturbance. Young seedlings occupy openings and compete well with developing herbaceous vegetation. Initial seedling densities can be low on naturally and artificially revegetated areas but can increase dramatically during community development. Stevens (25) monitored white-stem rubber rabbitbrush density and growth performance 2, 11, and 22 yr following chaining and seeding (mixture included white-stem rubber rabbitbrush, basin big sagebrush, and forbs and grasses) of a pinyon-juniper (*Pinus-Juniperus*) site in central Utah. Rabbitbrush densities increased to 3,174 plants/ha 11 yr after treatment, and decreased 40 % during the subsequent 11 yr.

Persistence appears to vary among rabbitbrush species and subspecies. Daubenmire (14) indicated that rubber rabbitbrush was short-lived for a woody plant, with less than 50 % survival 12 yr after a burn in a sagebrush steppe community in southeastern Washington. Harniss and Murray (13) reported that low rabbitbrush started to decline 12 yr after establishing on a burned sagebrush-grassland in southeastern Idaho. Measurement of annual growth rings on stems of rabbitbrush plants in 52 stands in northwestern Nevada indicated that plants persist from 10 to 15 yr (15). Persistence of four subspecies of low and rubber rabbitbrush was studied for 10 yr in a common garden in northwestern Nevada (26). All collections showed excellent initial establishment, but low rabbitbrush subspecies did not persist as long as rubber rabbitbrush subspecies. Rabbit utilization was highest on rubber rabbitbrush subspecies, but impacts of rabbit defoliation have been shown to be largely ameliorated by compensatory growth during the following growing season (27).

How competitive are rabbitbrush species and subspecies with other species? Plummer et al. (28) listed rabbitbrush and sagebrush among the undesirable competition to be removed to assure the success of introduced grass seedings. Frischknecht (29), on the other hand, reported that rubber rabbitbrush was not as competitive with crested wheatgrass (*Agropyron desertorum*) as big sagebrush. He observed that shallow lateral roots of big sagebrush were more abundant and thus more competitive with those of grasses than lateral roots of rubber rabbitbrush. Rooting profile data from an undisturbed sagebrush-grassland community in southeastern Idaho support this claim (30). Big sagebrush and low rabbitbrush root systems had respective lateral spreads of 50 cm and a trace (roots present, but not enough to determine spread) at 25 cm deep, and 100 cm and 20 cm at 40 cm deep. Big sagebrush root systems had greater lateral spread than low rabbitbrush root systems at deeper depths to 225 cm (low rabbitbrush roots penetrated to 190 cm). Miller (31) reported that Wyoming big sagebrush (*Artemisia tridentata* ssp. *wyomingensis*) had a greater impact on soil water use, at depths to 40 cm, by associated herbaceous vegetation than stickyleaf low rabbitbrush (*C. viscidiflorus* ssp. *viscidiflorus*) in a sagebrush steppe community in southeastern Oregon. Frischknecht (29) claimed that control of rubber rabbitbrush may not be justified for crested wheatgrass pastures, especially those receiving fall grazing.

Control/Management

Due to the resprouting ability of rabbitbrush, top removal practices such as mowing and burning are generally not effective by themselves. Some exceptions have been reported in the literature where burning has resulted in high rabbitbrush mortality. Daubenmire (14) observed that a fire in a sagebrush steppe community in southeastern Washington in July just after shrubs had been severely defoliated by grasshoppers proved fatal to nearly every rabbitbrush plant. Rubber rabbitbrush was also virtually eliminated from a sagebrush-grassland community in southern British Columbia after a July fire during the driest growing season on record (32). In both instances, reasons for the observed failures of resprouting by rabbitbrush were unknown. Robertson and Cords (33) used summer and fall burning in north central Nevada to precondition rubber and low rabbitbrush to be more susceptible to phenoxy herbicides applied during the following growing season.

Results from chemical control efforts have been extremely variable, depending on the species/subspecies of rabbitbrush targeted for control, site environmental conditions (particularly available soil moisture), phenological stage and physiological activity of the plant, type of herbicide and rate of application, and follow-up management. Chemical control efforts have focused primarily on simultaneous control of rabbitbrush and other undesirable shrub species, such as sagebrushes. When compared to big sagebrush, rabbitbrush species and subspecies are moderately tolerant to phenoxy herbicides. Thus, careful timing is required in order to consistently kill rabbitbrush in one application of herbicide.

Leaf tissue and apical meristems are always sensitive to the commonly applied herbicide 2,4-D, but plant mortality results only at the time when phenological and physiological conditions allow translocation of the herbicide in sufficient quantities to kill dormant buds on basal stems and crowns. Canopy reduction and mortality can be evident the first growing season, only to be followed by resprouting from buds that were not killed by herbicide.

Vegetative and reproductive phenologies of different rabbitbrush species/subspecies have been documented, and growth patterns have been related to phenoxy herbicide susceptibility. Young and Evans (34) recorded the phenology and growth of 100 stickyleaf low rabbitbrush plants in 14 stands in northwestern Nevada and divided vegetative development into four stages: (1) bud burst in late February/early March; (2) restricted growth from early March to early May (soil moisture available but low soil temperatures inhibited growth); (3) accelerated growth from mid/late May until as late as mid July (depending on available soil moisture); and (4) growth cessation in late July (lack of available soil moisture, onset of flowering). Optimum susceptibility to phenoxy and other herbicides occurs during the accelerated growth stage when plants are actively growing and translocating carbohydrates. The timing of the accelerated growth stage varies with plant age, associated plant competition, and environmental conditions. Accelerated growth may not occur during a growing season that has prolonged cool spring temperatures followed by low soil moisture due to drought conditions. Plants will essentially remain in a restricted growth stage and will not be susceptible to herbicides under these conditions. Hyder and co-workers (35,36), investigating the simultaneous control of sagebrush and rabbitbrush in southeastern Oregon, applied different rates of 2,4-D on different dates related to different phenological stages for low rabbitbrush. During a normal precipitation year, big sagebrush was susceptible to 2,4-D earlier and for a longer period (early May to mid-June) than low rabbitbrush (late May to mid-June). During a dry year, big sagebrush was susceptible to 2,4-D from mid-May to early June, and low rabbitbrush was not susceptible to 2,4-D at any application date during the growing season. Failure to recognize the low susceptibility of rabbitbrushes to 2,4-D during dry years has resulted in the control of relatively susceptible sagebrush and the release of rabbitbrush (37). Research (38,39) has shown that mixing picloram with 2,4-D can achieve better control of rabbitbrush in conjunction with sagebrush control by allowing more flexibility in the timing of herbicide application.

Recently released herbicides such as triclopyr and clopyralid have had variable effects on rabbitbrush species. Triclopyr, triclopyr + 2,4-D, and clopyralid + picloram did not provide adequate control of low or rubber rabbitbrush in a degraded big sagebrush community in southeastern Wyoming (40,41). Preliminary results from research in central Utah indicated that clopyralid could selectively suppress rubber rabbitbrush and big sagebrush (members of the Compositae) without adversely affecting bitterbrush (*Purshia tridentata*) and serviceberry (*Amelanchier alnifolia*) (members of the Rosaceae) (42). Final results from that study only discussed selective control of big sagebrush without adversely affecting associated rosaceous shrubs, indicating that rubber rabbitbrush must have resprouted after being initially suppressed by clopyralid (43).

Soil-applied herbicides such as tebuthiuron have also had variable effects on rabbitbrush species. Hairy low rabbitbrush (*C. viscidiflorus* ssp. *puberlus*) was successfully controlled by 1.3 kg/ha of tebuthiuron applied to a Utah juniper (*Juniperus osteosperma*) community in central Utah (44). In the same study, rubber rabbitbrush was not controlled by tebuthiuron at the same rate of application. In a sagebrush-bunchgrass community in southeastern Oregon, tebuthiuron applied at 2 kg/ha reduced the frequency of occurrence of low rabbitbrush from 11.5 to 3.3% while the 4 kg/ha rate virtually eliminated the shrub species (45). The 4 kg/ha application rate also eliminated most forbs and grasses in the treated area. Rubber rabbitbrush was not controlled by tebuthiuron at rates up to 1.7 kg/ha in a big sagebrush community in northeastern Nevada (46). The lack of control of this species is unfortunate because the chief advantage of soil-applied herbicides is that timing of application is not as critical as with foliar-applied herbicides.

The variable results from these control efforts reflect the variability in the susceptibility of rabbitbrush species and subspecies to different herbicides, rates and methods of herbicide application, and site environmental conditions (primarily available soil moisture). The tools are available for effective control, but control/management programs need to be more flexible. Plant development and site environmental factors need to be monitored more closely in order to seize the opportunities (favorable years) and avoid the hazards (dry years) (47), i.e., to avoid releasing rabbitbrush and creating more severe management problems than existed prior to treatment. Researchers and land managers have focused most of their attention on killing the target rabbitbrush species. More emphasis should be placed on the associated understory vegetation and the role it will play in moderating the subsequent establishment of rabbitbrush after implementing control treatments.

References

1. Weber DJ, DF Hegerhorst, TM Davis, and ED McArthur. In: *Proc. Fourth Utah Shrub Ecology Workshop, The Genus Chrysothamnus* (KL Johnson, Ed.), pp. 27-34. College of Natural Resources, Utah State Univ., Logan, 1987.
2. McArthur ED, AC Blauer, AP Plummer, and R Stevens. In: *Characteristics and Hybridization of Important Intermountain Shrubs III. Sunflower Family*, pp. 36-58. USDA Forest Service Res. Pap. INT-220, Intermountain Forest and Range Exp. Sta., Ogden, UT, 1979.
3. Hall HM and FE Clements. In: *The Phylogenetic Method in Taxonomy; The North American Species of Artemisia, Chrysothamnus, and Atriplex*, pp. 157-234. The Carnegie Inst. Wash. Publ. 326, Washington, D.C., 1923.
4. Anderson LC. In: *Proc. - Symp. on the Biology of Artemisia and Chrysothamnus* (ED McArthur and BL Welch, Comp.), pp. 29-43. USDA Forest Service Gen. Tech. Rep. INT-200, Intermountain Res. Sta., Ogden, UT, 1986.
5. McArthur ED and SE Meyer. In: *Proc. Fourth Utah Shrub Ecology Workshop, The Genus Chrysothamnus* (KL Johnson, Ed.), pp. 9-18. College of Natural Resources, Utah State Univ., Logan, 1987.
6. Hanks DL, ED McArthur, AP Plummer, BC Giunta, and AC Blauer. J. Range Manage. 28:144-148 (1975).
7. McArthur ED, DL Hanks, AP Plummer, and AC Blauer. J. Range Manage. 31:216-223 (1978).
8. McArthur ED, CF Tiernan, and BL Welch. Great Basin Nat. 39:81-87 (1979).
9. Leonard RL, ED McArthur, DJ Weber, and BW Wood. Great Basin Nat. 41:377-388 (1981).
10. Anderson LC. In: *Proc. - Symp. on the Biology of Artemisia and Chrysothamnus* (ED McArthur and BL Welch, Comp.), pp. 98-103. USDA Forest Service Gen Tech. Rep. INT-200, Intermountain Res. Sta., Ogden, UT, 1986.
11. Tueller PE and ED Payne. In: *Proc. Fourth Utah Shrub Ecology Workshop, The Genus Chrysothamnus"* (KL Johnson, Ed.), pp. 1-8. College of Natural Resources, Utah State Univ., Logan, 1987.
12. Roundy BA, JA Young, and RA Evans. Weed Sci. 29:448-454 (1981).
13. Harniss RO and RB Murray. J. Range Manage. 26:322-325 (1973).
14. Daubenmire R. Northwest Sci. 49:36-48 (1975).
15. Young JA and RA Evans. J. Range Manage. 27:127-132 (1974).
16. Young JA and RA Evans. J. Range Manage. 31:283-289 (1978).
17. McKell CM and WW Chilcote. J. Range Manage. 10:228-230 (1957).
18. Monsen SB and R Stevens. In: *Proc. Fourth Utah Shrub Ecology Workshop, The Genus Chrysothamnus* (KL Johnson, Ed.), pp. 41-50. College of Natural Resources, Utah State Univ., Logan, 1987.
19. Stevens R, KR Jorgensen, and JN Davis. Great Basin Nat. 41:274-277 (1981).

20. Stevens R, KR Jorgensen KR, JN Davis, and SB Monsen. In: *Proc. - Symp. on the Biology of Artemisia and Chrysothamnus* (ED McArthur and BL Welch, Comp.), pp. 353-357. USDA Forest Service Gen. Tech. Rep. INT-200, Intermountain Res. Sta., Ogden, UT, 1986.

21. Sabo DG, GV Johnson, WC Martin, and EF Aldon. USDA Forest Service Res. Pap. RM-210, Rocky Mountain Res. Sta., Ft. Collins, CO, 1979.

22. Khan MA, N Sankhla, DJ Weber, and ED McArthur. Great Basin Nat. 47:220-226 (1987).

23. Romo JT and LE Eddleman. J. Range Manage. 41:491-495 (1988).

24. Meyer SE, ED McArthur, and GL Jorgensen. Amer. J. Bot. 76:981-991 (1989).

25. Stevens R. In: *Proc. - Symp. on the Biology of Artemisia and Chrysothamnus* (ED McArthur and BL Welch, Comp.), pp. 278-285. USDA Forest Service Gen. Tech. Rep. INT-200, Intermountain Res. Sta., Ogden, UT, 1986.

26. Young JA, RA Evans, and BL Kay. J. Range Manage. 37:373-377 (1984).

27. Anderson JE and ML Shumar. J. Range Manage. 39:152-156 (1986).

28. Plummer AP, AC Hull, G Stewart, and JH Robertson. USDA Agric. Handbook 71 (1955).

29. Frischknecht NC. J. Range Manage. 16:70-74 (1963).

30. Reynolds TD and L Fraley. Environ. Exp. Bot. 29:241-248 (1989).

31. Miller RF. J. Range Manage. 41:58-62 (1988).

32. Johnson AH and RM Strang. J. Range Manage. 36:616-618 (1983).

33. Robertson JH and HP Cords. J. Range Manage. 10:83-89 (1957).

34. Young JA and RA Evans. Weed Sci. 22:469-475 (1974).

35. Hyder DN, FA Sneva, DO Chilcote, and WR Furtick. Weeds 6:289-297 (1958).

36. Hyder DN, FA Sneva, and VH Freed. Weeds 10:288-295 (1962).

37. Young JA, RA Evans, and PT Tueller. J. Range Manage. 29:342-344 (1976).

38. Tueller PT and RA Evans. Weed Sci. 17:233-235 (1969).

39. Evans RA and JA Young. J. Range Manage. 28:315-318 (1975).

40. Whitson TD and MA Ferrell. In: *1987 Research Progress Report*, pp. 49. Western Society of Weed Science, Boise, ID, 1987.

41. Whitson TD and MA Ferrell. In: *1988 Research Progress Report*, pp. 38. Western Society of Weed Science, Fresno, CA, 1988.

42. Whisenant SG. In: *Proc. - Symp. on the Biology of Artemisia and Chrysothamnus* (ED McArthur and BL Welch, Comp.), pp.115-121. USDA Forest Service Gen. Tech. Rep. INT-200, Intermountain Res. Sta., Ogden, UT, 1986.

43. Whisenant SG. Weed Sci. 35:120-123 (1987).

44. Clary WP, S Goodrich, and BM Smith. J. Range Manage. 38:56-60 (1985).

45. Britton CM and FA Sneva. J. Range Manage. 34:30-32 (1981).

46. Marion JH, EI Hackett, and JW Burkhardt. In: *Proc. - Symp. on the Biology of Artemisia and Chrysothamnus* (ED McArthur and BL Welch, Comp.), pp. 122-125. USDA Forest Service Gen. Tech. Rep. INT-200, Intermountain Res. Sta., Ogden, UT, 1986.

47. Westoby M, B Walker, and I Noy-Meir. J. Range Manage. 42:226-274 (1989).

35

Oakbrush: Classification, Ecology, and Management

G. Allen Rasmussen

Abstract

Oaks (*Quercus* spp.) are widespread with over 400 known species worldwide. They dominate or co-dominate in many plant communities. While they cause many problems for land managers (reduce herbaceous and browse forage production, restrict access, poison cattle, and reduce watershed characteristics), they are also valued by many. Oaks provide aesthetic values, wildlife habitat, fuel wood, watershed protection, and browse for livestock; in dry areas, oaks have improved herbaceous forage production. The control of oakbrush has often focused on single treatments and objectives. However, with the realization of the ecological role of oakbrush and management strategies that integrate both control techniques with long term land use objectives, improved management for the multiple uses of oakbrush communities has resulted.

Introduction

The oaks (*Quercus* spp.) are an important component of many ecosystems and their classification as a noxious weed depends on the perspective and objective of the individual land manager. They range from shrubs to large trees with the size of individuals regulated by environmental conditions and time since the last disturbance. Oakbrush generally refers to the shrub oaks but also includes the tree oaks when they or their resprouts dominate a community. The major characteristics of oaks that classify some of them as weeds are: reducing production of herbaceous plants (1-3) and browse species (4,5); restricting access and movement of livestock and recreationists at high densities; and they are toxic to cattle (6).

Oak can be toxic if it constitutes over 50% of cattle diets (6) and fatal if over 75% (7). All parts of an oak plant (leaves, stems, and acorns) contain sufficient tannins to be toxic. The tannins may reduce rumen efficiency (8) and are broken down into the toxic compounds of gallic acid and pyrogallol (9). On most ranges, oak toxicity to cattle only occurs when there is not sufficient feed from other sources. This generally occurs in the spring as the oaks begin leafing out. Tannin concentrations are also highest in the new growth. But oaks can also cause problems in the fall when cattle utilize the fallen acorns even though other forage is available. In areas where cattle are not consuming sufficient quantities of oak to be toxic (<50%), concern has been expressed that the high tannin levels will reduce the overall productivity of cattle because of reduced rumen efficiency (10). Generally, other livestock (11) as well as wildlife species do not show toxic effects from using oak.

Oakbrush types have been converted to other shrub types because they have lower forage values, particularly during winter months (12-14). These characteristics of oaks often result in managers trying to reduce the influence of oaks in the plant community, but many people look at the oakbrush type as a valuable commodity for aesthetics, fuel wood, wildlife habitat (game as well as nongame species), watershed values, and livestock feed (15,16). In most areas where the evergreen oaks occur, they are considered a valuable feed source during the winter for all classes of livestock including cattle (10), though managers must ensure sufficient feed from other sources is available for cattle. Oaks have even improved herbaceous plant production depending on the site, environmental conditions, and total oak canopy cover (16,17). On areas under a mature Gambel oak (*Quercus gambelii*) canopy on deep soils, herbaceous production was 66% greater (1650 kg/ha compared to 990 kg/ha) under the canopies compared to the interspaces (16). During periods of limited precipitation and at lower canopy covers, herbaceous production has also increased (18). Currently, the lack of oak regeneration in California has caused enough concern that traditional management strategies have been altered and research is being conducted to increase the regeneration potential to ensure the oakbrush type is not lost (19).

Oak Classification

The oak genus (*Quercus*) belongs in the Fagaceae family and Fagaidae subfamily. The *Quercus* genus first appeared in the Cretaceous period, with most modern species developing during the Pliocene and later (20). Currently, there are over 400 recognized species in the genus, but if the genus was reviewed carefully, the total number would probably be reduced. Several have even challenged the species concepts in the *Quercus* genus because of the high degree of variation within a species, frequent hybridization, and edaphic variation (21-23). The modern center of diversity is in southeast Asia (24). However, the primary distribution center of living *Quercus* species on the North American continent is approximately in the state of Arizona (20).

Species in the *Quercus* genus are woody and range in size from shrubs to tall trees. Oaks are moniceous with the staminate (male) flowers being pendulous, naked catkins, and the pistillate (female) flowers having six ovules (five abortive) enclosed in an involucre of flat scales. The fruit is a nut partially surrounded at the base by an involucre of flat to thickened scales which form a cup commonly referred to as an acorn (20). The genus is characterized by leaves that are simple, petioled, entire, lobed, toothed, alternate, and deciduous or evergreen. Leaf structure is highly plastic within individual trees and species varying with season, age, location on the plant, and environmental conditions (25). This makes leaf structure a tenuous taxonomic characteristic. The nut or acorn is the most reliable taxonomic characteristic but to compound the problem of identifying oaks, many species form fertile hybrids with each other (20). For example, the three naturally occurring *Quercus* species in Utah (the evergreen *Q. turbinella* and the two deciduous species *Q. harvardii* and *Q. gambelii*) form five recognized hybrids between them (26).

Quercus is divided into three subgenera; the white oaks (subgenus *Quercus* synonym *Lepidobalanus*), intermediate oaks (subgenus *Protobalanus*), and black oak or red oaks (subgenus *Erythrobalanus*). White oaks are found worldwide, while the black or red oaks are found only in the New World and intermediate oaks are confined to North America (21). In the United States, there are 54 species of white oaks, 26 black oaks, and 4 intermediate oaks (20). Tucker (21) characterized the differences between the subgenera by acorn and vegetative characteristics. The inner shell of the acorn on white oaks is glabrous, and tomentose on black and intermediate oaks. The acorns also mature within one year on white oaks and require two years for the black and intermediate oaks in most cases. The acorn cup scales on black oaks are thin and flat while they are thickened on the intermediate oaks. Vegetative characteristics are highly variable between the subgenera, though the bark on white oaks is usually light gray or brown with leaf lobes typically rounded and without bristle tips. The black oaks tend to have dark brown or blackish bark and leaf lobes that are generally pointed and bristle-tipped. White and intermediate oaks have greater growth rates and softer wood than the black oaks (20). Hybridization in the *Quercus* genus generally only occurs within species of the same subgenus (21). Recently, it has been proposed that the subgenera be further divided (27).

Ecology

Oaks are distributed over a wide variety of soil textures (sands to clays) and pH levels (acidic to calcareous) (28,29) as well as climatic zones. However, the majority of species are restricted to rather specific geographical areas (30). Shrub oaks are generally restricted to areas with high environmental stresses compared to the tree oaks, and the evergreen oaks tend to occupy relatively drier environments compared to deciduous oaks (31).

The anemophilous nature of the genus results in most of the oaks flowering early in the spring (32), making the catkins susceptible to freezing (33), and may limit the sexual reproductive success of the genus in colder environments. Seed dissemination of the genus is largely by animals and birds which are scatter hoarders (34). These animals will gather the seeds and hide them in various locations so they can be used later. But even these scattered seeds do not free them from predation because numerous animals continue to seek them out. The acorns are subjected to removal both before and after they drop (35-37). Predation losses have been in excess of 99% (36). Oak acorns have a high nutritional value and can be separated into subgenera by their lipid and tannin contents. The black oak (subgenera) acorns have lipid contents ranging from 18 to 25%, with tannin contents from 8-12%. White oak acorns have lower lipid (<10%) and tannin (<2.5%) contents (38). High tannin contents generally reduce the palatability of forages but the high lipid content seems to temper the tannin effects to the point that squirrels will select and spend more time foraging for black oak acorns (38).

Oak seedlings are also susceptible to drought periods during the summer and frost damage in the winter. They are susceptible to grazing by rodents, other wildlife and livestock (35,39,40). The greatest negative impact on oak seedling establishment by livestock tends to be soil compaction (35). However, the elimination of competition and the reduction of fine fuels has reduced fires, allowing many species to increase (41-43). Griffin (44) speculated that seedling recruitment of many oak species required the following special conditions to establish:

a good acorn production year, good seedling establishment conditions the following year, and low acorn and seedling predation conditions during this same period. The oaks are capable of some type of vegetative reproduction from rhizomes, lignotubers, dormant buds, or callus tissue (45,46). The lignotubers in Gambel oak are responsible for survival of the clone after a disturbance while the rhizomes are accountable for the expansion of the clone. Those oaks that are not rhizomatous generally can resprout readily after disturbances from dormant buds on the rootcrown and callus tissue. This ability for vegetative reproduction allows many of the oaks to maintain themselves in the stand despite their limited sexual reproduction. In most stands of the oakbrush type, oak is considered to be a higher seral stage than the herbaceous vegetation often managed for; many of these communities were originally savannahs where the density of oak has increased since disturbance types and frequencies have been altered.

Many oaks are often described as oakbrush and occur on a wide range of soils. Most of the oaks considered to be a problem tend to be found on shallow light textured soils. The most common are listed in Table 35.1.

TABLE 35.1 Oak species commonly referred to as oakbrush, their vegetative type of reproduction, and growth and leaf form.

Common Name	Scientific Name	Vegetative Regrowth[a]	Growth Form	Leaf[b] Persistence
Post Oak	Q. stellata	Resp	tree	d
Blackjack oak	Q. merlandica	Resp	tree	d
Shrub liveoak	Q. virginicus	Rhiz	tree/shrub	e
Sand shinnery oak	Q. harvardii	Rhiz	shrub	d
Gambels oak	Q. gambelii	Rhiz	shrub/tree	d
Scrub liveoak	Q. turbinella	Rhiz	shrub	e
California oaks				
Blue oak	Q. douglasii	Resp	tree	d
Coastal liveoak	Q. agrifolia	Rhiz	tree/shrub	e
Inter. Liveoak	Q. wislizenii	Rhiz	shrub	e
Scrub oak	Q. dumosa	Rhiz	shrub	e

[a] Resp=resprouts, Rhiz=rhizomes
[b] d=deciduous, e=evergreen

Post Oak, Blackjack Oak

Post oak and blackjack oak commonly occur together and have a wide distribution in the eastern portion of the United States (47). They reach their greatest densities on the Cross Timbers ecological region which runs through central Texas, eastern Oklahoma, and Kansas. The soil texture where they dominate tends to be shallow sandy loam to loam. As the clay content increases, these oak species decrease. Precipitation ranges from 46 to 115 cm. Of the two species, post oak is more competitive on dry sites although blackjack oak is more competitive on nutrient poor sites (43). These oaks originally occupied the area as savannah but with present disturbance levels of grazing and fire, their densities along with the associated understory have increased or "thicketized" to develop a closed forest in many areas. As these trees age, their resprouting capabilities tend to decrease. Blackjack oak tends to have a greater number of sprouts compared to post oak after top removal or disturbance (48).

Shrub Liveoak

Liveoak is found in the more southern portion of the United States along the coasts of the Atlantic and Gulf of Mexico (47). It is found on deep sands where it is considered a problem, though a subspecies (*Q. virginiana fusiformis*) occurs in central Texas on limestone-derived shallow clay loam soils. This evergreen tree often forms isolated mottes which are desirable shade and cover for wildlife and livestock and it is used as an ornamental. In other areas, it is valued as browse for sheep, goats, and wildlife (49), though it must be kept low enough for these animals to reach it. Shrub liveoak at high densities reduces access and herbaceous forage production (50).

Sand Shinnery Oak

It occurs on sandy soils primarily in New Mexico, western Texas, and Oklahoma (47). This rhizomatous shrub can reach 3 m in size on good soils, though normally it is about 1 m. Precipitation ranges from 30 to 50 cm. Sand shinnery oak generally occurs in monotypic stands that dramatically reduce forage production and has resulted in numerous cattle deaths in the spring when it is the only green forage available in many pastures. It is helpful in stabilizing sand dunes and provides habitat for upland game birds. It is found in limited areas in Arizona and Utah, but only in isolated areas on sands.

Gambel Oak

Gambel oak occurs in the Intermountain West in Utah, Colorado, New Mexico, and Arizona (47). Soil textures where Gambel oak dominates generally are loam to silt loams in a precipitation zone that receives 38 to 56 cm. Stem numbers increase rapidly after disturbances such as fire, though the number of stems, stand structure, and herbaceous species composition will return to preburned levels in approximately 18-20 yr (39). The dry summer growing season is often cited as the reason ponderosa pine (*Pinus ponderosa*) does not replace much of the gambel oak. Many areas now are being managed for multiple uses of wildlife, livestock, recreation, and watershed protection.

Scrub Liveoak

It has established in Arizona, southern Utah, southern Nevada, and southeastern California. It occurs on sandy loam to loam soils in a 20 to 38 cm precipitation zone. It is used extensively for winter livestock feed and is valued as wildlife habitat (10). It generally occurs on shallow rocky slopes. Problems are generally encountered when it dominates an area, reducing production of herbaceous plants and timber, access, and water yield from the watershed (51). Normally, it is found in association with other chaparral species (52).

California Oakbrush

The California oakbrush consist of several species. The most common are blue oak (*Q. douglasii*), coastal liveoak (*Q. agrifolia*), Englemann oak (*Q. englemannii*), interior liveoak (*Q. wislizeni*), and scrub oak (*Q. dumosa*). The distribution of these oaks are restricted to California for the most part (47). Blue oak and interior liveoak are codominants in the foothills and northern areas, and coastal liveoak and Englemann oak dominate in the south. Scrub oak is found in the chaparral communities. The oak woodlands are important for livestock production, recreation, fuel wood, and wildlife habitat. These numerous uses and the regeneration problems associated with blue oak and Englemann oak have created concern that the oak type may be lost (15). Research has shown that during dry periods, herbaceous production has been improved under the canopies of several oak species (53).

Management

Oaks have been "controlled" by a wide range of methods, including mechanical, herbicides, fire, and biological. These methods have met with varying degrees of success and most only have short-term effects. The method or technique selected depends on the growth form of the plant and the type of site the plants are on. Good reviews of the different techniques are found in Scifres (54) and Vallentine (55).

Use of most management treatments have tried to take advantage of the growth patterns by treating oak trees and shrubs when total nonstructural carbohydrates are low. However, many treatments effective on the mature oaks have not been as effective on the resprouts. This may in part be explained by the growth cycles of the mature plants and their resprouts. Growth of mature oak species typically exhibits an intermittent growth pattern where growth occurs in a single flush and is fairly predictable every year (56). The resprouts on at least two species exhibit a continuous growth pattern throughout the growing season (57). This complicates the timing of a treatment where there is no conspicuous low or translocation direction of total nonstructural carbohydrates.

The mechanical methods used on oaks include chaining, dozing, roller chopping, fuel wood cutting, root plowing, and shredding (54,55). Chaining, dozing, and fuel wood cutting are generally effective only on large shrubs or trees that are not rhizomatous, while shredding can only be used on smaller shrubs. All of the mechanical treatments need relatively gentle terrain. Mechanical treatments are not commonly used because the vegetative reproduction of the oak species allows them to come back quickly; there is considerable physical ground disturbance; and they are expensive.

Fire has been credited in helping to create and maintain many of North America's original oak savannah. In some areas, it is now being used to convert oak woodlands back to a savannah (43). Following single burn events, however, stem density and germination often increase (29,39,43,58). Multiple burns have been effective in reducing oak stem densities and improving herbaceous production, but only the smaller stems are removed. White (58), studying 13 years of annual burns, found that stems greater than 25 cm diameter at breast height were not removed. If the burn frequency was altered to allow the fine fuel to build up for four or five years, fire intensities might be high enough to topkill these large stems, though fire control might be more of a problem. In addition, prescribed burns would have to be continued to suppress the resprouts. In general, prescribed fire is effective in maintaining the density of shrubs in the community since it can remove seedlings. With the vegetative reproduction found in oaks, fire will rarely remove individuals from the community. Cost of using prescribed fire can be relatively cheap, although exact cost depends on factors such as deferment time, crew size, size of the burn, topography, and fine fuel loads (59).

Biological treatment of unwanted species often centers on the use of imported insects or increasing the number of native insects that naturally occur. Oaks are native and desired by many people, so the chances of finding an insect that would effectively control oaks and remain on a restricted site is unlikely. The use of livestock (goats) as biological control agents has been effective in reducing oak, improving browse, and herbaceous composition and production on an experimental basis (60,61). Large scale use of goats has not been practical in many areas because of herding problems and (more importantly) the lack of a market in areas where the goats could be used to manage the oak.

A wide variety of herbicides has been tested on oaks with varying degrees of success both from a biological and economic basis. The most commonly used herbicides currently are picloram (4-amino-3,5,6-trichloropicolinic acid) and tebuthiuron {N-[5-(1,1-dimethylethyl)-1,3,4-thiadiazol-2-yl]-N,N'-dimethylurea}. Herbicides and application rates depend on the species, soil, and precipitation (52,62-65). Forage has increased four- to nine-fold in treated areas compared to untreated, but the longevity of the treatments still tends to be short term. In some cases, other brush species increase to replace the oaks. Eastern redcedar (*Juniperus virginiana*) was found to replace post oak and blackjack oak after herbicide treatment with tebuthiuron or triclopyr (64). Foliar applied herbicides are generally more effective when leaves have fully developed or just prior to full leaf expansion, which allows them to take advantage of the total nonstructural carbohydrate translocation patterns (65). Costs and environmental concerns about the use of herbicides have reduced their use in recent years.

Most control treatments have only had short-term results because of oaks' vegetative reproduction capabilities. Furthermore, oaks tend to occupy a higher successional status in the community than managers are trying to maintain. This and the need to maintain oak in specific quantities for other purposes have required managers to move away from single treatment approaches to an integrated approach. The integrated approach generally incorporates several techniques which allow a manager to take advantage of each technique's strengths and to minimize their weaknesses. This also requires the manager to include the costs involved and maintain the long term integrity of the resource (66). In one such approach, herbicides were used to initially reduce the composition of post oak and blackjack oak in the community. Prescribed burning was then used to improve accessibility and forage quality and to reduce other shrubs that were increasing on the area (67). This approach worked well, although forage production began to decline after several years of annual burns. To stabilize the forage production level, the fire frequency could be changed to three to five years or more, depending on plant community structure.

The use of an integrated approach on oakbrush has great potential as long as the manager has a good knowledge of how each management technique works on a particular species of oak and the associated community and thoroughly understands the objectives on the area. Integrated plans are long-term (set up for several years) but they must be dynamic to allow the manager to change details of the plan, depending on the circumstances that develop. The manager can then achieve long-term, workable, economical solutions which will meet his objectives and maintain the resource.

References

1. Kufeld RC. Colorado Div. Wildl. Tech. Pub. 34 (1983).
2. Parker VT and CH Muller. Amer. Midl. Natur. 107:69 (1981).
3. Sears WE, CM Britton, DB Wester, and RD Pettit. J. Range Manage. 39:403 (1986).
4. Kufeld RC. J. Range Manage. 28:216 (1977).
5. Riggs RA and PJ Urness. J. Range Manage. 42:354 (1989).
6. Dollahite JW, GT Housholder, and BJ Camp. J. Amer. Vet. Med. Assoc. 148:908 (1966).
7. U.S. Department of Agriculture. ARS Agric. Inf. Bull. 327 (1968).
8. Provenza FD and JC Malecheck. J. App. Ecol. 21:831 (1984).
9. Sandusky GE, CJ Fosnaugh, JB Smith, and R Mohan. J. Amer. Vet. Assoc. 171:627 (1977).
10. Ruyle GB, RL Grumbles, MJ Murphy, and RC Cline. Rangelands 8:124 (1988).
11. Villena F and JA Pfister. J. Range Mange. 43:116 (1990).
12. Riggs RA, PJ Urness, and KA Gonzalez. J. Range Manage. 43:229 (1990).
13. Bidwell T. Personal communication, Oklahoma State Univ., 1990.
14. Welch BL, SB Monsen, and NL Shaw. In: *Proc. Res. and Manage. of Bitterbrush and Cliffrose in Western North America* (AR Tiedemann and KL Johnson, Comp.), pp. 173-175. USDA Forest Serv. Gen. Tech. Rep. INT-152 (1982).
15. Huntsinger L and LP Fortmann. J. Range. Manage. 43:147 (1990).
16. Bowns J. In: *Proc. Third Utah Shrub Ecology Workshop* (KL Johnson, Ed.), pp. 29-32. Utah State Univ., Logan, 1985.
17. Holland VL. In: *Ecology, Management and Utilization of California Oaks* (TR Plumb, Tech. Coord.), pp. 314-318. USFS Gen. Tech. Rep PSW-44 (1980).
18. Duncan DA and JN Reppert. USFS Misc. Pap. PSW-46 (1960).

19. Bartolome JW, MC Stroud, and HF Heady. J. Range Manage. 33:4 (1989).
20. Trelease W. *The American Oaks*. Memoirs National Academy Sci. 20 (1922).
21. Tucker JM. In: *Ecology, Management and Utilization of California Oaks* (TR Plumb, Tech. Coord.), pp. 19-29. USFS Gen. Tech. Rep PSW-44 (1980).
22. Burger WC. Taxon 24:45 (1975).
23. Van Valen. Taxon 25:233 (1976).
24. Kaul RB. Amer. J. Bot. 72:1962 (1985).
25. Blue MP and RJ Jensen. Amer. J. Bot. 75:939 (1988).
26. Welsh SL, ND Atwood, LC Higgins, and S Goodrich. *A Utah Flora*. Great Basin Naturalist Memoirs No. 9, Brigham Young Univ., Provo, 1987.
27. Nixon KC and WL Crepet. Amer. J. Bot. 72:934 (1985).
28. Muth GJ. In: *Ecology, Management and Utilization of California Oaks* (TR Plumb, Tech. Coord.), pp. 19-29. USFS Gen. Tech. Rep PSW-44 (1980).
29. Rasmussen GA and HA Wright. J. Range Manage. 42:295 (1989).
30. Axelrod DI. Ann. Mo. Bot. Gard. 70:629 (1983).
31. Rundel PW. In: *Ecology, Management and Utilization of California Oaks* (TR Plumb, Tech. Coord.), pp. 43-54. USFS Gen. Tech. Rep PSW-44 (1980).
32. Kaul RB, EOC Abbe, and LB Abbe. Biotropica 18: (1986).
33. Neilson R and LH Wullstein. J. Amer. Bot. 67:426 (1980).
34. Stapanian MA and CC Smith. Ecology 65:1387 (1984).
35. Borchert MI, FW Davis, J Michaelsen, and LD Oyler. Ecology 70:389 (1989).
36. Sork VL. Ecology 65:1020 (1984).
37. Sork VL, P Stacey, and JE Averett. Oeclogia 59:49 (1983).
38. Smallwood PD and WD Peters. Ecology 67:168 (1986).
39. Harper KT, FJ Wagstaff, and LM Kunzler. USDA For. Ser. Gen. Tech. Rep. INT-179 (1985).
40. Griffin JR. Ecology 52:862 (1971).
41. Curtis JT. *The Vegetation of Wisconsin*. Univ. Wisconsin Press, Madison, 1959.
42. Madany MH and NE West. Ecology 64:661 (1983).
43. Penfound WT. Ecology 49:1003 (1968).
44. Griffin JR. In: *Terrestrial Vegetation of California* (MR Barbour and J Major, Eds.), pp. 383-416. John Wiley and Sons, New York, 1977.
45. Muller CH. Madrono 11:129 (1951).
46. Tiedemann AR, WP Clary, and RJ Barbour. Amer. J. Bot. 74:1065 (1987).
47. Little EL Jr. *Atlas of United States Trees, Vol. 1*. USDA For. Ser. Mis. Pub. No. 1146 (1971).
48. Scifres CJ and RH Haas. Texas Agric. Exp. Sta. MP-1136 (1974).
49. Bryant FC, MM Kothmann, and LB Merrill. J. Range Manage. 33:410 (1980).
50. Meyer RE and RW Bovey. Weed Sci. 28:51 (1980).
51. Brown TC, PF O'Connell, and AR Hibbert. USDA For. Ser. Res. Pap. RM-127 (1974).
52. Johnsen TN Jr. Weed Sci. 35:810 (1988).
53. Bartolome JW. Rangelands 7:122 (1987).
54. Scifres CJ. *Brush Management: Principles and Practices for Texas and the Southwest*. Texas A&M Univ. Press, College Station, 1980.

55. Vallentine JF. *Range Development and Improvements*. Brigham Young Univ. Press, Provo, 1971.
56. Zimmermann MH and CL Brown. *Trees - Structure and Function*. Springer-Verlag, New York, 1971.
57. Engle DM, CD Bonham, and LE Bartel. J. Range Manage. 36:363 (1983).
58. White AS. Ecology 64:1081 (1983).
59. Rasmussen GA, GR McPherson, and HA Wright. J. Range Manage. 41:413 (1988).
60. Davis GG, LE Bartel, and CW Cook. J. Range Manage. 28:216 (1975).
61. Riggs RA and PJ Urness. J. Range Mange. 42:354 (1989).
62. Scifres CJ, JW Stuth, and RW Bovey. Weed Sci. 29:270 (1981).
63. Meyer RE, RW Bovey, LF Bouse, and JB Carlton. Weed Sci. 31:639 (1983).
64. Engle DM, JF Stritzke, and FT McCollum. Okla. Agric. Exp. Sta. MP-119 (1987).
65. Engle DM and CD Bonham. J. Range Mange. 33:390 (1980).
66. Scifres CJ. J. Range Mange. 40:482 (1987).
67. Bernardo DJ and DM Engle. J. Range Mange. 43:242 (1990).

36

Mesquite: Classification, Distribution, Ecology, and Control

Pete W. Jacoby, Jr., and R. James Ansley

Abstract

Mesquites are woody legumes which inhabit arid and semiarid regions throughout the southwestern U.S.A., Mexico, and South America. The two principle species found in the United States, honey mesquite (*Prosopis glandulosa*) and velvet mesquite (*Prosopis velutina*), occur in a variety of growth forms ranging from decumbent shrubs to medium-sized trees. Increases in mesquite density since the late nineteenth century have been attributed to man's influence, either from suppression of natural fires or dissemination of seed by domestic livestock. Mesquite is viewed as a noxious plant because dense stands impede livestock operations. However, mesquite in lower densities (i.e., as a savanna) has been increasingly viewed as a resource which must be managed. Management for a mesquite savanna rather than total eradication may necessitate initial stand reduction using herbicides or mechanical methods followed by maintenance of the stand with prescribed burning. Control methods employed to reduce dense stands should emphasize achievement of root mortality rather than only a suppression of aerial growth because of mesquite's potential for coppicing.

Classification and Distribution

Mesquite (genus *Prosopis*) is a thorny shrub or tree of the legume family which is found in the southwestern U.S.A. Forty-four species occur naturally in arid and semiarid areas of North and South America, northern Africa, and eastern Asia. However, introductions have been made into other parts of the world. Forty of the above species are New World species and 31 of these are indigenous to South America. Morphological diversity and pattern of flavonoid chemistry suggest South America as the area of origin for mesquite (1).

The two species of concern in the U.S. are honey mesquite and velvet mesquite. Honey mesquite is found mainly in Texas, New Mexico, and northern Mexico with plants occurring in California, Oklahoma, Kansas and Louisiana (Figure 36.1). Velvet mesquite occurs mainly in southern Arizona, but is found in California and northern Mexico.

Leaves on honey mesquite are alternate, deciduous, long-petioled, bipinnately compound of 2 (occasionally 3 or 4) pairs of pinnae; with 12-20 leaflets. Leaflets are glabrous, linear, acute or obtuse at apex, 3-5 cm long, and 0.5-1 cm wide. Velvet mesquite leaves are bipinnately compound with mostly four pairs of pinnae. Leaflets are pubescent and smaller than those on honey mesquite. Because of ecological similarities, further discussion will encompass both species unless otherwise indicated.

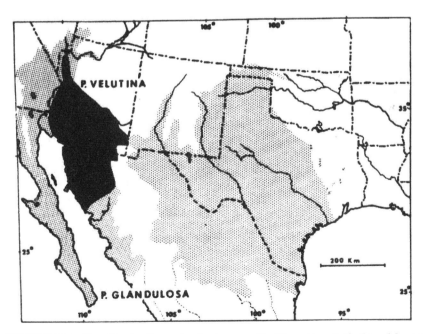

FIGURE 36.1 The distribution of velvet mesquite (*Prosopis velutina*) and honey mesquite (*Prosopis glandulosa*) in North America [from Simpson and Solbrig (2)].

Ecology

Establishment of mesquite is by seed which are borne in pods. Survival of seed varies with site. Seed deposited in the soil on an Arizona site were able to survive and remain viable for up to 10 years, especially when seed were within the pod (3). Longevity of seed in areas with higher rainfall is thought to be much shorter than that of southern Arizona. We have found that a very low percentage of seed survive more than three months in north central Texas. Most are destroyed by insects or fungi or are consumed by rodents (4).

Unlike many legumes, mesquite pods do not split open at maturity (1). This feature may contribute to the distribution of seed away from the parent tree in that foraging animals may ingest seed in the process of consuming the mesocarp which is high in sugars (5). In feeding trials, 91%, 79%, and 16% of mesquite seeds passed undamaged through the digestive tracts of horses, cows, and sheep, respectively (6). Germination is enhanced when seed are scarified by passage through animal digestive tracts. Seed deposited directly to the soil in this manner have an immediate source of nutrients in the dung which enhances seedling survival (7-9). Many species of wildlife consume and/or distribute seed in a similar fashion—including coyotes, peccary, rodents, and lagomorphs. Seed recovered in fossil dung indicate that mesquite was consumed by the Shasta ground sloth which became extinct at the end of the Pleistocene (29).

Major seedling establishment usually occurs following drought when competing plant cover and vigor are reduced or in areas where continuous overgrazing places similar pressures on competing plants (10-13). Mesquite is a heliophyte which germinates in response to openings in canopy cover (14). Standing herbaceous biomass has been found to inhibit germination of honey mesquite (15).

Survival of mesquite seedlings has been observed to occur principally during early spring and late fall when soil moisture is favorable. Seedlings are established by about 10 days after germination (9). A substantial proportion of the carbohydrate in the embryo is devoted to establishing the radical (5). If the seed is buried more than 3 cm, a weak root system may develop at the expense of developing the hypocotyl. Emerging seedlings are susceptible to grazing and are killed if clipped below the cotyledons (16).

Seedlings develop as single or few-stemmed plants unless the top is removed. Plants are able to coppice or resprout from a series of dormant buds located near the zone of stem/root differentiation. This bud zone is termed the lignotuber in plants having similar sprouting characteristics and seems appropriate for this species.

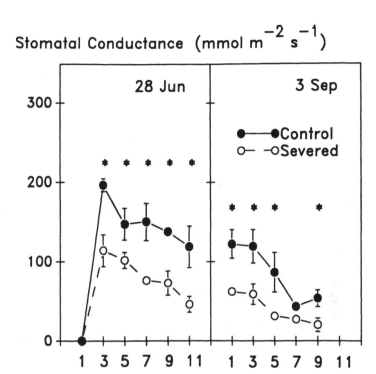

Hours Post Sunrise

FIGURE 36.2   Diurnal stomatal conductance of lateral root-severed honey mesquite and unsevered control trees. Vertical bars indicate ±1 standard error. Asterisks indicate significant difference between treatments ($P \leq 0.05$) [from Ansley et al. (21)].

Established mesquite trees can develop both lateral and tap roots. Mesquite has a reputation as a deep-rooted phreatophyte which avoids drought (5,17). This characterization is based primarily on research in the Sonoran desert of California, an area of 70 mm annual precipitation, but which has unlimited water occurring at about 5 m depth (18,19). In higher rainfall areas, mesquite rely heavily on shallow lateral roots which extend as much as 15 m from the plant (20). In a study in north Texas, severing the lateral roots of large mesquite trees significantly reduced transpiration and stomatal conductance when compared to unsevered trees (Figure 36.2) (21).

The annual growth cycle of mesquite begins during April and May with a period of leaf emergence and twig elongation. This process is completed within six weeks, followed by a period of radial stem growth. Aerial vegetative growth subsides by mid-June with the onset of summer drought but new cohorts of leaves may be produced later in the growing season if moisture is abundant (22). Less is known about dynamics of root growth. However, downward net flow of carbohydrates following leaf and stem elongation suggests that root growth may follow these processes (23,24).

Flowering begins shortly after leaf development. Mesquite have been reported to develop flowers up to four times during a growing season (5). During flowering, trees are covered with thousands of blooms. However, fewer than 3% of flowers initiate fruit development and only one-half to one-third of these actually produce fruits.

Although the geographical distribution of mesquite has remained stable, densities within stands have increased since the late nineteenth century. This trend has been attributed to man's influence, either through suppression of natural fires or dissemination of mesquite seed by the herding and migration of domestic livestock (7,8,25,26). Movement of cattle containing seed in their rumen may also have introduced considerable variation into the mesquite stands which dominate much of the southwestern rangelands today.

Control of Mesquite

There are four recognized methods for the control of mesquite: biological, mechanical, chemical, and fire. Biological agents and fire are credited as factors which limited densities of mesquite on rangelands before development of the ranching industry and settlement of the southwest. Mechanical and chemical methods have been employed by man during a period with plentiful petroleum supplies. There is some movement toward the use of fire as a method of mesquite suppression as petroleum-based strategies escalate in cost.

Biological agents have not been employed successfully for the control of mesquite. Insects, rodents, and lagomorphs exert a biological influence on mesquite; however, the closest resemblance to control occurs in form of consumption of flowering and fruiting parts, and to a lesser extent, the destruction of young seedlings. Foliage may be reduced by leaf-consuming grasshoppers, walkingsticks, beetles, and moth larvae (27). Defoliation by insects is short-lived and poses more of a problem to herbicidal control efforts using foliar-applied herbicides than an advantage as a biological control method.

Small twigs of mesquite may be pruned by the twig girdler (*Oncideres cingulata* or *Oncideres rhodosticta*) (28,29). Girdling of small twigs may temporarily reduce canopy size but ultimately has the effect of pruning and results in increased foliage density (30). Moisture-stressed mesquite appear to be more susceptible to twig girdlers than less-stressed plants (31).

Mesquite can be killed if the root is severed immediately below the bud zone (6). The root plow has been an effective tool to accomplish mesquite control by this method. A horizontally supported blade is pulled behind a large crawler tractor at a depth of 30-36 cm, effectively destroying mesquite plants in its path. Root plowing is expensive and usually destroys the grass cover. The roughened soil surface may inhibit access for animals until settling occurs. Follow-up measures such as smooth chaining, reseeding, and weed control with herbicides may be required (32). The advantages of rootplowing are a high rate of efficacy and a treatment life which may last 20 yr or more. Rootplowing is usually recommended only on deep soils having a high potential for subsequent forage production.

Chaining is another widely used mechanical technique for controlling mesquite. By this method, a large anchor chain is pulled between two crawler tractors and the trees are pulled down. A two-way operation is usually performed to increase the number of trees uprooted. Soil moisture is critical to the success of this operation, as well as the forementioned rootplowing, and soil must be moist below the surface and fairly dry on the immediate surface. Chaining when conditions are dry will only break the plants off at the base and promote coppicing. Chaining when the surface is too moist will damage the grass component.

Chaining was most widely used following the spraying of herbicides on virgin stands of mesquite to fell standing dead stems. Multi-stemmed regrowth from surviving and new plants require many years for stems to achieve sufficient size to accommodate successful chaining (33). Small plants with supple stems will not provide enough physical resistance to cause uprooting. Noxious plants such as pricklypear and other cacti may be spread by chaining, rootplowing, and other forms of mechanical treatment.

Grubbing of individual plants is a viable option when density is not great. Grubbing blades attached to front-end loaders or small crawler tractors is made even more efficient with a hydraulic attachment on the blade to improve ease of removal (33). Grubbers are limited by tree size and best employed to maintain mesquite densities at desired levels.

Herbicides have been a widely used form of mesquite control since the early-1950s (6). Until its removal from use in the mid-1980s, 2,4,5-T [(2,4,5-trichlorophenoxy)acetic acid] was the product of choice, and most current recommendations made for mesquite control with herbicides are based on research with this product. While this chemical provided excellent suppression of top growth, few trees were actually killed in a typical 2,4,5-T application. Surviving trees resprouted as multi-stemmed shrubs which required periodic treatment every five to seven years for suppression.

Other herbicides used for mesquite control include clopyralid (3,6-dichloro-2-pyridinecarboxylic acid), triclopyr [[(3,5,6-trichloro-2-pyridinyl)oxy]acetic acid], picloram (4-amino-3,5,6-trichloro-2-pyridinecarboxylic acid), dicamba (3,6-dichloro-o-anisic acid), hexazinone [3-cyclohexyl-6-(di-methylamino)-1-methyl-1,3,5-trizaine-2,4(1H,3H)-dione], and tebuthiuron [N-[5-(1,1-dimethylethyl)-1,3,4-thiadiazol-2-yl]-N,N'-dimethylurea]. The latter two products are soil-applied, root-absorbed products with limited use on rangelands. Picloram and dicamba are normally applied in a mixture with another herbicide. Triclopyr yields efficacies similar to 2,4,5-T and mainly produces suppression through destruction of top-growth (34).

Clopyralid (sold in the rangeland market as Reclaim) is the most effective herbicide for mesquite control and has the potential for high kill rates when properly applied (35). Clopyralid is not only significantly more effective than other herbicides, but is equally effective during the entire growing season (36). Clopyralid also produces a predictable dosage response which provides opportunities for a landowner to determine anticipated levels of efficacy at given rates.

Resistance of mesquite to herbicides may occur through several modes of action. Because most herbicide applications are made aerially with foliarly absorbed herbicides, condition of the leaf surface may strongly affect herbicide efficacy. Accumulation of epicuticular wax during the growing season may create a barrier to absorption and penetration of herbicides. For this reason, oils or other adjuvants may be required to achieve efficacy, especially later in the growing season.

In our research of wax accumulation in mesquite, we found evidence of considerable variation within a stand of mesquite trees (37). Some trees accumulated lighter or heavier wax loads than the population mean and were observed to maintain this ability from year to year (Figure 36.3). This intraspecific variation may cause a variability of response to herbicides. Intraspecific variation can be considered a source of resistance to herbicides when timing decisions are made on the assumption that all mesquite are in a similar physiological status at the time of herbicide application.

Other forms of resistance to herbicides are imposed by the growth form of mesquite. We found in examination of thousands of treated mesquite that number of stems was positively correlated with increasing resistance to foliarly applied herbicides (Figure 36.4) (38). Improper coverage can contribute to incomplete kill of mesquite plants and this may be a partial explanation for increased herbicide resistance in multi-stemmed plants because of their greater foliage density than few-stemmed plants. Plant height was not found to influence whether a plant was killed or not (38,39).

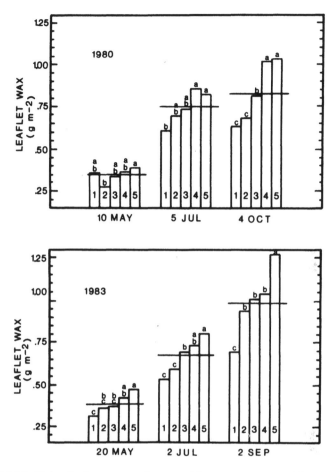

FIGURE 36.3 Epicuticular wax on leaves of honey mesquite. Numbers within bars refer to individual trees. Similar letters above bars indicate no significant differences between means within a date. Horizontal lines indicate the mean of five trees within a date [from Jacoby et al. (37)].

Herbicide effectiveness is dependent on translocation of the chemical to the mesquite root system. Because herbicides are transported in the phloem with carbohydrates (40), it is desirable for net flow of carbohydrate translocation to be downward to the roots at the time of herbicide application.

Mesquite density in pristine times was probably held in check to some degree by the action of fire (7,25). Fire is known to have occurred periodically on rangelands prior to settlement, although the frequency is not well established (41). Fires are presumed to have been ignited naturally by lightning and, because such storms occur principally during

372

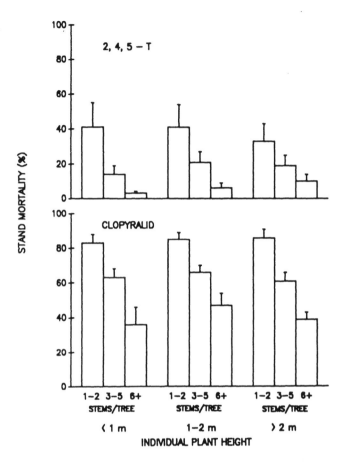

FIGURE 36.4 Influence of stem number and plant height on resistance of honey mesquite when treated with 0.56 kg/ha of 2,4,5-T or clopyralid [from Jacoby et al. (38)].

warmer months, fires probably occurred mainly during summer months when grasses were dry and highly combustible. Seedlings and juveniles have been observed to be susceptible to fire before the age of 2-3 yr (25,42). After this time, mesquite becomes increasingly fire tolerant.

Research employing fire as a control method for mesquite has been conducted mainly during the winter months when fire can be more easily controlled and intensities are low enough to minimize damage to forage grasses (41). Fires during winter generally do not achieve intensities sufficient to cause top kill on larger trees, although smaller trees may be suppressed. Conversely, fires during summer months have been reported to cause light to moderate mortality of established mesquite (43). The need for research on summer burning is recognized but few studies have been reported.

Studies have shown that mesquite having dead stem tissue can be ignited, and this can cause plant mortality (44). Dead stem material can be created by using herbicides several years prior to burning. Combination of techniques such as burning with herbicides or mechanical treatment may produce advantages over single treatment approaches.

Management of Mesquite

Mesquites are classified as pests by most landowners owing to their thorny branches, their increased density on rangeland, and their perceived competition with forage grasses. Many studies have shown that forage production increases following control of mesquite (45-50). However, response is highly variable and dependent on many factors such as density of mesquite prior to treatment, effectiveness of treatment, and climate (50).

Although mesquite is widely regarded as a noxious plant because of its interference with livestock production, mesquite has an emerging image as a *resource* which should be managed. Unfortunately, decades of control attempts have destroyed many mature stands of mesquite which contained single to few-stemmed trees. These trees were desirable in that they occupied far less surface area than multi-stemmed growth forms which resulted from destruction of aerial tissue and subsequent resprouting. While the visibility factor for livestock gathering was enhanced, one can only speculate as to the benefits which were lost through this action. Livestock shade is given consideration in many feedlot operations, but range livestock are not afforded this benefit when the trees are removed.

The decision on whether mesquite will continue to be viewed as a noxious plant or managed as a resource will be made on both economic realities and environmental consciousness. Complete elimination of mesquite has been a goal that few landowners have achieved, and the concept of complete removal is questionable both economically and environmentally (25). Mesquite has many benefits to the ecosystem when maintained at controlled densities (i.e., as a savanna). Such benefits include nitrogen fixation, livestock shade, habitat for nesting birds, and the potential as firewood or wood products (25). Mesquite has the potential to produce commercial hardwood in some regions with higher rainfall. In lower rainfall areas, shrubby growth forms of mesquite may have other benefits, such as wildlife habitat. A mesquite savanna offers a pleasant landscape and may improve the value of a property over either an unmanaged woodland or a treeless grassland (51,52).

Some landowners are beginning to select a level of mesquite density desired for their properties. Recent research suggests that mesquite savannas can be sustained as long as the herbaceous understory is maintained at sufficient densities to outcompete mesquite seedlings (7,10,15). Management for a mesquite savanna rather than total eradication may necessitate initial stand reduction using herbicides or mechanical methods, followed by maintenance of the stand with prescribed burning. Methods employed to reduce dense stands should emphasize achievement of root mortality rather than only suppression of aerial growth to avoid coppicing. Maintenance treatments with prescribed fire may be necessary to kill mesquite seedlings without damaging desired savanna trees (52).

References

1. Burkhart A and BB Simpson. In: *Mesquite—Its Biology in Two Desert Ecosystems* (BB Simpson, Ed.), pp. 201-215. US/IBP Synthesis Series No. 4, Dowden, Hutchison and Ross, Inc., Stroudsburg, Pennsylvania, 1977.
2. Simpson BB and OT Solbrig. In *Mesquite—Its Biology In Two Desert Ecosystems* (BB Simpson, Ed.), pp. 1-25. US/IBP Synth. Ser. No. 4, Dowden, Hutchinson and Ross Inc., Stroudsburg, Pennsylvania, 1977.
3. Tschirley FH and SC Martin. J. Range Manage. 13:94-97 (1960).
4. Cabral I and PW Jacoby. Abstr. No. 244, 43rd Ann. Mtg., Soc. Range Manage., Denver, CO (1990).
5. Mooney HA, BB Simpson, and OT Solbrig. In: *Mesquite—Its Biology in Two Desert Ecosystems* (BB Simpson, Ed.), pp. 26-41. US/IBP Synth. Ser. No. 4, Dowden, Hutchinson and Ross, Inc., Stroudsburg, Pennsylvania, 1977.
6. Fisher CE, CH Meadors, R Behrens, ED Robinson, PT Marion, and HL Morton. Texas Agric. Exp. Sta. Bull. 935, Texas A&M Univ., College Station (1959).
7. Archer S. Naturalist 134:545-561 (1989).
8. Brown JR and S Archer. Vegetatio 73:73-80 (1987).
9. Haas RG, RE Meyer, CJ Scifres, and JH Brock. Mesquite Res. Monogr. 1, Texas Agric. Exp. Sta., College Station, pp. 10-19, 1973.
10. Brown JR and S Archer. Oecologia 80:19-26 (1989).
11. Ueckert DN, LL Smith, and BL Allen. J. Range Manage. 32:284-287 (1979).
12. Van Auken OW and JK Bush. Bull. Torr. Bot. Club. 114:393-401 (1987).
13. Van Auken OW and JK Bush. Amer. J. Bot. 75:782-789 (1988).
14. Bush JK and OW Van Auken. Southwest. Nat. 32:469-473 (1987).
15. Bush JK and OW Van Auken. Bot. Gaz. 151:234-239 (1990).
16. Scifres CJ and RR Hahn. J. Range Manage. 21:296-298 (1971).
17. Nilsen ET, PW Rundel, and MR Sharifi. Oecologia 50:271-276 (1981).

18. Jarrell WM and RA Virginia. J. Arid Environ. 18:51-58 (1990).
19. Nilsen ET, MR Sharifi, PW Rundel, WM Jarrell, and RA Virginia. Ecology 64:1381-1393 (1983).
20. Heitschmidt RK, RJ Ansley, SL Dowhower, PW Jacoby, and DL Price. J. Range Manage. 41:227-231 (1988).
21. Ansley RJ, PW Jacoby, and GJ Cuomo. J. Range Manage. 43:436-442 (1990).
22. Nilsen ET, MR Sharifi, PW Rundel, and RA Virginia. Oecologia 69:95-100 (1986).
23. Fick WH and RH Sosebee. J. Range Manage. 34:205-208 (1981).
24. Scifres CH, RW Bovey, CE Fisher and JR Baur. Mesquite, Res. Monog. 1, Texas Agric. Exp. Sta., College Station, pp. 24-32, 1973.
25. Fisher CE. In: *Mesquite—Its Biology in Two Desert Ecosystems* (BB Simpson, Ed.), pp. 177-188. US/IBP Synth. Ser. No. 4, Dowden, Hutchinson and Ross, Inc., Stroudsburg, Pennsylvania, 1977.
26. Hastings JR and RM Turner. *The Changing Mile: An Ecological Study of Vegetation Change With Time in the Lower Mile of an Arid and Semiarid Region.* Univ. of Arizona Press, Tucson, 1965.
27. Ueckert DN. J. Range Manage. 27:153-155 (1974).
28. Polk KL and DN Ueckert. Ann. Entomol. Soc. Amer. 66:411-417 (1973).
29. Rogers CE. J. Kan. Entomol. Soc. 50:222-228 (1977).
30. Whitford WG, DJ DePree, and RK Johnson Jr. J. Arid Environ. 1:345-350 (1978).
31. Ansley RJ, CH Meadors, and PW Jacoby. Southwest. Entomol. 15:469-474 (1990).
32. Wiedemann HT and BT Cross. Rangeland Resources Research Cons. Prog. Rep. 3665, Texas Agric. Exp. Sta., College Station, pp. 109-11 (1980).
33. Fisher CE, HT Wiedemann, CH Meadors, and JH Brock. Mesquite, Res. Monog. 1, Texas Agric. Exp. Sta., College Station, pp. 46-52 (1973).
34. Jacoby PW and CH Meadors. Weed Sci. 31:681-685 (1983).
35. Jacoby PW, CH Meadors, and MA Foster. Weed Sci. 29:376-378 (1981).
36. Jacoby PW, RJ Ansley, and CH Meadors. J. Range Manage. 44:56-58 (1991).
37. Jacoby PW, RJ Ansley, CH Meadors, and AH Huffman. J. Range Manage. 43:347-350 (1990).
38. Jacoby PW, RJ Ansley, CH Meadors, and GJ Cuomo. J. Range Manage. 43:36-38 (1990).
39. Jacoby PW, CH Meadors, and RJ Ansley. J. Range Manage. 43:33-35 (1990).
40. Bovey RW and RE Meyer. In: *Proc. Brush Management Symp.* (K McDaniel, Ed.), pp. 45-52. Texas Tech Univ. Press, Lubbock, 1983.
41. Wright HA and AW Bailey. *Fire Ecology—United States and Southern Canada.* John Wiley and Sons, New York, 1982.
42. Wright HA, SC Bunting, and LF Neuenschwander. J. Range Manage. 29:467-471 (1976).
43. Cable DR. J. Range Manage. 18:326-329 (1965).
44. Britton CH and HA Wright. J. Range Manage. 24:136-141 (1971).
45. Cable DR. J. Range Manage. 29:286-289 (1976).
46. Gonzales CL and JD Dodd. J. Range Manage. 32:305-309 (1979).
47. McDaniel KC, JH Brock, and RH Haas. J. Range Manage. 35:551-557 (1982).

48. Jacoby PW, CH Meadors, MA Foster, and FS Hartmann. J. Range Manage. 35:424-426 (1982).
49. Bedunah DJ and RE Sosebee. J. Range Manage. 37:483-487 (1984).
50. Dahl BE, RE Sosebee, JP Goen, and CS Brumley. J. Range Manage. 31:129-131 (1978).
51. Fisher CE, GO Hoffman, and CJ Scifres. In: *Mesquite: Growth and Development, Management, Economics, Control, Uses*, Res. Monog. 1, pp. 5-9. Texas Agric. Exp. Sta., College Station, 1973.
52. Jacoby PW. In: *Ecology and Management of the World's Savannas* (JC Tothill and JJ Mott, Eds.), pp. 223-228. Australian Acad. Sci., Canberra, 1985.

37

Saltcedar (Tamarisk): Classification, Distribution, Ecology, and Control

Gary W. Frasier and Thomas N. Johnsen, Jr.

Abstract

Saltcedar or tamarisk (*Tamarix* spp.) is a woody phreatophyte found along the drainage ways of many river systems in the United States, especially in the West. The shrubby tree was introduced into the United States as an ornamental in the early 1800s, but escaped cultivation in the mid to late 1800s and invaded many river systems, often replacing the native vegetation. In many, places the trees form dense stands which congest river channels, creating potential flood hazards. There have been claims that saltcedar evapotranspires considerably more ground water than the displaced native vegetation, thereby affecting the total flow of the river. Cited saltcedar benefits include nesting areas for doves and flowers which are a source of pollen and nectar for honey bees. Saltcedar has been classified as one of the 10 worst noxious weed in the United States. Its growth characteristics make it very difficult to control by either mechanical or chemical means. With a favorable supply of groundwater, there will probably not be a natural reversal in saltcedar stand compositions. Any type of saltcedar control must be part of a complete riparian habitat management program which evaluates merits of control, environmental effects, and public acceptance.

Introduction

Saltcedar or tamarisk (*Tamarix* spp.) is a woody phreatophyte, originally introduced into the United States from the Middle East as an ornamental shrub. Some benefits attributed to saltcedar are possible nesting sites for doves and a source of pollen and nectar for honey bees. It escaped from cultivation and is now found along the riparian drainageways of many river systems, especially in the western U.S. In places, trees have become so dense that they congest the river channels, creating potential flood hazards. There have been claims that the plant evapotranspires considerably more ground water than the displaced native vegetation which may affect the total flow of the river. Saltcedar's growth characteristics and persistence makes it difficult to control by either mechanical or chemical means. Robinson (1) stated: "...saltcedar is without doubt the outstanding problem phreatophyte of the Southwest because of its aggressive nature and thirst for water." He listed it as one of the 10 worst noxious weeds in the United States.

Description and Classification

Saltcedar is a phreatophytic shrub or small tree that may grow up to 10 m tall. Its slender upright or spreading branches covered with a smooth reddish-brown bark that becomes furrowed and ridged with age. The branches reach the ground and form a narrow or rounded crown. The small, scalelike, appressed leaves are deciduous. The flowers are crowded into numerous clusters 2 to 5 cm long on the ends of the twigs and are present throughout the growing season (2). The leaves secrete salt as a white bloom and may be covered with a salty fluid during periods of high humidity (3). The seed capsule is reddish brown which splits into three to five parts with many tiny hairy seeds (4).

The taxonomy of saltcedar is uncertain. It is a deciduous plant in the genus *Tamarix*, one of four genera in the Tamaricaceae, the tamarisk family (5). Four species have been commonly referred to as saltcedar: *Tamarix gallica* (4,6); *T. pentandra* (7-9); *T. ramosissima* (10,11); and *T. chinensis* (12). Most land managers use the common names of saltcedar or tamarisk for all of these indistinct species. The name "saltcedar" is believed to come from the small scale-like leaves that resemble cedars and the salty residue that collects on the foliage (13). The common name "tamarisk" is often confused with "tamarack," a coniferous tree.

Distribution

Saltcedar is believed to be a native to southern Europe, northern Africa, and eastern Asia (5) and is reported to have grown for centuries in the Jordan River Valley (13). The specific time when the plant was introduced into the United States is not firmly known. Horton (14) reported that saltcedar was introduced in the eastern United States in the 1820s. Seed and nursery catalogs listed the species for distribution as a landscape ornamental plant as early as 1856. It is believed that the plant escaped from these early plantings into the river basin areas where it has become a noxious weed problem (9). A saltcedar herbarium record was made in 1884 on the San Jacinto River in Texas and a specimen was collected in Fairmont Park, Philadelphia, in 1887. Other records of collections were infrequent until 1915, when the name saltcedar came into popular usage.

The extent of saltcedar infestations was not considered a major problem in the period prior to 1915. Turner (15) referenced reports of saltcedar spread in the Canadian River flood plain of Oklahoma from 1920 to 1935, along rivers in Utah from 1935 to 1955, and along the Arkansas River in Colorado from 1936 to 1957. The invasion of saltcedar was most noticeable in the valleys of central New Mexico (Figure 37.1) (16).

Similar results were documented for the upper Gila River in Arizona (15). It now grows along the lower elevation drainageways of many major rivers throughout the lower half of the United States and, at present, is found across the midsection of the United States from the central valleys of California to the southeastern coastal plains (17) (Figure 37.2). Saltcedar is found in nearly every watercourse in the southwestern states at elevations below 1830 m. It is rapidly spreading into drainageways in Oregon, Wyoming, western Nebraska, and Kansas, and is becoming a serious problem along the tributaries of the Missouri River and perhaps in the entire Missouri River system (18). Robinson (19) estimated that a total of 364,000 ha in the United States were infested by saltcedar in 1960 and the area was expected to expand to 526,000 ha by 1970.

Ecology

Saltcedar is a rapid-spreading phreatophyte on moist riparian areas along river channels and is dependent upon groundwater for growth and survival. In 1957, it was considered one of the 10 worst unwanted water-using plants, with an annual rate of water consumption greater than any other phreatophyte (1). It was believed to be the main reason for low

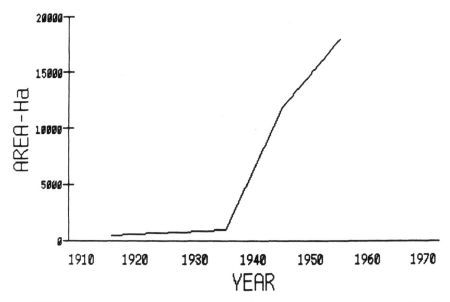

FIGURE 37.1 Increase in the extent of saltcedar in the Rio Grande river valley of New Mexico from Bernardo to San Macial (16).

water deliveries in the Rio Grande River to the Elephant Butte Reservoir in southern New Mexico as required by the Rio Grande Compact (20). Water use by saltcedar has received much research attention with extensive studies incorporating sophisticated techniques such as energy budgets and lysimeters (21-23) but it is still not certain if saltcedar uses more water than the vegetation it replaces (24).

Saltcedar spreads by seeds and resprouts vigorously from roots if the top growth is damaged or removed. The seeds are shed throughout the growing season (2) and usually germinate within 24 hr after moistening. Seeds have been known to germinate while floating on the water during May and June in Arizona. The germinated seeds float to the shore and become established on the saturated soil as the water recedes.

Saltcedar seedlings are capable of surviving long periods (several weeks) of submergence (25) and are tolerant to saline soils (26). Early seedling growth is slow, with plants reaching only 10 cm after 60 days. During the early seedling stage, the soil must be kept moist as most young seedlings can be killed by one day of drought. Older plants grow rapidly and are alkali-tolerant and drought-resistant (4). After the plants become established, saltcedars can withstand severe droughts of several months or more. Saltcedar can also be readily established from cuttings if planted in moist soil (9).

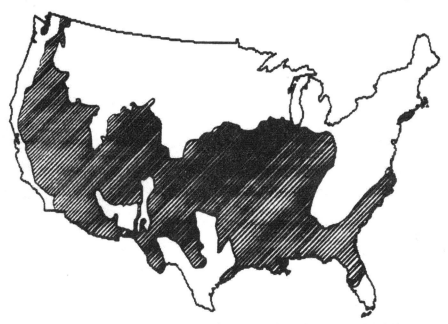

FIGURE 37.2 Extent of saltcedar invasion in the United States by 1938 (17).

The saltcedar will invade a well-lit bare, moist seedbed, often invading an area as a few scattered seedlings. It then rapidly spreads into other adjacent, favorable areas. Saltcedar may grow in association with mesquite, willows, and cottonwoods. However, after the saltcedar has become firmly established in large numbers, the native plants are almost completely excluded due to shading and salt accumulation under the plant canopy. Everitt (27) indicated that river peak discharges in the spring favored cottonwoods but peak discharges in the summer favored saltcedar.

Saltcedar has changed the plant successional stages in some areas and will ultimately dominate many riparian communities (28). Once saltcedar becomes established, even large changes in soil moisture will not completely eliminate it (15). As long as there is an abundance of available ground water, there will probably not be a reversal in saltcedar stand composition.

A dense cover of saltcedars can have a major impact on the hydrologic balance of an area. Saltcedar stands can increase hazard of floods in an area by choking or reducing the width of the normal channel, obstructing flood waters, and impeding river flow. Along the Brazos River flood plain in Texas, saltcedar dominated about 37% of the area (7000 ha) and was estimated to be using over 93 million cubic meters of water annually (29).

While it is commonly believed that saltcedar is an important habitat component for nesting white-wing and mourning dove, dense stands of saltcedar along the lower Colorado river were also found to have a low value for other avian populations (30) and other animals. Some saltcedar stands are used extensively by bee keepers. This is frequently done to avoid pesticides sprayed on crops and to maintain the hives when other plants are not producing pollen or nectar. The honey produced from saltcedar is generally sold to the baking industry (31).

Replacement of native plants and changing habitats for animals are the main effects of saltcedar invasion into riparian habitats. This invasion is a result of many factors including (1) overgrazing, (2) land clearing, (3) changing seasonal river flows, (4) lower water tables, and (5) factors such as woodcutting and recreational uses.

Control

The sprouting characteristics of saltcedar, its aggressive invasion of open areas, ease of establishment, and its dense stands make it very difficult to control. During the past 40-50 yr, many control methods have been investigated with variable degrees of success. Control methods investigated include burning, mowing, chopping, disking, root plowing, herbicides, and combinations of the methods (32-37). Usually only temporary suppression of plant growth was obtained, even with repeated treatments. The most successful control was a combination of root plowing followed by burning the debris and repeated spraying of herbicides on the regrowth. Root plowing 35 to 60 cm deep with a cutting blade equipped with fins to pull roots and buried stems out of the ground was generally effective but destroyed the grass and other vegetation in the area. Root plowing must be done while the soil is dry and conditions are such that the plant material will rapidly dry. This leaves the soil in an unprotected state subject to wind and water erosion.

There are few herbicides available for use on saltcedar. Tebuthiuron is effective and currently labeled for spot treating saltcedar as a soil application (38). Some newer herbicides, hexazinone and imazapyr, show promise but need further testing on saltcedar (39,40). Herbicide spraying of saltcedar during the growing season is restricted by potential drift onto sensitive crops adjacent to treatment areas. Most chemical treatments have to be repeated to maintain plant suppression. Injecting long residual life herbicides at 35 to 60 cm in the soil with a finless root plow killed 90% or more of the saltcedar with one treatment without destroying the cover of grasses and other shallow-rooted plants (41). Long-lived herbicide residues can possibly contaminate ground water in areas with shallow water tables and may affect the water quality for downstream water users. Individual trees and small stands of trees may be most suited to herbicidal control. However, currently available, effective herbicides are not selective for saltcedar alone, but kill other plant species as well. Before any herbicides are used, current registrations and limitations should be determined.

Changing the ground water level may kill the saltcedar as well as other riparian vegetation. While the trees are considered drought resistant, a dropping water table along the Gila River in Arizona reduced saltcedar stands (42). Extended inundation of one or two years will also kill the trees (43).

Grazing by livestock of seedlings and fresh sprouts can, in some places, repress saltcedar regrowth and seedlings (44). The use of insects and diseases has been suggested as possible control measures for saltcedar, but these techniques have not been fully tested (45). A leafhopper reduced saltcedar growth by 76% in limited tests in New Mexico (46). There are several species of insects from the eastern Mediterranean and Pakistan areas which attack saltcedar and may have potential in some area as biological control agents (47). The advantages of introducing biological control agents are that the method is usually self-perpetuating and achieves most objectives of multiple use of riparian sites (47).

Controlling saltcedar is a controversial subject (48-50). Those favoring saltcedar control cite: (1) removal of barren disclimax saltcedar stands to restore the original riparian flora and fauna, (2) increasing channel flow by reducing excessive evapotranspirations losses, (3) reducing flooding hazards from vegetative restricted channels, and (4) reduction of salt build-up in soils. Those opposing control cite: (1) saltcedar thickets used for nesting areas of white-wing and mourning doves in Arizona, (2) saltcedar as a source of pollen and nectar for honey bees, (3) potential hazards of increased soil erosion following saltcedar removal, and (4) potential damage to ornamental *Tamarix*. Even the proposed forms of biological control will face many of the same environmental objections used against other saltcedar control methods. There are also potential problems of restricting biological agents to the applied sites and eliminating them if they become established outside the control areas. Several proposed saltcedar control projects have been halted or delayed by court orders on challenges by opponents to vegetation control.

After saltcedars are killed, other vegetation must be established to protect the soil resource and to prevent or retard saltcedar re-invasion. Establishing a canopy cover on treated areas with seeded grasses and planted cottonwood cuttings could reduce the chances of saltcedar successfully re-invading an area but this is costly. Saltcedar control must be part of a complete riparian habitat management program. All factors, such as merits of control, environmental effects, and public acceptance, must be considered.

References

1. Robinson TW. In: *Symposium on Phyreatophytes*. Pacific Southwest Regional Meeting, Amer. Geophy. Union, Sacramento, CA, 1958.
2. Horton JS. The Southwestern Naturalist 2(4):135-139 (1957).
3. Decker JP. Forest Science 7(3):214-217 (1961).
4. Little EL Jr. Agric. Handbook No. 9, USDA Forest Service, Washington, DC, 1950.
5. Baum BR. Israel Acad. Sci. Humanit., 1978.
6. Benson L and RA Darrow. Bio. Sci. Bull. No. 6, Vol. XV No. 2, Univ. of Arizona, Tucson, 1944.
7. Little EL Jr. Agric. Handbook No. 41, USDA Forest Service, Washington, DC, 1953.
8. McClintock E. J. Calif. Hort. Soc. 12:76-83 (1951).
9. Reynolds HG and RR Alexander. In: *Seeds of Woody Plants in the United States*, pp. 794-795. Agric. Handbook No. 450, USDA Forest Service, Washington, DC, 1974.
10. Baum BR. Baileya 15:19-25 (1963).
11. Botherson JD, JG Carman, and LA Szyska. J. Range Manage. 37(4):362-364 (1984).
12. Horton JS and JE Flood. The Southwestern Naturalist 7(1):23-28 (1962).
13. Bowser CW. In: *Symposium on Phyreatophytes*. Pacific Southwest Regional Meeting, Amer. Geophy. Union, Sacramento, CA, 1958.
14. Horton JS. USDA Forest Service, Rocky Mountain Forest and Range Exp. Sta. Res. Note RM-16, Fort Collins, CO, 1964.
15. Turner RM. US Geol. Survey Prof. Paper No. 655-H, US Govt. Printing Office, Washington, DC, 1974.
16. Thompson CB. In: *Symposium on Phyreatophytes*. Pacific Southwest Regional Meeting, Amer. Geophy. Union, Sacramento, CA, 1958.
17. Van Dersal WR. USDA Misc. Pub. No. 303, Washington, DC, 1938.
18. Timmons FL and DL Klingman. Amer. Assoc. Advance. Sci., pp. 157-170 (1960).
19. Robinson TW. US Geol. Survey Prof. Paper 491-A (1965).
20. Fletcher HC and HB Elmendorf. In *Yearbook of Agriculture 1955*, pp. 423-429. US Govt. Printing Office, Washington, DC, 1956.
21. Culler RC, RL Hanson, RM Myrick, RM Turner, and FP Kipple. US Geol. Survey Prof. Paper 655-P (1982).

22. Gay LW. In: *Riparian Ecosystems and Their Management: Reconciling Conflicting Uses*. Rocky Mountain Forest and Range Exp. Sta. Tech. Rep. RM-120, Fort Collins, CO, 1985.

23. Weeks EP, HL Weaver, GS Campbell, and BD Tanner. USGS Prof. Paper 491-G (1978).

24. Van Hylckama TEA. USGS Geol. Surv. Prof. Paper 491-E (1974).

25. Horton JS. In: *Symposium of Hannoversch-Munder*. International Sci. Hydrol. Assoc. Pub. No. 48 1:76-83 (1959).

26. Carman JG and JD Botherson. Weed Sci. 30:360-364 (1982).

27. Everitt BL. Environ. Geol. 3:72-84 (1980).

28. Campbell CJ and WA Dick-Peddie. Ecology 45:4592-5020 (1964).

29. Busby FE Jr and JL Schuster. J. Range Manage. 24:285-289 (1971).

30. Cohan DR, BW Anderson, and RD Ohmart. In: *Symp. Proc. Strategies for Protection and Management of Floodplain Wetlands and Other Riparian Ecosystems*." Callaway Gardens, GA, Gen. Tech. Rep. WO-12, USDA Forest Service, Washington, DC, 1979.

31. Waller GD and R Schmalze. *Gleanings in Bee Culture*. USDA-ARS, Tucson, AZ, 1976.

32. Horton JS. USDA Forest Service, Rocky Mountain Forest and Range Exp. Sta. Res. Note 50, Fort Collins, CO, 1960.

33. Arle HF, GN Bowser, and GN McRae. In: *1962 Res. Rep. Western Weed Control Conf.*, 1962.

34. Hughes EE. In: *Chemical Control of Range Weeds, Chemical Plant Control Subcom., Range Seeding Comm. USDA-USDI*, pp. 36-37. US Govt. Printing Office, Washington, DC, 1966.

35. Campbell CJ. USDA Forest Service, Rocky Mountain Forest and Range Res. Exp. Sta. Res. Note RM-76, Fort Collins, CO, 1966.

36. Bracher B. Cattleman 58(10):46-47,87 (1972).

37. Peterson P, JO Evans, and CJ Hurst. Proc. Western Weed Sci. 33:119 (1980).

38. Temple AJ and BR Murphy. In: *Res. Highlights Noxious Brush and Weed Control Range and Wildlife Management 1981*, pp. 57-58. Texas Tech Univ., Lubbock, TX, 1981.

39. Shrader T. In: *1985 Res. Prog. Rep. Western Soc. Weed Sci.*, pp. 347-348. Phoenix, AZ, 1985.

40. Beardmore RA and TE Nishimura. In: *1989 Res. Prog. Rep. Western Soc. Weed Sci.* Honolulu, HI, 1989.

41. Hollingsworth EB, PC Quimby Jr, and DC Jaramillo. J. Range Manage. 32(4):288-291 (1979).

42. Horton JS. J. Soil and Water Conserv. 27(2):57-61 (1972).

43. Warren DK and RM Turner. J. AZ Acad. Sci. 10:135-144 (1975).

44. Gary HL. USDA Forest Service, Rocky Mountain Forest and Range Res. Exp. Sta. Note RM-51, Fort Collins, CO, 1960.

45. Watts JG, DR Liesner, and DL Lindsey. Bull. New Mexico Agric. Exp. Sta. 650 (1977).

46. Liesner DR. MS Thesis, New Mexico State University, Las Cruces, 1971.

47. DeLoach JC. In *Proc VII Int. Symp. Bio. Control Weeds*. Rome, Italy, 1989.

48. Bristow B. In *Proc. 12th Annual Arizona Watershed Symp.*, pp. 41-43. Phoenix, AZ, 1968.

49. Pugh CA. In: *Proc. 12th Annual Arizona Watershed Symp.* Phoenix, AZ, 1968.

50. Winter NA. In: *Proc 12th Annual Arizona Watershed Symp.* Phoenix, AZ, 1968.

38

The Importance, Distribution, and Control of Dyers Woad (*Isatis tinctoria*)

John O. Evans

Abstract

Dyers woad infests crops and waste areas in several western states, but its spread and invasion of western range and forest land is of extreme concern to land managers and weed scientists throughout the region. Dyers woad appears to be especially adapt to the physical and environmental conditions of the intermountain states since it currently does not exist as a threatening weed in the eastern United States where it was initially introduced as a cultivated crop. Dyers woad produces abundant seed, each of which is encased in a large winged silicle that appears to be wind disseminated and, in concert with birds and animals, is responsible for initial infestations of the species. Dyers woad can regenerate itself from a rudimentary root after being cut off at the ground surface by mechanical operations or handpulling. Of the many control strategies available, the most important in many situations is hand removal because the plant is particularly easy to identify from considerable distances. Care must be taken to remove the plant well down into the root itself.

Introduction

Dyers woad (*Isatis tinctoria*) is a rapidly spreading mustard weed that was intentionally introduced into North America from Europe. It occurs mainly in northern Utah, southern Idaho, and western Wyoming in the intermountain region, but it also occurs to a limited degree in northern California, Oregon, and Nevada. Its principle direction of spread appears to be to the north and east, and it is presently invading northern Wyoming and Montana.

Dyers woad has been extensively cultivated as a commercial dye crop in Europe and in the United States (1). Perhaps it came west with the early pioneers as a textile dye or even as an ornamental. The first occurrence of the weed in Utah appears to be in the early years of the twentieth century in Box Elder County. To date this area supports extremely dense populations of dyers woad, suggesting near ideal conditions for this weedy species.

Importance and Distribution

Botanically, it is a member of the mustard family and mature woad plants produce hundreds of purple-black pods or silicles, each containing a single seed. Occasionally, a silicle may contain two seeds but a predominate number of pods contain a single seed under typical environmental conditions of the western United States. Some plants have been observed to produce over ten thousand seeds in one year. Similar to other mustards, this species is a prolific seed producer and, since the fruit may be wind disseminated, it is capable of rapid distribution in the intermountain states.

In the intermountain region, dyers woad germinates in the fall or early spring. The seedlings develop rosettes which produce large taproots during the first growing season. Seedlings arising in the spring will bolt and flower the following spring whereas fall-germinating seedlings overwinter as small rosettes and usually require the following growing season to develop sufficient below-ground support and reserves to sustain the flowering plants. Undisturbed dyers woad plants behave as biennials or winter annuals and require portions of two growing seasons to produce seed. A perennial behavior by the species is elicited by mowing, handpulling, or breaking the bolting stalk above ground. Dyers woad has a distinctive blue-green vegetative color. The oblong to lanceolate leaves are connected to the stem by a petiole and are three to four inches long. The upper leaves are distinct in that they have no petiole and clasp the stem with ear-like projections. All leaves have a slight pubescence and are nearly entire, but occasionally toothed. It has a large fleshy taproot from which it may reproduce asexually.

Dyers woad is readily identified in the blossom stage by its characteristic bright yellow flowers. Perhaps this single feature has focused unusual concerns about the weediness of this mustard compared with other equally troublesome but less noticeable ones. The bright flowers also provide opportunities to locate dyers woad invasions of previously uninfested fields and may help in handpulling isolated plants to prevent dyers woad migration.

As the seeds mature, the pods take a blue-black color and dangle conspicuously from the one to three foot tall umbrella-shaped panicle. The pods do not release the seed, but fall to the ground intact. Seeds are spread by wind, animals, and perhaps by man. Seeds have no dormancy, but are contained in the pod which has a water soluble germination inhibitor which prevents germination of dyers woad and other plant seeds. It also has an inhibiting effect on root elongation of competing species. Some dyers woad seeds germinate in the presence of this inhibitor and have a competitive advantage over other seedlings. Some woad seeds remain until the inhibitor is leached from the pod before they germinate. This insures a prolonged supply of germinable seed in the soil. It has not been determined how long seeds are viable in the soil.

Dyers woad is an important range weed since it competes favorably with shrubs and browse species characteristic of rangelands (2,3). It is especially well suited to dry, rocky soils common to many steep hillsides throughout the western United States, and infestations are frequently observed in rugged, inaccessible mountain terrain. It also exists along roadsides, rights-of-way, wasteland, and disturbed sites. It rarely exists in cultivated fields that are mechanically tilled on an annual basis. For example, it is not a common weed in grain fields nor in rowcrops. It routinely invades forage crops such as nonirrigated grass pastures and dryland alfalfa fields not tilled.

Recent estimates by the Bureau of Land Management project annual increases in dyers woad infestation of approximately 14% (4) and the U.S. Forest Service reports that dyers woad has increased several fold on intermountain forests during the past three decades (5). Dewey et al. (6) employed satellite remote sensing to measure potential expansion of dyers woad in the Cache National Forest in Utah. They associated plant communities and terrain where dyers woad currently exists with similar plant cover and topography to suggest that perhaps present infestations may increase more than 100-fold unless effective control strategies are enforced.

Dyers Woad Control

Several control strategies have been developed for dyers woad that should help prevent the spread of this aggressive weed. In recent years, new herbicides have appeared that are selective against mustards and compliment the existing impressive arsenal of herbicides available for agrosystems impacted by this plant. Dyers woad is sensitive to many 2,4-D formulations and this chemical continues to be the favored herbicide provided the crop or setting is within phenoxy herbicide registrations. Advantages of 2,4-D for dyers woad control include plant sensitivity to this compound from the early rosette growth stage continuing to the early blossom stage, and it is one of the least expensive materials available.   Recently, the sulfonylurea herbicide registrations have provided additional possibilities for dyers woad control since they are active against dyers woad rosettes in particularly low dosages and express excellent safety with respect to monocot plants often associated with dyers woad.   Further, they appear to offer additional control opportunities in that they significantly inhibit seed production by dyers woad when applied during the flowering process.

Control in Cultivated Land

Dyers woad is commonly found in areas where small grains are grown.   Unlike alfalfa and rangelands, small grain plantings provide a situation conducive to dyers woad control.

Of the several methods of control possible, the most important in many cases is rogueing. Rogueing is simply the process of going into the field and removing the undesirable plants by hand. It is most convenient to wait until the plant bolts and flowers since the distinct yellow flowers and umbrella-shaped stalk make it easy to identify. The plant can then by easily removed by handpulling or digging. Dyers woad has a thick, fleshy taproot and must be removed below the crown of the plant and well down into the root.

Rogueing is very effective in hard to reach spots such as ditchbanks, rock piles, and ravines or draws bordering grain fields. These sites can often be a great source of seed, and only through rogueing can this seed source be removed. Rogueing is also effective in the fields themselves. Grain fields with upwards of 100 plants per acre can be maintained successfully by hand weeding.

In fields where dyers woad infestations are more severe, cultivation and herbicide programs can be used to advantage. Small grains are cultivated often enough that tillage itself should control the greatest extent of the problem. In spring wheat plantings, a thorough spring cultivation will satisfactorily control plants in either the bolting or seed production stages. Since dyers woad acts as a biennial or winter annual in small grains, it must pass through a cold temperature period to be able to produce seed. Early spring cultivation destroys rosettes that have passed through the cold period. It is possible that new dyers woad seedlings may germinate and grow in the grain during the summer, but these plants cannot produce seed until the next growing season. In order to remove the competitive effect of these immature weeds, herbicides should be used.

The use of selective herbicides is an important component of an effective dyers woad control program in small grains. The control of woad in winter wheat and barley during the spring requires the use of herbicides. Woad as well as other broadleaf weeds can be readily controlled in wheat by treatment with 2,4-D (amine or low volatile ester), bromoxynil, and dicamba.

Dyers woad is a serious problem in alfalfa, especially many areas of dryland alfalfa. As more and more acres of alfalfa become infested with dyers woad, its control becomes more critical. The need for control of dyers woad is due to two main factors. First, while woad may not decrease the tonnage harvested with alfalfa, it does result in a lower protein content. Second, the ease with which the weed is spread by both forage and seed alfalfa practices makes it essential that alfalfa plantings be kept free from dyers woad infestations. It is a relatively simple process for mature, seed-bearing dyers woad plants to be cut and baled with alfalfa. This baled hay is often shipped great distances where it is used as livestock feed. Animals can further aid the dissemination process to even more remote areas. This process can spread seed to rangelands where the effect of dyers woad competition is much more severe and much more difficult to control than in alfalfa fields.

Dyers woad can also be a serious problem for alfalfa grown for seed. Woad seeds are enclosed in a fruit or silicle that is much larger than alfalfa seed, and at first glance it appears that this fruit would be separated from the alfalfa seed during harvesting. However, if the fruit is broken and the seed escapes, it is much closer in size to alfalfa seeds and can be mixed into the alfalfa seed. Contamination of seed results in lower prices for the seed as well as severe problems in the new alfalfa plantings.

One of the most important tools in controlling dyers woad in alfalfa is that of cultivation. Infested fields should be cultivated by some method in the spring with implements such as flex-tine harrows. Herbicides provide yet another alternative for woad control in alfalfa, but they should be used only where infestations warrant their use. At this time there are a few selective compounds which aid in the control of dyers woad. These herbicides are not, however, 100% effective against dyers woad, but they do help control other broadleaves in the crop. The alkylphenoxy herbicide 4-(2,4-DB) is selective in alfalfa, but it is rather weak in control of dyers woad and tends to burn the plant back and then allow it to regrow. This is especially true in poor stands of alfalfa.

Metribuzin is another herbicide that can be used for woad control. It provides a quick knockdown, but it does not maintain effective control as long as desired. Hexazinone is also an effective chemical or dyers woad control. As with 2,4-DB, both metribuzin and hexazinone are much more effective when used in vigorous healthy stands of alfalfa.

Control on Rangeland

There are three major techniques that can be used to control dyers woad in these areas: rogueing, mowing, and herbicide application. Rogueing or hand removal of individual weeds is probably one of the simplest, yet most essential, methods of dyers woad control. Rogueing is most effective in areas along the fringe of the major infestation and in those areas where the weed has been introduced far from any other major infestation. To be effective, it is generally necessary to wait until the woad bolts and flowers before attempting the rogueing operation. The distinctive yellow flowers make it easy to identify and locate all of the plants in an area. Once the plants have been identified, they can be removed by pulling or digging them up with a hoe or shovel. The important thing to remember is that there are only 4-6 wk from the time of flowering until the seeds are mature. It is essential that the plants be removed as soon as possible after flowering to prevent the possibility of some slipping by and going to seed.

The importance of hand rogueing cannot be overstressed, especially in those areas which have a light infestation of dyers woad. Land managers and other should be trained to identify the weed. As they make surveys in May and June, they can remove any small, isolated patches of dyers woad.

Excellent control of dyers woad can be obtained by spraying with 2,4-D in the rosette stage. As the plant enters early bud and blossom stages, 2,4-D often cannot kill it quick enough to prevent seed production. Combining 2,4-D with some other herbicides shows more promise to immediately stop dyers woad growth and seed production. The use of 2,4-D should be confined to those areas where adjacent properties will not be damaged by drift or spray.

Dyers woad typically enters an area by moving along highways, railroads, or canals. Apparently dyers woad seed can be spread by vehicles or railcars where it is dropped onto suitable sites. As the seeds germinate, new plants readily grow and produce seed and spread to neighboring fields or are picked up once again by passing vehicles and thus continue the cycle. Because roadsides and railways are such effective avenues of seed dispersal, it is extremely critical that any woad growing in these areas be destroyed or removed. It is especially important not to allow the plants to produce seed.

One of the most exciting discoveries with regard to stopping the advance of dyers woad is the impact that a native rust pathogen, *Puccinia thlaspeos*, has on this noxious weed. Fruit and seed production are completely prevented on almost all infected plants. Studies are underway to determine optimum conditions for the pathogen and whether rust spores can be hand disseminated to remote dyers woad locations. Recent surveys reveal that the rust is naturally spreading to new dyers woad infestations and significantly slowing the growth and reproduction of many dyers woad plants.

References

1. King LJ. Proc. N.E. Weed Control Conf. (New York) 21:589-593 (1967).
2. Young JA and RA Evans. Weed Sci. 19(1):76-78 (1971).
3. Farah KO, AF Tanaka, and NE West. Weed Sci. 36:186-193 (1988).
4. Northwest Area Noxious Weed Control Program Environmental Impact Statement. USDI, Bureau of Land Managaement, Portland, OR, 1985.
5. Intermountain Region Noxious Weed and Poisonous Plant Control Program Environmental Impact Statement. USDA, Forest Service, Intermountain Region, Ogden, UT, 1986.
6. Dewey SA, KP Price, and D Ramsey. Weed Tech. 5:in press (1991).

39

Medusahead: Importance, Distribution, and Control

William H. Horton

Abstract

Medusahead (*Elymus caput-medusae*) is a serious weed problem across large areas of rangeland in several western states. It continues to invade additional areas and poses a dilemma for land managers and owners, increasing soil erosion and seriously reducing both quantity and quality of forage resources. Attempts to control the annual medusahead have generally resulted in failure. Its vigorous competitive nature and seed carry-over makes establishment of more desirable perennial vegetation very difficult. However, when perennial grass plantings are successful in medusahead infestations, some of the more vigorous perennial species maintain dominance with proper management.

Introduction

Medusahead (*Elymus caput-medusae*) is a member of the grass tribe Hordea and the only annual plant found in the genus *Elymus*. Medusahead has its origins in areas bordering the Mediterranean Sea where it is also considered a weedy species. A detailed discussion of medusahead origin can be found in Major (1).

The first known collections of medusahead in the United States were made near Roseburg, Oregon, in 1884. Shortly after the turn of the century additional collections were made in Washington in 1901 and California in 1903. Idaho ranchers observed the plant in Gem County,

Idaho, in the 1930s (2). The first collections in Idaho were made in Payette by J.F. Pechanec in 1944 (3). In 1988, medusahead was discovered in Box Elder County, Utah, about four miles south of Tremonton on marginal clayey soil areas used mainly for early spring sheep grazing. The following year it was collected from several areas in the foothills south and west of Avon in southern Cache County, Utah. By the summer of 1990, it was evident medusahead was increasing and invading additional sites in these areas, even though private landowners agree the plant had been present on several sites for a number of years without noticeable aggression.

The rapid spread of medusahead across vast acreages of western rangeland has presented a serious problem to ranchers and wildlife and range managers. During a twenty-year period (1944 to 1964), medusahead spread from an almost unheard of species to a major problem on more than 700,000 acres of rangeland in the 10-20 inch precipitation zone in southwestern Idaho (2). It has progressively dominated numerous additional acres in western and southern Idaho and continues to be a major problem on rangelands in California and Oregon.

Torell et al. (2) identified medusahead as the worst weed in Idaho because of its vigorous competitive abilities and low forage values. Hironoka (4) reported a 50 to 80% reduction in livestock carrying capacity on areas infested with medusahead.

Distribution and Ecology

The distribution of medusahead can be extended beyond its natural range of adaptation when competition from other existing species is reduced or removed. Medusahead is not normally found above 4500 feet nor in areas having less than 9 to 12 inches of annual precipitation. However, it can survive in deep soils of swales or catch basins in lower rainfall areas. If adjacent drier areas are disturbed or misused, medusahead can rapidly move onto these sites (5).

Medusahead's life cycle and adaptation is somewhat similar to cheatgrass. Large areas in Idaho once dominated with monocultures of cheatgrass now host monotypic stands of medusahead.

Cheatgrass is palatable from six to eight weeks in the spring and again in the fall and winter after the seed and awns have fallen from the plant. In comparison, medusahead is only moderately palatable for three to five weeks in the spring. Many of the barbed awns do not drop from the plant until very late and the silica content (6) is more than double that of cheatgrass, which may partially explain its avoidance by livestock. It appears that grazing animals do not prefer medusahead. However, they will utilize some of the material if other feed is not available.

Medusahead depends on more rainfall than cheatgrass and on heavier soils that support moisture later into the season for seed set and ultimate survival (1). Therefore it would not appear that medusahead will replace cheatgrass on all sites now occupied by cheatgrass (7).

The most influential site characteristic contributing to medusahead domination is the type of soil development in the A and B soil horizons and topographic position. Generally, a higher content of montmorillonite clay in the upper soil profiles and more mesic topography greatly increase the chance for medusahead domination (8).

Well aerated A and B soil horizons generally promote cheatgrass dominance, whereas finer clayey and silty-clay soils encourage medusahead invasion. Even though surface soils may appear to be loamy and aerated, if the B horizon is clayey, the site is still very susceptible to medusahead establishment. Dahl (7) indicates that it is common for cheatgrass to dominate aerated mounded areas while medusahead infests the heavier soils of the intermounds.

Several factors favor medusahead in competition with cheatgrass. Medusahead produces more reproductive seeds per plant than cheatgrass. Medusahead is able to withstand extreme crowding, with populations of 1000 plants per square foot possible. Both cheatgrass and medusahead seeds have an after-ripening dormancy of 100 days or longer. However, cheatgrass matures two to three weeks earlier than medusahead, and greater amounts of seed have germinated by early September. In contrast, many medusahead seeds do not germinate in the fall but over-winter and germinate in the early spring. Lack of palatability in medusahead reduces grazing pressure which in turn leads to larger available seed sources to promote dominance. There are 21 diseases known to affect cheatgrass, while medusahead is essentially disease free (9).

Sharp and Tisdale (3) reported that five inches of accumulated medusahead litter is not unusual on many sites in southwestern Idaho. Areas with heavy litter are avoided by livestock, especially sheep. Increased litter also discourages revegetation efforts and the establishment of more beneficial species.

Control

The control of medusahead so that more preferred forages can be planted is a major concern. Past research has explored various combinations of burning, dalapon application, discing, and plowing with varying degrees of success. Medusahead control for only a single year is questionable due to excessive weed seed carryover into spring, which will normally out-compete more desirable seeded species (2).

Clipping or early spring grazing of medusahead seriously depletes seed development (10). Intensive spring grazing before seed set may be an effective management tool used over several seasons to deplete seed sources sufficiently so successful perennial seedings can be accomplished (7).

Research efforts to establish perennial grasses on burned and unburned medusahead arid rangeland sites with a treatment of herbicide, followed by a minimum-till drill planting, was conducted in southern Idaho (Horton, unpublished data). The unburned area resulted in total failure to establish perennials. It became evident that revegetation efforts in monotypic stands of medusahead would require at least a prerequisite burn to reduce litter and weed competition sufficient to establish perennial grasses.

On the burned site, seedings were made in the late fall (mid-October). In these trials, heavy late summer precipitation led to excellent germination of medusahead seeds which had not been destroyed by the earlier burn. Medusahead competition was reduced further by applications of 1.2 liters/ha of the herbicide Roundup to seedlings prior to minimum-till seeding of perennial grasses in the late fall. The following spring, perennial seedlings of more vigorous grass varieties such as crested wheatgrass and Russian wildrye were able to germinate and establish with greatly reduced competition from medusahead.

The use of minimum-till planting became an important factor in accurate seed placement, moisture retention in arid conditions, and reduced crusting problems on heavy, fine soils typically associated with medusahead sites.

Conclusion

Prevention and control of medusahead invasion should be major objectives of users and land managers. The best safeguard against the spread of medusahead is to retain a healthy perennial plant community. Competition from perennial species have considerable importance in limiting the dominance and invasion of medusahead (7). It appears that any effort to control this species should include the establishment of competitive perennials.

References

1. Major J. *Proc. Calif. Section, Amer. Soc. Range Manage. Ann. Mtg.*, pp. 35-39 (1960).
2. Torell PJ, LC Erickson, and RH Haas. Weeds 9:124-131 (1961).
3. Sharp LA and EW Tisdale. Res. Note 3, For., Wildlife, and Range Exp. Sta., Univ. of Idaho, Moscow, 1952.
4. Hironaka M. J. Range Manage. 14(5):263-267 (1961).
5. Hironaka M. PhD Diss., Univ. of Wisconsin, 1963.
6. Bovey RW, D LeTourneau, and LC Erickson. Weeds 9:307-311 (1961).
7. Dahl BE. PhD Diss., Univ. of Idaho, 1966.
8. Baver LD. *Soil Physics, 3rd Ed.* John Wiley and Sons, Inc., New York, 1956.
9. Klemmedson JO and JG Smith. Botan. Rev. 30:226-262 (1964).
10. Heller TH. MS Thesis, Univ. of Idaho, Moscow, 1962.

40

Dalmatian Toadflax, Yellow Toadflax, Black Henbane, and Tansymustard: Importance, Distribution, and Control

Don W. Morishita

Abstract

Several weed species are considered to be of minor importance as rangeland weeds because of the number of acres they infest. Although they are not found in extremely large numbers, control of Dalmatian toadflax (*Linaria genistifolia* ssp. *dalmatica*), yellow toadflax (*Linaria vulgaris*), black henbane (*Hyoscyamus niger*), tansymustard (*Descurainia pinnata*) or flixweed (*Descurainia sophia*) is important in range management because of their potential to spread and create a detrimental impact on wildlife, livestock, and man. Dalmatian and yellow toadflax are perennial weeds that reproduce by seeds and creeping roots. Black henbane is an annual or biennial, and tansymustard and flixweed are annuals. All of these species have the ability to produce large numbers of seed that can spread by wind or animals. Seed from these weeds also can lie dormant in the soil for several years. One factor these weed species have in common is that they establish themselves most readily on disturbed and/or poorly maintained soils. All of these weeds are somewhat to very opportunistic and are generally found along roadsides, waste areas, and poorly managed pastures and rangeland. These weeds can be found throughout most of North America. Dalmatian and yellow toadflax will grow in more humid areas than black henbane, tansymustard, or flixweed. Tansymustard and flixweed tend to favor the less humid climates. Black henbane would be considered the most dangerous plant to animals because of its toxic properties. However, because of its foul odor, it is not readily grazed and probably does not cause any more of a problem to animals than the other weed species. Dalmatian and yellow toadflax are the most difficult of these weeds to control because of their perennial life cycle. Black henbane, tansymustard, and flixweed can be effectively controlled with chemicals as well as by hoeing or cultivation where possible.

Introduction

The invasion of weeds on rangeland can reduce the productivity and aesthetic value of that land. If those invading weeds are noxious, they present a more potentially serious threat to wildlife, livestock, and man. There are several noxious and less threatening weed species found on rangeland that are considered to be of minor importance. The primary reason for this is because of the relatively low number of acres these particular weeds infest compared to more commonly occurring or aggressive range weeds, such as the knapweeds (*Centaurea* spp.), leafy spurge (*Euphorbia esula*), dyers woad (*Isatis tinctoria*), etc. Nevertheless, weeds such as Dalmatian toadflax, yellow toadflax, black henbane, tansymustard, and flixweed can present serious localized problems in range and pasture management. This paper provides a brief review of the description, importance, distribution, and control of four weed species found in rangeland.

Dalmatian Toadflax and Yellow Toadflax

Description

Dalmatian and yellow toadflax are members of the Schophulariaceae (figwort) family. Both weeds are perennial herbs that reproduce by seeds and horizontal or creeping rootstocks (1). Dalmatian toadflax grows 0.8 to 1.5 m tall and yellow toadflax grows 0.2 to 0.8 m tall. The stems of both species are erect. Dalmatian toadflax is a more robust plant with more branching near the top of the plant than yellow toadflax. Leaves of both species are alternate, but crowded and sometimes appearing opposite. Dalmatian toadflax leaves are broad, ovate to ovate-lanceolate or sometimes lanceolate near the lower part of the plant. The leaves at the base are sessile and cordate clasping. Yellow toadflax leaves are soft, linear or linear-lanceolate, sessile, and pale-green. They are generally 2.5 cm long by 2 to 4 mm wide. The flowers of both species are bright yellow. Dalmatian toadflax racemes are elongate and loosely flowered. Yellow toadflax racemes have numerous flowers in a more compact spike. The pedicels are short to almost nonexistent. Yellow toadflax seeds are flattened and winged while Dalmatian toadflax seeds are sharply angular, slightly winged. Seeds of both species are 1 to 2 mm long (1). Dalmatian and yellow toadflax flowers from May to August and produces seed from July to September (2).

Importance

Dalmatian toadflax is a native of the Mediterranean region from Yugoslavia to Iran (3). It was grown as a ornamental in Europe in the late 1500s and introduced into North America around 1900 (4). Yellow toadflax is of Eurasian origin. It, too, was grown as an ornamental and introduced into North America in the mid 1800s (5).

The seedlings of toadflax are considered ineffective competitors for soil moisture with established perennials and winter annuals. Average stand life of Dalmatian toadflax once it is established is 3 to 13 yr (3). A mature Dalmatian toadflax plant can produce up to 500,000 seeds. Seed dispersal is primarily by wind, but it is also dispersed by livestock. Estimated seed longevity in the soil is 10 yr. Over 90% germination has been obtained with 2- to 3-yr old seeds in the laboratory (6). Under field conditions, some mature seeds of Dalmatian toadflax will germinate immediately after fall harvest, but most germinate the following spring (7).

Kingsbury (8) reported that neither species is poisonous. There are reports, however, that yellow toadflax contains a poisonous glucoside and may be harmful to livestock (9). Both species are considered to be unpalatable so reports of livestock poisoning are rare.

Distribution

Dalmatian and yellow toadflax are found throughout North America. Much of their distribution can be attributed to their long-time usage as an ornamental and casual collecting as a wild flower (10). Dalmatian toadflax is most common in the western United States and has a tolerance to low temperatures and coarse-textured soils. Yellow toadflax is more common throughout eastern North America, but also is found in many areas in the western United States.

The habitat of both species is similar. They can be found in fields, roadsides, waste areas, and overgrazed pastures and rangeland (2).

Control

Control measures for both species are very similar. Successful control has been obtained using intensive clean cultivation. This requires at least two years with eight to ten cultivations in the first year and four to five cultivations the second year (9). This also includes planting competitive perennial and winter annual grasses (3).

No widely effective biocontrol agents are currently available for controlling these weeds. Two beetles, *Chrysomela rossia* and *Chrysolina gypsophilae*, have been evaluated by Canadian researchers as biocontrol agents, but both species had too wide of host range to be used in North America (11,12). *Calophasi lunula*, a moth, and *Brachyterolus pulicarius*, a beetle, are currently being evaluated by the University of Idaho and Montana State University (13).

Herbicides currently registered for Dalmatian and yellow toadflax control include dicamba, 2,4-D, and picloram. Picloram, at 2.25 kg/ha, controlled Dalmatian toadflax for two years. Combinations of picloram at lower rates with fluroxypyr also were effective (14). Yellow toadflax control was best with picloram applied at 2.25 kg/ha. However, control was not long term (15).

Studies on the timing of herbicide application for perennial weed control (such as the toadflax species) indicate the best time for herbicide application is when carbohydrate reserves in the underground portions are lowest (16). It was found that reserve carbohydrates of Dalmatian toadflax were at their highest levels in the fall at the end of the growing season, and at the lowest point at the beginning of flowering (summer).

Black Henbane

Description

Black henbane is a member of the Solanaceae (nightshade) family. It is also known as fetid nightshade, insane root, and hog's bean (17). It is an annual or more frequently biennial. The plant is coarse, strongly-scented (due to alkaloids), viscid, and hairy. Plants range from 0.2 to 1 m tall with alternate leaves that are 5 to 20 cm long, rather narrow, irregularly lobed or divided, sessile, with the upper leaves clasping the stem. The flowers are borne in leaf axils and terminal spikes. The calyx is bell-shaped, 2 to 2.5 cm long, and enlarges as the fruit matures. Petals are fused into a tube with five spreading lobes that are greenish-yellow with dark purple veins (18). Seeds are kidney-shaped to oval, light brownish-gray, pitted, and about 1.5 mm long (19). Black henbane flowers from June to September and produces seed from July to October (16).

Importance

Black henbane is a native of Europe. It was introduced into North America as a medicinal and ornamental plant (20). It is commercially important in the pharmaceutical industry for medicinal purposes.

The major alkaloids that have been isolated from the leaves and flowering tops of black henbane are hyoscyamine, hyoscine (scopalamine), and atropine. Medicinal uses of this species date back to the 10th century (21). It has been used as a sedative and narcotic in cases of maniacal excitement, sleeplessness, and nervous depressions. It has frequently been used as a relief for coughs, especially in early stages of bronchitis and asthma. It also has been used internally as a laxative and carminative. External applications of the alkaloids have been used for neuralgia, rheumatism, painful glandular enlargements, ulcers, and hemorrhoids (22).

Establishment of black henbane usually occurs on disturbed or heavily grazed sites. Though not reported to be a highly competitive aggressor, it is capable of invading pastures and rangeland (20). Seed remain in the soil up to five years (23).

The alkaloids found in black henbane are very toxic to livestock, and all parts of the plant are poisonous. Because of the foul odor of the plant, livestock will seldom graze it and few cases of livestock poisoning have been reported.

Distribution

Since its introduction into North America, it has spread throughout much of the United States, especially in the Northeast, Midwest, and Rocky Mountains. Its typical habitat is pastures, rangeland, roadsides, and fence rows (18,19). Under commercial production in India, black henbane responds very favorably to increasing levels of soil nitrogen (24).

Control

The most commonly recommended method of control is cutting, hoeing, or digging of isolated plants before seed production (19,20). Because of the seed longevity in the soil, cultural control practices must be maintained annually. Since this species is usually found on disturbed or overgrazed soils, prevention of establishment may be by proper land management and by seeding desirable plant species. No biological methods of black henbane control have been reported in the literature.

Herbicides registered for black henbane control include dicamba at 0.28 to 1.7 kg/ha, 2,4-D at 1.12 to 1.24 kg/ha, and picloram at 0.28 to 0.56 kg/ha. Dicamba and 2,4-D have been effective for season-long control, while picloram provides residual control of germinating seeds the following year (25).

Tansymustard and Flixweed

Description

Tansymustard and flixweed are members of the Brassicaceae (mustard) family. These two species are often confused with each other. They are similar in appearance, but have certain vegetative and floral characteristics that distinguish them from each other.

Tansymustard is an annual with erect stems growing 0.3 to 0.6 m tall and branches above. Flixweed is also an annual but sometimes can behave as a biennial. It grows from 0.3 to 1 m tall and branches more above than tansymustard. Leaves of both species are alternate. Tansymustard leaves are divided into small segments, are smooth or often minutely gray and hairy. The lower leaves are 3 to 10 cm long and the upper leaves are smaller. Flixweed leaves are more finely divided, often fern-like, and 2 to 10 cm long, covered with stellate branched hairs which often gives the leaves a grayish color. Tansymustard flowers in a long terminal raceme whereas flixweed flowers in a dense terminal raceme (1). Both species have yellow to cream colored corollas. The most distinguishing floral characteristic that sets these two species apart is the fruit of each species. Tansymustard siliques are clavate with less than 20 seeds per pod which produce about 12,000 seeds per plant (26). The seeds are elliptic to oblong and less than 1 mm long (27). Flixweed siliques are narrowly linear, with usually greater than 20 seeds per pod and produce over 75,000 seeds per plant (28). The seeds are bright orange, about 0.7 mm long, and oblong to ellipsoid (27). Both species flower from March to August and fruit from June to October.

Importance

Tansymustard is native to the United States and flixweed was introduced from Eurasia (2,28). Both species tend to establish themselves on disturbed soils such as roadsides and move into fields and rangeland. There are several varieties of tansymustard. These differ mainly in the length of seed pods, lobing of leaves, and hairiness of plant stems (29). As an introduced species, flixweed has spread very rapidly, and it is estimated that 80% of the *Descurainia* population in eastern Washington is flixweed (5). Flixweed has been listed as one of the main weeds encountered during a five-year survey in the semiarid region of southwest Saskatchewan (28).

Flixweed has an estimated seed longevity of up to three years (28). No report was given for tansymustard seed longevity in the soil. Both mustards, because of their small size, are spread by wind, animals, and man.

Tansymustard has been reported to be an important component of the diet of cattle on desert grassland range while it was small (30). However, it has been reported to cause a poisoning in cattle referred to as "paralyzed tongue" (31). This occurs when tansymustard is consumed in large quantities at the flowering stage of growth. Treatment of such poisoning has been reported to consist of intravenously administered doses of ethanol. The efficacy of the treatment is dramatic in most cases with a single injection relieving all signs except those involving visual acuity. Some ranchers, in medicating their own stock, have given an undiluted bolus of "mainline Everclear" without fatal results (32).

Distribution

Tansymustard and flixweed are found throughout the United States and into Canada, primarily in the less humid areas (2,28). Their habitats are similar in that they are found in cultivated fields, roadsides, waste areas, and overgrazed pastures and rangeland. Studies on flixweed have shown that flixweed flowered earlier when root temperatures wee maintained at a constant 5 C rather that at 12, 20, or 27 C (2). In its native Europe, flixweed is primarily found on open, warm, nutrient-rich sandy or stony soils. However, in North America, it is not restricted to such soils.

Control

Fall-germinating tansymustard and flixweed can be mechanically controlled via cultivation in the fall or early spring as long as the rosettes are small. In wet periods, cultivation may not kill the plants because of their root system. Prevention of overgrazing of pastures and rangeland is an effective control practice. There are no specific biological control organisms reported for tansymustard or flixweed.

Herbicides registered for tansymustard and flixweed control include 2,4-D at 0.56 to 1.12 kg/ha, metsulfuron methyl at 14 to 21 g/ha, and chlorsulfuron at 53 to 158 g/ha. These herbicides and others will effectively control both species when applied according to the directions on the label.

References

1. Hitchcock CL and A Cronquist. *Flora of the Pacific Northwest*, 1978.
2. USDA Agricultural Research Service. *Selected Weeds of the United States*, 1970.
3. Robocker WC. Washington State Univ., Coll. of Agric. Tech. Bull. No. 79 (1974).
4. Alex JF. Canad. J. Bot. 40:295-307 (1962).
5. Gaines XM and DG Swan. *Weeds of Eastern Washington and Adjacent Area*. Camp-Na-Bor-Lee Assoc., Inc., 1972.
6. Robocker WC. Weed Sci. 18:720-725 (1970).
7. Lange AW. Weeds 6:68-70 (1958).
8. Kingsbury JM. *Poisonous Plants of the United States and Canada*. Prentice-Hall, Inc., Englewood Cliffs, NJ, 1964.
9. Parker R and D Peabody. PNW Extension Publication 135 (1983).
10. Hurst SJ. Newsletter of the Assoc. of Official Seed Analysts 55:61-72 (1981).
11. Rizza A and P Pecora. Ann. Entomol. Soc. Amer. 73:95-99 (1981).
12. Rizza A and P Pecora. Ann. Entomol. Soc. Amer. 77:182-187 (1984).
13. McCaffrey J. Personal communication.
14. Ferrell MA and TD Whitson. West. Soc. Weed Sci. Res. Prog. Rep. 40:72 (1988).
15. Dewey SA, R Michelson, and PW Foote. Proc. West. Soc. Weed Sci. 39:118-119 (1986).
16. Robocker WC. Weed Sci. 20:212-214 (1972).
17. Georgia AE. *A Manual of Weeds*, pp. 373-374. The Macmillan Publishing Co., 1927.
18. Dennis LJ. *Gilkey's Weeds of the Pacific Northwest*, p. 250. Oregon State Univ. Press, 1980.
19. Alberta Environmental Centre. *Weeds of Alberta*, pp. 152-153. Alberta Agriculture, 1983.
20. Lorenz RJ and SA Dewey. In: *The Ecology and Economic Impact of Poisonous Plants on Livestock Production* (LF James, MH Ralphs, and DB Nielsen, Eds.), pp. 309-336. Westview Press, Boulder, CO, 1988.
21. Kuma A, A Sharma, AK Singh, and OP Virmani. Curr. Res. Medicinal and Aromatic Plants 6:197-211 (1984).
22. Mital SP and RK Saxena. Indian J. Hort. 23:11-12 (1978).
23. Roberts HA. J. Applied Ecol. 23:639-656 (1986).
24. Pareek SK, MA Kidwai, RK Saxena, and R Gupta. Indian J. Agron. 33:249-252 (1988).
25. Edwards G. Personal communication.

26. Alley HP and GA Lee. *Weeds of Wyoming*, p. 89. Univ. of Wyoming Ext. Bull. 498 (1979).
27. Southern Weed Science Society. *Weed Identification Guide*. Southern Weed Science Society, Champaign, IL, 1990.
28. Best KF. Canad. J. Plant Sci. 57:499-507 (1977).
29. Welsh SL and JL Reveal. Great Basin Naturalist 37:279-365 (1977).
30. Hakkila MD, JL Holechek, JD Wallace, DM Anderson, and M Cardenas. J. Range Manage. 40:339-342 (1987).
31. Muenscher WC. *Poisonous Plants of the United States*. Macmillan Publishing Co., New York, 1975.
32. Staley E. Bovine Practitioner 11:35 (1976).

41

Cheatgrass

James A. Young

Abstract

Cheatgrass (*Bromus tectorum*) was introduced to the sagebrush (*Artemisia*) ranges of western North America largely during the 20th century. It has become a significant portion of the forage base for the range livestock industry in the Intermountain Area. The introduction of cheatgrass has changed secondary plant succession on most of the sagebrush ranges. The dominance of cheatgrass is closely related to the increased frequency and changed periodicity of wildfires that occur following establishment of this annual grass. Cheatgrass communities are open to invasion by other weed species that may not support grazing.

Introduction

In the Intermountain Area, cheatgrass is generally recognized by ranchers, land managers, and members of the general public as the annual grass *Bromus tectorum*. Older ranchers may refer to the plant as broncograss or six-weeks grass, but most recognize cheatgrass as a plant that furnishes the bulk of the livestock forage harvested by grazing animals from sagebrush ranges in the valleys and foothills of the Intermountain Area.

Range ecologists recognize cheatgrass as the species that has revolutionized secondary succession on sagebrush rangelands during this century, effectively delaying or blocking the establishment of seedlings of native herbaceous species. Wildlife managers recognize cheatgrass as the major component in the diet of upland game birds, especially chukar partridge (*Alectoris chukar*). All wildland managers recognize cheatgrass as the source of fine fuel that leads to disastrous wild fires on rangelands. The general public associates cheatgrass with repeated wildfires that destroy homes and other property and threaten human lives.

Nature of Cheatgrass

Cheatgrass is not a noxious weed in the legal sense. As far as I know, it is not listed as a noxious weed by the laws pertaining to weeds in any state. The Weed Science Society of America (WSSA) uses the officially recognized common name of downy brome for *Bromus tectorum* (1). WSSA uses cheat as the common name for *Bromus secalinus*. Cheatgrass is a major weed in winter wheat (*Triticum aestivum*) in the Pacific Northwest and on the Great Plains (2). In the Palouse wheat country of the Pacific Northwest, a population density of about 105 cheatgrass plants per m^2 will produce an average reduction in wheat yield of 27%.

Origin of Cheatgrass

Cheatgrass is not native to North America. Currently, the species is found in all of the adjacent 48 states of the United States except the coastal southeast. Cheatgrass was probably introduced independently into the United States several times in the 19th century. The source of the introductions could be about any part of the temperate world where domestic livestock are produced. The origins of cheatgrass probably are in Central and Southwestern Asia along with the origins of agriculture. Flannery (3) reported on an excavation in Iran of what are judged to be some of the oldest known examples of agriculture settlements in the world. The carbonized seeds found in association with the primitive domesticated animals included *Bromus tectorum*. Apparently since that time, cheatgrass has been the shadow of herds of domestic animals wherever man has spread agrarian cultures.

Introduction to the Intermountain Region

The timing of the introduction of cheatgrass to the sagebrush areas of the Great Basin is generally given as near 1900. Sereno Watson (4) of the 40th Parallel Survey did extensive botanical collections across the northern Great Basin. His collections included about 68 species of

Poaceae in 35 genera. Watson deemed nine of the taxa to be introduced, but *Bromus tectorum* was not included. In 1902, P.B. Kennedy (5) made a survey of range conditions in the sagebrush/bunchgrass area of northeastern Nevada. His trip started at Beowawe in Eureka County on the Humboldt River and ended across the Tuscarora Mountains in Elko County. He found no cheatgrass in the area. A previous survey of sheep winter and summer ranges north of Reno, Nevada, conducted by Kennedy and Doton (6) had also failed to identify cheatgrass. The trained botanist David Griffiths (7) collected plants and observed range conditions from Winnemucca, Nevada, to Burns, Oregon, during the same time period and failed to collect cheatgrass.

According to Mack (8), the first collections of cheatgrass in the Intermountain Region were from Spence's Bridge, British Columbia, in 1889; Ritzville, Washington, in 1893; and Provo, Utah, in 1894. The collections in the Pacific Northwest were in areas of wheat production and suggest that introduction may have occurred through contaminated seed. This was also the time period when steam powered threshing machines or combines became widely used in the Pacific Northwest (9). Custom operators traveled from farm to farm with both types of machines, often without properly cleaning residual wheat and weed seeds from them (10).

A rancher remembered seeing cheatgrass in Elko County, Nevada, in 1905 after an area had been heavily grazed the previous season by sheep from California (11). Mack (8) reports that during the period 1898 to 1904, collections of cheatgrass were made through the Intermountain Region. As Aldo Leopold (12) so aptly expressed it, "... the spread was often so rapid as to escape recording. One simply woke up one fine spring to find the range dominated by a new weed..."

The first collection of cheatgrass from western Nevada for which there is a voucher specimen was made by A.A. Heller in 1912 from the Reno area (Herbarium of the University of Nevada of Reno, Nevada, Heller #10488). Billings (13) suggested that specimens were collected in Elko County, Nevada, as early as 1905 to 1910. He also suggested from apparent herbarium notes that P.B. Kennedy reported cheatgrass along the railroad at Reno as early as 1906. Billings reported an old resident of Verdi, Nevada, told him of first noticing cheatgrass along a railroad siding where stock cars had previously been cleaned. The movement of cattle from the vast holdings of Miller and Lux in California to western Nevada was common during this time period (14). However, it should be noted that cheatgrass is a very rare species in the annual dominated flora of California (15).

Spread of Cheatgrass

Railroads provided a transportation mechanism for dispersal of cheatgrass; but perhaps of more importance, steam locomotives provided a continued source of ignition of wildfires along railroad rights-of-way which aided establishment and maintenance of cheatgrass stands. From the 1890s until World War II, the range sheep industry had a tremendous impact on rangelands. The industry consisted of flocks with no base property, known as tramp sheep; and flocks operating from ranches with great distances between seasonal ranges. Both types of range sheep operations were seasonally nomadic, moving for vast distances from winter, through summer and fall ranges, back to wintering areas annually (16,17).

The caryopses of cheatgrass, with their villose covering and sharp awn, are ideally adapted for dispersal by wool or hair. The nomadic nature of the sheep industry probably aided in the rapid spread of cheatgrass in the Intermountain Area.

Relation to Wildfires

Obviously, wildfires or prolonged, excessive, season-long grazing are not requirements for the spread of cheatgrass, but they are necessary for the seral dominance of cheatgrass. Fire had to be a part of the ecosystems that evolved in the sagebrush/bunchgrass region, but the advent of cheatgrass dominance of depleted understory vegetation, following excessive grazing, changed the seasonal occurrence and frequency of wildfires. Degraded sagebrush rangelands become segregated into subcanopy and shrub interspace environments (18). The predominance of litter accumulation, and subsequent nitrification occurs under the canopies of the sagebrush plants. The shrubs mine the interspaces of soil moisture and nutrients. If this condition persists long enough, the soils of the interspaces versus the subcanopy area reflect the large differences (19). The quality of the surface soil to serve as a seedbed becomes altered with vesicular crusting of the interspace soils.

Cheatgrass matures in July, while the native herbaceous species it replaced matured in late August. This makes it possible for wildfires to occur much earlier in the season, both prolonging the fire season and producing fires when the physiological stage of some species of native perennial species make them susceptible to injury by burning (20). The abundance of fine herbage produced by cheatgrass on years with sufficient moisture provides a fuel that is readily ignited. On degraded sagebrush ranges with minimal cheatgrass, there must be sufficient wind to bend flame heights to ignite the adjacent shrub for spread of the fire. An abundant, continuous cover of cheatgrass leads to rapid spread of wildfires so that under conditions of high temperatures, low humidity, and wind, the fires are very difficult to suppress.

The impact of wildfire on degraded sagebrush ranges infested with cheatgrass has long been described (21), but the effects are not totally understood. The dominant, woody sagebrush plants have most of their aerial portions consumed by the fire, and the plants do not sprout from crown or root buds. Most of the litter and surface soil-borne seedbank of cheatgrass seeds is found under the canopies of the sagebrush plants (22). The woody biomass of the shrub, plus litter accumulations, provides sufficient fuel to elevate temperatures high enough and long enough to destroy the cheatgrass seeds. Some of the seeds of cheatgrass in the interspace are consumed by the fire, but many survive even though the cheatgrass herbage is completely consumed.

The season following the fire, cheatgrass plants establish in reduced numbers in the interspace areas, and the subcanopy areas are bare except for seedlings of a few species of herbaceous annual species. The density of cheatgrass plants in the interspaces before burning may average 10,000 plats per m^2, while the post-burn population may be only 10 plants per m^2 (22). Despite the reduction in density, the post-burn population may equal or exceed the preburn herbage and seed production. Cheatgrass responds dynamically to the increased environmental potential available after the fire.

Influence of Cheatgrass on the Livestock Industry

On average to above average moisture years, cheatgrass probably results in more forage available for use by livestock than existed under pristine conditions. This is especially true on the more arid portions of the region, as in the shadscale (Atriplex confertifolia) zone where cheatgrass has recently expanded at the expense of relatively few native herbaceous species. In terms of serving as the forage base for the livestock industry, one of the major problems with cheatgrass is consistent production from year to year. Reliable forage production is vital in an area where cow and calf operations are predominant rather than stocker-type operations.

Charles Fleming was among the first to realize that cheatgrass was becoming a significant portion of the forage base for the livestock industry (23). He enumerated the disadvantages of cheatgrass as a forage species: (a) short green feed period, usually six weeks; (b) even with fall germination, the plants overwinter as rosettes that produce little harvestable herbage until mid-spring; (c) tremendous variation among years in forage production depending on available moisture for plant growth; (d) at maturity the awns reduce livestock preference and can be injurious to livestock; (e) especially on years of above average forage production, smut (*Ustilago*) may infest many seed florets; and (f) dry forage is often lost to wildfires.

Possibly the longest continuous records of cheatgrass forage production are those gathered by Burgess L. Kay at the Likely Table in northeastern California (24). The violent swings in cheatgrass production he recorded ranged from no harvestable forage (actually limited germination, much less growth) to production far in excess of what the rancher was capable of utilizing with a cow and calf operation. These changes in production were erratic in occurrence and very difficult to predict with any precision. Comparable perennial dominated ranges at the same location also had wide swings in forage production, but the perennials always produced some harvestable forage, even on the driest years.

On some desert winter ranges, cattle have been observed selecting cheatgrass seeds from the plants. The seeds persist in the inflorescence much longer on desert ranges than on upland sagebrush sites. On the Fish Springs Ranch in northwestern Nevada, virtually all the weaner calves were treated one season for lumpy-jaw, which is often attributed to damage from cheatgrass awns. The pathology of the organisms responsible for lumpy-jaw may be of more importance in producing the symptoms than the potential damage from cheatgrass awns.

On most public rangelands that are burned in wildfires, grazing is excluded for up to two years after the fire. If a portion of a pasture is burned, the entire pasture may be rested for that period. This can have a very significant influence on individual ranch operations and even on regional areas on bad fire years.

Influence on Plant Succession

In the Columbia Basin, Daubenmire (25) and Harris (26) clearly demonstrated that cheatgrass could invade perennial bunchgrass plant communities that had been protected from grazing and fire for as long as 50 years. Perhaps competition between established native perennials and cheatgrass is minimal, with the cheatgrass existing on environmental potential that no native species evolved to exploit, or which varies so greatly among years that no perennial species could exploit it.

R.L. Piemeisel (27,28) was probably the first scientist to recognize that secondary succession in disturbed sagebrush communities was preempted by a seral continuum of alien annual species that reached its peak with cheatgrass dominance. George Stewart and a young A.C. Hull were among the first to report the significance of cheatgrass to range management (29).

Perhaps the most significant paper in the chain of research which enumerated the dominant role of cheatgrass in succession in the sagebrush region was published by Robertson and Pearse (30). They pointed out, based on experimental evidence, that stands of cheatgrass closed sites to the establishment of seedlings of perennial herbaceous species. Piemeisel's research suggested this was the case based on observations of secondary succession, but Robertson and Pearse based their conclusions on experimental results. Raymond Evans, in a long series of experiments, was able to illustrate that competition for available soil moisture was the critical factor in cheatgrass competition (31,32). Richard Eckert Jr. worked with Evans and concurred on the importance of soil moisture as the limiting factor in cheatgrass competition and also researched the role of nitrate as the catalyst for plant growth once moisture was available in sagebrush ecosystems (33).

It has long been suspected that cheatgrass also suppressed the establishment of the seedlings of native woody species, but experimental evidence for landscape dominant species such as big sagebrush has not been available. In an attempt to release seedlings of bitterbrush (*Purshia tridentata*) from competition with cheatgrass, we applied the herbicide atrazine at the rate of 1.1 a.i. kg/ha over one year old bitterbrush seedlings. The treatment effectively controlled cheatgrass without harming the shrub seedlings, but it prompted a tremendous establishment of seedlings of big sagebrush and low rabbitbrush (*Chrysothamnus viscidiflorus*) (unpublished research ARS/USDA, Reno, NV). Because we were monitoring soil moisture in these experiments, we have evidence that controlling the cheatgrass initially conserved soil moisture compared to the control where no seedling establishment of the shrubs occurred. It also influenced the litter, surface soil structure, and nitrate content of the soil forming the seedbed (34-37). All of these factors are known to influence the potential of seedbed to support germination. It would appear that in the case of this experiment, cheatgrass does influence the successful establishment of seeds of woody species.

Closed Versus Open Community

The late Fritz Went often spoke of how desert annual communities were open to invasion (38). When I questioned him about cheatgrass, he found no conflict with this statement, despite the obvious closing of cheatgrass communities to seedlings of native perennial grasses. Dr. Went suggested it was only a matter of time and the introduction of the right species. Obviously, he was correct as indicated by the following examples: (a) medusahead (*Taeniatherum asperum*), on selected fine textured soils in the sagebrush zone; (b) bur buttercup (*Ranunculus testiculatus*), extremely widespread in the sagebrush zone; and (c) forage kochia (*Kochia prostrata*), introduced on research plots. Medusahead can virtually exclude cheatgrass on soils with considerable moisture holding capacity (39). Under drought conditions in northeastern California, I have seen medusahead disappear and be replaced by cheatgrass which in return is replaced by medusahead when precipitation returned to normal. Bur buttercup spread across the Great Basin as if cheatgrass was not present (40). In terms of the production of large quantities of viable seeds, persistence of seedbanks, periodicity of germination and emergence, soil moisture, and mineral nutrition requirements, cheatgrass must nearly overlap with the seedlings of native perennial grasses. Cheatgrass occupies the same environmental potential at the same moments in time as the seedlings of the native perennials. When two or more biologic organisms occupy the same time and space, interference occurs. The species with the most biologically effective, inherent physiologic systems are successful in these cases of interference, and in the case of sagebrush ecosystems, the winner is cheatgrass.

If an organism is introduced to a cheatgrass community that has the inherent potential to move the areas of biological interference in time and/or space, it can be successful in terms of establishment, persistence, and reproduction. Bur buttercup illustrates these concepts as an extreme ephemeral that completes its life cycle before cheatgrass initiates growth in the spring. Bur buttercup interference with cheatgrass is minimal because they actively grow and reproduce in different periods of time.

Based on what has already happened in portions of the Columbia Basin that formerly supported sagebrush/bunchgrass plant communities, Professor Ben Roché, Jr., of Washington State University has suggested that the dominance of cheatgrass in sagebrush ecosystems is a transitory event. Perennials will return to dominance, but they will be alien noxious weeds such as leafy spurge (*Euphorbia escula*) and the knapweed group (*Centaurea* spp.). Perhaps the perennial with the inherent potential to dominate the more arid portions of the Intermountain Area has not yet been introduced. This is perhaps the most important issue of this conference.

Competitive Nature of Cheatgrass

In the half century since it has become generally recognized how competitive cheatgrass is and how difficult it is to restore native species, many scientists have undertaken to identify the characteristic that makes the species so successful [see Klemmedson and Smith (41) for the last comprehensive literature review].

To illustrate this approach, let us examine one group of closely related characteristics, seed germination and dormancy. Seeds of cheatgrass are generally highly viable (>85%) at maturity. There have been reports of after-ripening requirements, but generally seeds of cheatgrass produced in the sagebrush zone will germinate soon after maturity without pretreatment. Germination is more rapid and higher at low, moderate, and high incubation temperatures than the seeds of any other native or exotic perennial grass species adapted to the sagebrush zone (42). Cheatgrass seeds are dispersed to the surface of seedbeds and are primed for simultaneous germination with the first moisture event following the summer drought. Considering the relative seed production of cheatgrass and native perennial grass species, it is obvious that this germination-dormancy system assures an overwhelming abundance of cheatgrass seedlings with the first rain.

The seeds of cheatgrass are not particularly adapted by size, protective coverings, or speed of germination for germination on the surface of seedbeds (36,37). Seeds that fall into favorable microtopographic situations or are covered by litter will germinate. Seeds placed in unfavorable seedbed conditions that are exposed to repeated cycles of partial germination and drying acquire a dormancy (43). This dormancy can last for as long as three years. Once cheatgrass has established on a site and gone through a couple of cycles of germination, growth, seed production, and dispersal, there are always twice as many viable, but dormant, seeds in the seedbank as there are established plants in the community. The seeds in the seedbank lose their dormancy slowly over time, but the germination of dormant seeds can be enhanced by enrichment of the seedbed with a nitrate source (44).

Seed germination is just one of the numerous ecophysiologic characteristics where similar, complex, and interacting webs of adaptive characteristics have been enumerated for cheatgrass. Perhaps no single characteristic is responsible for the inherent competitiveness of cheatgrass. The breeding system of the species that combines self-pollination with environmentally induced outcrossing may be the key to the success of the species because it allows for rapid evolution in changing environments (45).

Influence of Grazing Systems

Since the 1960s, virtually all grazing allotments on public lands in the sagebrush zone have been placed under a grazing management system to favor perennial grasses by allowing some type of deferment of grazing until after the perennials have had the opportunity to reproduce. This deferment may not be annual, but rotated yearly so that once in every three years the grazing is deferred. In retrospect, these systems are ideally adapted for the propagation of cheatgrass. The only thing better would be to eliminate grazing entirely. If current grazing management practices are continued as we cycle through years where cheatgrass growth is abundant to superabundant, there will be wildfires in the sagebrush grasslands at an unprecedented scale. Grazing management to favor native perennial grasses is doomed to failure unless there are sufficient perennial grasses remaining in the ecosystem to dampen the dynamics of cheatgrass. We are managing cheatgrass ranges as if they were perennial grass dominated ranges, not managing cheatgrass.

References

1. Anon. Weed Sci. 32:1-91 (1984).
2. Young JA, RA Evans, WO Lee, and DG Swan. Farmers Bull. No. 2278, USDA GPO, Washington, DC (1984).
3. Flannery KV. In: *The Domestication and Exploitation of Plants and Animals* (PJ Voke and GW Dimbelby, Eds.), pp. 73-100. Aldine Publ. Co., Chicago, IL, 1969.
4. Watson Sereno. *Botany, U.S. Geological Exploration of the Fortieth Parallel, Vol. 5.* GPO Washington, DC, 1871.
5. Kennedy PB. Nev. Agric. Exp. Sta. Bull. 55 (1903).
6. Kennedy PB and SB Doton. Nev. State Univ., Agric. Exp. Sta. Bull. 51 (1901).
7. Griffiths D. Bureau of Plant Industry Bulletin 15, USDA, 1902.
8. Mack RN. Agro-Ecosystems 7:145 (1981).
9. Young JA and JD Budy. *Endless Tracks in the Woods.* Crestline Publ., Saratoga, FL, 1989.
10. Young JA. Agric. History 62:123-130 (1988).
11. Young JA, RA Evans, and J Major. J. Range Manage. 27:127-132 (1972).
12. Leopold A. The Land 1:310-313 (1941).
13. Billings WD. In: *The Earth Intravistion: Patterns and Processes of Biotic Impoverishment* (GM Wadwell, Ed.), Chapter 12. Cambridge Univ. Press, New York, 1990.
14. Young JA and BA Sparks. *Cattle in the Cold Desert.* Utah State Univ. Press, Logan, UT, 1985.
15. Heady HF. In: *Terrestrial Vegetation of California.* (MG Barbour and J Major, Eds.), pp. 491-514. John Wiley and Sons, New York, 1977.
16. Sawyer BW. *Nevada Nomads.* Harlan-Young Press, San Francisco, 1971.
17. Harris LR. PhD Thesis, Yale Univ. Dept. Anth., New Haven (1974).
18. Charley JL and NE West. J. Ecol. 63:945-964 (1975).

19. Eckert RE Jr, RR Peterson, MK Wood, WH Blackburn, and JL Stephens. Nev. Agric. Exp. Sta. Bull. TB-89-01, Univ. Nevada, Reno (1989).

20. Wright HA and JD Klemmedson. Ecology 46:680-688 (1965).

21. Kearney TH, LJ Briggs, and HT Shautz. Utah. Agric. Res. 1:365-417 (1914).

22. Young JA and RA Evans. J. Range Manage. 31:283-289 (1978).

23. Fleming CE, MA Shipley, and MR Miller. Nev. Agric. Exp. Sta. Bull. 159 (1942).

24. Young JA, RA Evans, RE Eckert Jr, and BL Kay. Rangeland 9:266-270 (1987).

25. Daubennaire R. Ecology 21:55-64 (1946).

26. Harris GA. Ecol. Monogr. 89-111 (1967).

27. Piemeisel RL. USDA Agric. Tech. Bull. 654 (1938).

28. Piemeisel RL. USDA Cir. No. 739 (1945).

29. Stewart G and AC Hall. Ecology 30:58-74 (1949).

30. Robertson JH and CK Pearse. Northwest Sci. 19:58-66 (1945).

31. Evans RA. Weeds 9:216-223 (1961).

32. Evans RA, RE Eckert Jr, and BL Kay. Weed Sci. 18:154-167 (1970).

33. Eckert RE Jr, CJ Klemp, JA Young, and RA Evans. J. Range Manage. 23:445-447 (1970).

34. Eckert RE Jr and RA Evans. J. Range Manage. 20:35-41 (1967).

35. Evans RA, RE Eckert Jr, BL Kay, and JA Young. Weed Sci. 17:166-169 (1969).

36. Evans RA and JA Young. Weed Sci. 18:697-703 (1978).

37. Evans RA and JA Young. Proc. WSSA, Dallas, TX, p. 115 (1971).

38. Went RW. Sci. Amer. 192:68-75 (1955).

39. Hironka M and EW Tisdale. Ecology 4:810-812 (1963).

40. Buchanan BA, KT Harper, and NC Frischknecht. Great Basin Naturalist 38:90-96 (1978).

41. Klemmedson JO and JG Smith. The Botanical Review, April-June, pp. 226-262 (1964).

42. Young JA and RA Evans. Agric. Res. Results ARR-W-271 November, Agric. Res. Ser. USDA, Oakland, CA (1982).

43. Young JA, RA Evans, and RE Eckert Jr. Weed Sci. 17:20-26 (1969).

44. Evans RA and JA Young. Weed Sci. 23:351-357 (1975).

45. Young JA and RA Evans. Weed Sci. 24:186-190 (1975).

42

Tansy Ragwort (*Senecio jacobaea*): Importance, Distribution, and Control in Oregon

Eric M. Coombs, Thomas E. Bedell, and Peter B. McEvoy

Abstract

Tansy ragwort (*Senecio jacobaea*, Asteracaeae) is one of the most serious noxious weed problems in western Oregon. The pyrrolizidine alkaloids in ragwort are toxic to cattle and horses and less so to sheep. Livestock losses associated with ragwort poisoning were estimated at over $4,000,000 annually in the 1970s. Tansy ragwort infested over 300,000 km^2 in 17 counties in western Oregon in 1987. Traditional control practices in pastures, including chemical, mechanical, and cultural control, have provided localized results. Biological control of ragwort was implemented on range and forest lands where other weed control methods were ineffective or too costly. Since 1960, three natural enemies of ragwort (the cinnabar moth, ragwort flea beetle, and ragwort seed fly) were introduced from Europe as biological control agents. In 1974, an intensive biological control program was initiated by the Oregon Department of Agriculture. By 1987, the density of ragwort in western Oregon was reduced by over 90%. Data from intensive research studies and regional surveys demonstrated that biological control was an effective method of reducing tansy ragwort infestations in western Oregon.

Introduction

Tansy ragwort is a biennial or short-lived perennial weed that is native to Europe. The flowers of tansy ragwort are yellow and daisy-like, usually with 13 ray flowers. The leaves are light to dark green and deeply lobed. Single plants, which range from 0.4 - 1.5 m tall, are capable of producing over 150,000 seeds, which can remain viable in the soil for three years or longer. Its native range extends from Norway south through Asia Minor and from Great Britain east to Siberia. Ragwort infests the northwest coast of the United States, the maritime provinces of Canada, New Zealand, Australia, and in Argentina (1). It was first recorded in western North America in British Columbia in 1913 (2). Tansy ragwort was first collected in Oregon in 1922 (3). Tansy ragwort has since become widely established throughout Oregon west of the Cascades. New infestations in eastern Oregon are becoming more frequent despite intensive eradication efforts. In 1987, it was estimated that in the 17 western Oregon counties, over 300,000 km^2 were infested with tansy ragwort (4).

The toxic properties of tansy ragwort, along with its ability to invade and displace desirable vegetation in pastures, have characterized ragwort as a serious noxious weed. Tansy ragwort causes chronic toxicity in cattle, horses, and swine when pyrrolizidine alkaloids are converted to pyrroles in the liver and results in hepatic dysfunction or failure (5). Death can occur when animals have ingested three to seven percent of their body weight in ragwort. Sheep and goats are less susceptible to poisoning caused by the ingestion of ragwort (6).

In the 1970s, it was estimated that losses to the livestock industry in western Oregon exceeded $4,000,000 annually (4). This figure estimates only the direct loss due to animal mortality and does not include weight loss, decline in animal condition, health care, and costs associated with traditional weed control methods including herbicides. Some ranchers suffered annual losses of five to ten percent of their herds and some dairy operations were forced out of business. As a result of these losses, the Oregon State Legislature classified tansy ragwort as a noxious weed, mandated control measures, and required the Oregon State Department of Agriculture to start an intensive biological control program in 1975.

Control

The Oregon Department of Agriculture in cooperation with Oregon State University initiated an intensive integrated tansy ragwort control program in 1975 which included chemical, physical, cultural, preventative, and biological methods of weed control (7-9).

In 1975, tansy ragwort was classified as a noxious weed by legislative action. State laws and local county regulations describe warranted actions against tansy ragwort infestations. Regulations may vary from complete eradication of any ragwort detected in a county, to supporting an intensive biological control program. Guidelines that require cleaning logging equipment when it is moved from ragwort-infested sites to uninfested sites have been established. In some counties, weed supervisors cite landowners who allow ragwort to go to seed.

Chemical Control

The use of herbicides to control tansy ragwort has been successful in some locations (10). Treated areas must not be grazed until the label requirements for reentry by livestock have been satisfied. The use of 1.12 - 2.24 kg/ha of 2,4-D LV ester or amine has been used in pasture situations and works best before flowers appear. Treatments in the fall after seedling germination are also effective. Retreatment is often required for plants that germinate from the ragwort seed bank in the soil. Picloram (Tordon), a restricted use herbicide in Oregon, has been used during the flowering stage at 0.28 kg/ha on rangelands and permanent pastures, but should not be used near sensitive crops. Dicamba (Banvel) at 1.12 kg/ha can also be used up to the flowering stage. Two combination herbicides [2,4-D + dicamba (Weedmaster) at 2.17 kg/ha and triclopyr + 2,4-D (Crossbow) at 1.26 - 1.68 kg/ha] have also been successful in controlling tansy ragwort. The use of herbicides requires specific conditions, equipment, and licensing to be used in range and pasture lands. Labeling changes and local regulations vary, so always consult local authorities. Read all labels and comply with all appropriate regulations before using herbicides.

Physical Control

Physical methods of controlling tansy ragwort by physically or mechanically removing or altering the plants have had variable success. In areas where tansy ragwort occurs as occasional scattered plants, hand pulling may be a useful method of control. In 1984 a court order imposed a ban on the use of herbicides on federal lands in Oregon and Washington, and hand pulling of tansy ragwort in satellite infestations was used more frequently. Successful control occurred primarily where plants were only recently established. In areas where tansy had been established, the seed bank was sufficient to warrant periodic retreatment.

Also, pulling can cause micro-disturbances where tansy seeds can germinate.

Mowing ragwort removes the upper parts of tansy plants and prevents them from going to seed, but plants may become short-lived perennials. After mowing, regrowth of plants may occur, particularly if soil moisture conditions are conducive. Seed viability of late flowering plants appears to be lower.

Cultural Control

Various cultural practices have been utilized to control tansy ragwort infestations. Competitive plantings of aggressive grasses in disturbed areas can reduce seedling survival and establishment. Type of livestock utilized in pastures may assist in controlling ragwort. Sheep grazing in heavily infested pastures has been successful in western Oregon. Grazing cattle and horses in pastures that are heavily infested with tansy ragwort is not recommended in the spring when the ragwort rosettes are small and often accidentally ingested while feeding on grass. When ragwort plants are tall and in flower, cattle and horses seem to be able to avoid consuming ragwort, unless the pasture is severely overgrazed. Heavy grazing during early fall rains tends to promote ragwort seedling establishment.

Using fertilizers and irrigation in conjunction with managing the class, season, and intensity of livestock has been shown to be successful in controlling tansy ragwort (7).

Preventative

In eastern Oregon in Wallowa county, a hay quarantine on western Oregon hay went into effect in 1983. Hunters bringing hay from western Oregon were required to exchange their hay for locally grown hay. As part of this program, the Oregon Department of Fish and Wildlife has included a section on tansy ragwort in the hunting proclamation, informing western Oregon hunters to avoid moving contaminated hay into eastern Oregon. Since the start of the hay quarantine, the number of new tansy ragwort sites in Wallowa county has decreased 85%, despite increased efforts in the ragwort detection program (Figure 42.1). The success of the biological control program on tansy ragwort in western Oregon was becoming evident at the time the hay quarantine was imposed. With the drastic reduction of tansy ragwort in western Oregon, hay taken into eastern Oregon may have been less likely to cause new infestations.

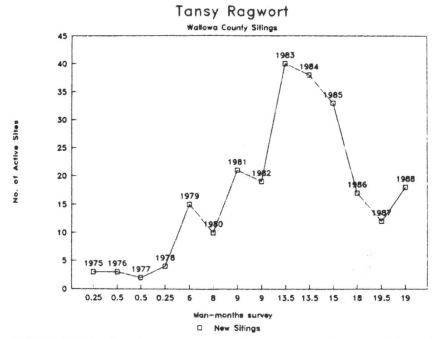

FIGURE 42.1 Number of new tansy ragwort sites detected each year in Wallowa County, Oregon. In 1983, a hay embargo was enforced to prevent hay from western Oregon from entering the county.

Biological Control

Biological control is the most successful method of controlling tansy ragwort in western Oregon. Classical biological control involves the introduction and management of selected natural enemies of a pest. We do not include sheep grazing as a form of biological control because sheep are not natural enemies nor are they host specific to ragwort, but rather as a method of cultural control. In the 1970s, economic losses caused by tansy ragwort poisoning were having a major impact on the livestock industry in the western part of the state. Similar problems were also occurring in northwestern California and western Washington. In western Oregon, some ranchers were losing from five to ten percent of their herds each year to ragwort poisoning. By 1975, when the state biological control program was funded, cooperation with USDA-ARS and other states was already in progress. The Oregon Department of Agriculture in cooperation with Oregon State University and USDA-ARS began an interdisciplinary program that emphasized the introduction, collection, distribution, and study of the biological agents of tansy ragwort. Three insects were cleared for use as biological control agents in the Pacific Northwest by the USDA-ARS. After host specificity studies, approval was given for the USDA-ARS to introduce a moth, a flea beetle, and a seed fly on tansy ragwort.

The cinnabar moth (*Tyria jacobaeae*, Lepidoptera:Arctiidae) was first released in Oregon in 1960 (11). The moth, originally from France, was originally released in California in 1959 (12). The ragwort flea beetle (*Longitarsus jacobaeae*, Coleoptera:Chrysomelidae) was first released in Oregon in 1971 (13). The flea beetle was originally released in California from Italy in 1969 (14). The ragwort seed fly (*Pegohylemyia seneciella*, also referred to as *Hylemyia seneciella*, Diptera:Anthomyiidae) was first released in Oregon in 1966, and re-released in 1976 because the first release was thought to have failed (4). The seed fly was introduced from Italy (15). Since 1975, all three biological control agents have been widely distributed throughout the range of tansy ragwort in Oregon (16).

Cinnabar Moth. The striking red and black adult moths emerge in the late spring from overwintering pupae. Mated females lay eggs in small groups on the underside of the basal leaves of ragwort. Starting at the terminal buds, the larvae consume floral parts and leaves, defoliating the plant, often leaving a bare stalk. Larvae will move to nearby plants after stripping a plant. Plants defoliated by cinnabar larvae may reflower if adequate moisture becomes available in the late summer, and may even become short-lived perennials. After five instars, the larvae pupate in soil debris. The moth has only one generation a year. A more in-depth biology of the cinnabar moth is presented by Dempster (17).

Collection of larvae for distribution usually occurs when the larvae are nearly full grown. The larvae can be shaken into a paper sack and given some plant material for food. If the larvae are small, the stalk can be clipped below the group of caterpillars and placed into a sack. The paper sack must be taped shut and corners and seams taped, or larvae will escape during transit. Releases are generally made with 1000 larvae which are gently dumped out onto tansy plants at the new site. Care should be taken to insure that the insects do not become overheated or too damp while being stored and transported. The cinnabar moth alone is not an effective biological control agent at most sites. It is usually released with the flea beetle and the seed fly.

Flea Beetle. The tiny golden ragwort flea beetle has one generation per year. This insect is unique in that both the larvae and the adults feed on the plant. The adults emerge in the early summer, feed somewhat, and enter a reproductive diapause until the fall when they begin to lay eggs. Eggs are laid on the plants and the larvae tunnel into the leaf petioles and the roots. Extensive damage to roots can often lead to plant mortality. The adults feed on the leaves, making a characteristic "shot hole" appearance. The life cycle and biology of the ragwort flea beetle is presented by Frick and Johnson (18).

Collection of the ragwort flea beetle generally begins in the fall when they are mating. Adults become more active following rains. The insects are collected with a gas-powered vacuum apparatus which sucks the adults off the rosettes. The adult flea beetles are then sorted and counted and put into paper cups in groups of 250-500. Some tissue paper is added to give the beetles a place to crawl around and also to absorb extra moisture. Some leaf material is also added for food. The flea beetles are then shipped to various areas around the state and released onto rosettes in infested areas.

The ragwort flea beetle has proven to be the most effective biological agent against tansy ragwort in Oregon (19). Intensive studies at research plots studied by scientists at Oregon State University coupled with regional surveys at 42 sites in western Oregon (Figure 42.2) conducted by

FIGURE 42.2 Map of Oregon showing major tansy ragwort infestation area (dotted line), several infestations in the northeastern counties (open circles), and the 42 regional survey sites (solid dots).

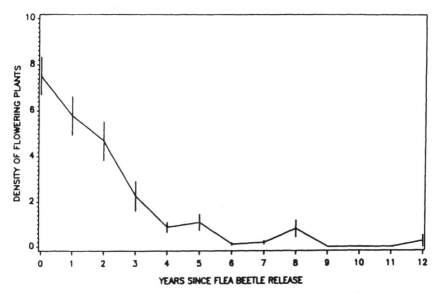

FLOWERING PLANT DENSITY ($\bar{x}$ ± SE) PER m^2

OREGON DEPARTMENT OF AGRICULTURE SURVEY
EXCLUDING MONROE SITE

FIGURE 42.3 Average density of flowering tansy ragwort plants m^2 from 50 0.25 m^2 plots at 41 of the 42 regional survey sites in western Oregon, based on the number of years since release of ragwort flea beetles (one site at Monroe was eliminated after a conifer nursery was planted on the study site).

the Oregon State Department of Agriculture have shown 90% reductions in ragwort populations six years following the release of the flea beetle (Figure 42.3). At several sites, no flowering ragwort was observed in the plots nine years after the release of the beetles. Research has shown that the flea beetle can effectively control low densities of ragwort. The cinnabar moth in contrast is more dependent on ragwort density in order to have much of an impact.

Seed Fly. The ragwort seed fly looks similar to the common house fly. The adults emerge in the spring from pupae that overwinter in the soil. Adult flies are often found walking around on ragwort seed heads when they are at the bud stage. Females lay eggs between the florets in the flowers. The larvae hatch and tunnel into the involucre and feed on tissues and cause a reduction of the seeds. There is usually one larva per flower. A flower infested with a seed fly larva often exhibits a frothy exudate emanating from the flower disc, or the disc may be discolored and appear darker than normal. After the larva completes its cycle within the flower, it emerges and drops to the ground and pupates in the soil . The biology of the seed fly is covered by Cammeron (20).

Collection and distribution of the ragwort seed fly is accomplished by transplanting flowering plants infested with larvae to new locations. This is now unnecessary in most cases because of the fly's ability to widely distribute itself. The seed fly has been found in satellite infestations in eastern Oregon over 200 km away from any known release sites.

Summary and Conclusions

The integrated control of tansy ragwort has been one of the most successful weed control programs for improving infested pastures and rangelands in Oregon. The interdisciplinary application of integrated pest management methods is saving over $4,000,000 a year in direct losses of livestock due to tansy ragwort poisoning. Biological control is the primary method of weed control that keeps this toxic pest in check. The cinnabar moth, flea beetle, and seed fly have been widely distributed throughout the ragwort-infested areas in western Oregon through an intensive collection and distribution program implemented by the Oregon Department of Agriculture. Research studies have demonstrated that the flea beetle is the primary driving force of the regional reduction of tansy ragwort, exceeding 90% in most areas.

The success of the biological control program is not without problems. In eastern Oregon, where eradication programs are ongoing, tansy ragwort continues to persist. Attempts to establish the biological agents in the eastern counties have failed, with the exception of the seed fly. A search for additional biotypes or cold hardy strains may be required if biological control is going to be implemented in the eastern part of the state. This problem is also manifested by the difficulty of maintaining the biological agents at elevations exceeding 1000 m.

In recent years, periodic resurgences of ragwort have occurred at numerous locations. In most cases, by the time the problem was evident to land managers, the biological agents were beginning to control the outbreak. Most of the sporadic reinfestations are manifestations of a management problem that existed two years before the observed resurgence. Apparently, the biological agents are able to maintain a presence in the community so that when outbreaks occur, there are sufficient numbers of agents to eventually control them.

Studies are needed to determine an economic threshold level, the amount of ragwort to warrant control for rangeland and pastures. Experienced observers generally consider a threshold level of one to two percent cover in tansy ragwort a threat to susceptible species of livestock.

428

In some instances, after the biological control agents have controlled ragwort infestations, heavily grazed sites may become dominated by other noxious weeds. An understanding of the ecological principles of vegetation management is crucial in maintaining pastures and rangeland in proper condition for livestock use.

References

1. Harper JL. Herbage Abstracts 28:151-157 (1958).
2. Harris P, ATS Wilkinson, LS Thompson, and ME Neary. In: *Proc. IV Intl. Biol. Contr. Weeds* (TE Freeman, Ed.), pp. 163-173. Univ. Florida, Gainsville, 1976.
3. Isaacson DL. MS Thesis, Oregon State Univ., 1971.
4. Brown RE. In: *Proc. VII Intl. Symp. Biol. Contr. Weeds* (ES Del Fosse, Ed.), pp. 299-305. Inst. Patol. Veg. Rome, Italy, 1988.
5. Maddocks AR. *Chemistry and Toxicology of Pyrrolizidine Alkaloids.* Academic Press, London, UK, 1986.
6. Muth OH. J. Amer. Vet. Med. Assoc. 153:310-312 (1968).
7. Bedell TE, RE Whitesides, and RB Hawkes. *Pasture Management for Control of Tansy Ragwort.* Pac. NW Coop. Ext. Bull. PNW-210, Oregon State Univ. Ext. Serv., Corvallis, 1981.
8. Isaacson DL and DT Ehrensing. *Biological Control of Tansy Ragwort.* OR Dept. Agric. Weed Contr. Bull. No. 1, 1977.
9. Hepworth HM and LO Guelette. *Tansy Ragwort.* Pac. NW Ext. Bull. PNW-175, 1977.
10. Burrill LC, RD William, R Parker, SW Howard, C Eberlein, and RH Callihan. *Pacific Northwest Weed Control Handbook*, p. 276. Ext. Serv. OR, WA State Univ., and Univ. ID, 1990.
11. Isaacson DL. Entomophaga 18:291-303 (1973).
12. Hawkes RB. J. Econ. Entomol. 61:449-501 (1968).
13. Isaacson DL. In: *Proc. IV Intl. Symp. Biol. Contr. Weeds* (TE Freeman, Ed.), pp. 189-192. Univ. Florida, Gainsville, 1976.
14. Frick KE. California Agriculture 24:12-13 (1970).
15. Frick KE. J. Econ. Ent. 62:1135-1138 (1969).
16. Coombs EM, RE Brown, and RB Hawkes. Proc. 36th Ann. Symp. OR Soc. Weed Sci. pp. 33-40 (1987).
17. Dempster JP. Adv. Ecol. Res. 12:1-36 (1982).
18. Frick KE and GR Johnson. Ann. Ent. Soc. Amer. 66:358-367 (1973).
19. McEvoy PB, CS Cox, RR James, and NT Rudd. In: *Proc. VII Intl. Symp. Biol. Contr. Weeds* (ES Del Fosse, Ed.), pp. 53-64. Inst. Patol. Veg., Rome, Italy, 1989.
20. Cammeron E. J. Ecol. 23:266-322 (1935).

43

Other Toxic Pyrrolizidine Alkaloid-Containing Plants: Importance, Distribution, and Control

M.H. Ralphs and D.N. Ueckert

Abstract

Plants containing pyrrolizidine alkaloids (PA) are the most important group of poisonous plants in terms of poisoning to both livestock and humans worldwide. In addition, many are considered noxious weeds. PAs primarily affect the liver, causing chronic hepatic failure. Death may occur after three weeks to several years later. Three plant families account for the majority of livestock poisoning in the U.S. *Senecio* species (Compositae family) cause the greatest livestock loss in the western U.S: tansy ragwort (*S. jacobaea*) in the northwest and Canada; threadleaf groundsel (*S. douglasii* var. *longilobus*) in the southwest; and Riddell's groundsel (*S. riddellii*) in the midwest. Seed from *Crotalaria* species (Leguminosae family) contaminate livestock feed in the southeast. Several weeds from the Boraginaceae family also contain high concentrations of PA.

Introduction

Pyrrolizidine alkaloids (PA) are the most significant group of secondary plant compounds in terms of adverse effects on humans and their economic activities (1). It is estimated that 3% of the world's flowering plants contain PA (2). Over 160 PA have been isolated from 332 species in 63 genera and 13 families (1). Poisoning occurs when livestock consume sufficient quantities of plants containing high concentrations of PA or are fed hay contaminated with these plants or grain contaminated with their seeds. Humans have been poisoned by consuming plants containing PA as green vegetables, herbal teas, and from foods contaminated with seeds of PA plants. They may also ingest PA indirectly through milk products from livestock consuming PA plants and from honey produced by bees feeding on the plants, although the levels of PA detected in such foodstuffs are low.

Toxicity

Pyrrolizidine alkaloids contain the basic nitrogen atom in the bicyclic 5-membered pyrrolizidine ring system. All toxic PA have a 1,2 double bond in one of the five-membered rings and esterification of the substituent hydroxyl groups (3). PA are not toxic in either the free-base or N-oxide form as they occur in the plant. They become toxic when the free-base is converted to highly reactive pyrroles by mixed function oxidase enzymes in the liver.

Since the conversion to a pyrrole occurs in the liver, this organ is primarily affected. Liver cells are damaged and are unable to reproduce. As the cells senesce or die, they are not replaced and the animal suffers a slow insidious death from chronic liver failure (4). Another syndrome of the liver is veno-occlusive disease. The small branches of hepatic veins become occluded, leading to ascites and edema (fluid accumulation in the abdominal cavity and interstitial space between cells, respectively), reduced urinary output, and subsequent urea poisoning (1).

Some PA (particularly in *Crotalaria*) also produce changes in the lungs and heart, resulting in pulmonary arterial hypertension and right ventricular hypertrophy that proceed to congestive right heart failure (1). PA can also damage the central nervous system, kidneys, visceral organs, and any organ to which the reactive pyrroles may be transported.

Many PA are also carcinogenic. Pyrroles are bidentate alkylating agents which crosslink with RNA, DNA, and other proteins resulting in carcinogenic and mutagenic activity. The high incidence of liver cancer in humans in South Africa has been attributed to low levels of PA in the diet (1).

The acute or chronic nature of poisoning depends upon the particular PA in the plant, their toxicity and relative concentration, the amount eaten, and rate at which the plant is ingested. There appear to be thresholds of tolerance in the liver, or degradation in the rumen, that must be exceeded before PA become dangerous (5). Above the threshold level, PA plants ingested for 15 to 20 days or longer may lethally damage the liver, and the animal may die three weeks to 18 months later from liver failure. Acute poisoning is rare because of the low palatability of plant species containing high PA concentrations. However, if animals are forced to eat large quantities in a short time, necrosis in the liver and severe hemorrhage cause death within one to two days (6).

Young animals are most susceptible to PA because of increased cellular metabolic activity and higher levels of mixed function oxidase enzymes in the liver (7). Cattle and horses are most susceptible to PA, while sheep are almost 100 times more resistant (7). Clinical signs of poisoning in cattle (8) include:

- Slight disinterest in food; if manger-fed with other animals, the animal may mouth its food without eating much.
- Loss of interest in surroundings, stands off by itself, and is reluctant to move; may stand much of the day with its head lowered.
- Signs of discomfort evidenced by occasional kicking at the belly.
- Gradual weight loss or failure to gain weight; the hair coat becomes dull and rough.
- Diarrhea or constipation, depending on the type of diet fed.
- Behavioral changes, most typically belligerence or stubbornness.
- As intestinal edema develops, the animal may strain severely, sometimes causing prolapsed rectum.
- In terminal stages, the animal may pace incessantly, bumping into fences and other objects; at this stage, it may appear to be blind, weak from lack of food, and unsteady in the hindquarters when walking; the rear knees become springy.
- As further weakness develops, the animal becomes recumbent, lying on its brisket and straining; it usually dies without struggle.
- The liver is enlarged, discolored, and fibrotic.

Plant Families

PA poisoning occurs from three principal plant families: Compositae, Leguminosae, and Boraginaceae (9). In the western U.S., *Senecio* species (Compositae) family are the most important group. Houndstongue (*Cynoglossum officinale*, Boraginaceae) is also becoming more of a problem.

In the southeast, *Crotalaria* (Leguminoseae) poses significant problems in that its seeds contaminate grains and soybeans.

Compositae

The *Senecio* species are commonly called groundsel, ragwort, or butterweed. The name *Senecio* is derived from the Latin "senex" (old man), referring to the prominent white bristles comprising the pappus on the yellow flowers. Over 3000 species exist, ranging from annual forbs to shrubs and vines. Of the 112 species that occur in North America, only eight have been documented as poisonous to livestock (6), but many other species also contain PA. Information on the relative toxicity of selected species is presented in Table 43.1.

Tansy ragwort (*S. jacobaea*) has caused the greatest economical loss from the standpoint of both livestock poisoning and as an introduced noxious weed in northwestern U.S. and Canada. Snyder (10) estimated that economic loss due to cattle poisonings in Oregon exceeded $1.2 million in 1971. Forage losses from tansy ragwort invading pastures exceeded $600,000 (11). Tansy ragwort is described in detail by Coombs, Bedell, and McEvoy in this volume. It can be controlled by applications of 2,4-D during spring or by dicamba during late spring-early summer (4).

Threadleaf or wooly groundsel (*S. douglasii* var. *longilobus*) has caused extensive livestock losses in west Texas, New Mexico, and Arizona.

TABLE 43.1 Toxicity of *Senecio* species fed to 220 kg calves for 20 days (18).

Species	PA Lethal Dose	PA Level in Plant	Lethal amount of plant Dry	Lethal amount of plant Fresh[4]
	mg/kg body wt	%	----- g/day -----	
Tansy ragwort	2.3[1]	0.31	161	806
Threadleaf groundsel	10-13[2]	2.19	118	594
Riddell's groundsel	15-20[3]	6.40	63	312
Riddell's groundsel (high PA)	15-20[3]	18.00	22	111

[1] from Johnson (8)
[2] from Johnson and Molyneux (5)
[3] from Johnson et al. (7)
[4] Based on dry matter content of 20%

Cattle losses in the Trans Pecos and Davis Mountains of Texas were 5 to 30% during 1927-1933 (12-14), and threadleaf groundsel continues to be a serious threat to cattle and horses (15). It has also been implicated in human poisoning from its use in medicinal herbal teas (1).

Threadleaf groundsel is an evergreen perennial subshrub. The plants have several branching stems arising from the base that become woody upon maturity. Stems and leaves are covered with whitish wooly hair. It occurs from Nebraska and Wyoming southward into the western half of Texas and westward into New Mexico, northern Mexico, and Arizona. It occurs principally on loamy to clay soils in the midgrass and shortgrass prairies, southern mixed prairie, desert grasslands, Chihuahuan Desert, and pinyon-juniper vegetation types.

Threadleaf groundsel is usually rare on grasslands in good condition (4). It increases with retrogression caused by overgrazing and drought. It is classified as an increaser or invader species and is frequently a major component of early seral stages on areas such as road and pipeline rights-of-way, livestock bedgrounds, and sacrifice areas.

The palatability of threadleaf groundsel to cattle is apparently quite low. Sheep and goats are less affected by it and will readily graze it. Grazing with a mixture of cattle, sheep, and goats has been proposed as a potential method for controlling this species in western Texas (14). Acute poisoning occurs when cattle are crowded in pastures where it is abundant and availability and/or quantity of desirable forages is limited. Chronic poisoning in cattle is most common and can occur when threadleaf groundsel is still green and other forage is dormant or scarce, or during snowstorms.

Cattle losses can be reduced, but probably not eliminated, by adhering to good grazing and management practices and specific weed control methods. Hungry animals should never be turned into pastures where threadleaf groundsel is abundant, even though there is adequate availability of desirable forage. Stocking rates should always be carefully balanced against the availability of desirable forage. Grazing management should allow the desirable forages to increase in vigor, abundance, and variety, and stocking with a combination of cattle, sheep, and goats should be used if possible. Localized dense infestations of threadleaf groundsel may be controlled with broadcast sprays of 2,4-D plus picloram (0.9 plus 0.2 kg a.e./ha), picloram alone (0.28-2.2 kg/ha), and metsulfuron at rates as low as 17.5 g a.i./ha (4).

Riddell's groundsel (*S. riddellii*) was recognized early in this century as causing "walking horse disease" in Nebraska and has been systematically eradicated in many areas of the Midwest (16). It is a deciduous suffrutescent half-shrub that dies back to the ground each fall. It resembles threadleaf groundsel in stature and appearance, but the stems and leaves are glabrous and bright green in color. It occurs predominately on sandy soils in a belt from western Nebraska southward through western Kansas and eastern Colorado into Oklahoma, western Texas, and New Mexico. It presents serious problems in only a few localized areas, but because of its extremely high toxicity, it should be considered dangerous. Concentrations of PA as high as 18% of the plant's dry weight have been reported (17). Although the particular PA in Riddell's groundsel is the least toxic of the species in Table 43.1, the high concentration in the plant makes it very danger-ous, requiring as little as 111 g/day of fresh plant for 15-20 days to be lethal to a 220 kg calf. In comparison, the extremely toxic alkaloids in tansy ragwort occur at low concentrations in the plant and would require a dose of 806 g/day to be lethal (18).

Common groundsel (*S. vulgaris*) is an introduced species which infests pasturelands of the Central Valley of California. In dry years, it presents potential problems as a contaminant of forage. It contains four PA at concentrations averaging 0.25% (19).

Broom groundsel (*S. serra* var. *spartioides*) occurs in mountainous areas in the sagebrush, pinyon-juniper, mountain brush, and aspen zones. It occurs from Wyoming to South Dakota and south to New Mexico and California. Concentrations of PA average 2.13% and range from 0.28 to 7.88% (19). *S. integerrimus* is another mountain plant that is toxic and occurs from British Columbia south to California. *S. plettensis* occurs on limy soils throughout the eastern U.S. and Canada, and *S. glabellus* occurs on wet soils in the south U.S.

Leguminosae

Crotalaria species are common throughout the southeast. *C. sagittalis* occurs in bottomlands throughout the area and is the only native species of the genus known to be poisonous. Showy crotalaria (*C. spectabilis*) is a winter annual that was introduced for use as a cover crop and green manure. It is unpalatable to livestock; cattle would not eat it in common garden trials due to its high alkaloid levels (20). However, its seed contaminates harvested grain and soybeans and has resulted in large losses of chickens and pigs. The potential threat is widely recognized and grain contaminated with showy crotalaria seed is severely discounted. Average PA content of showy crotalaria leaves and seeds were 0.24% and 4.08%, respectively (19).

Boraginaceae

Houndstongue (*Cynoglossum officinale*) is increasingly being recognized as a plant of concern. Knight et al. (21) first reported PA toxicity in horses fed hay contaminated with houndstongue. Bagley et al. (22) reported that 14 horses died between 1972 and 1974 and one in 1984 from consuming houndstongue-contaminated hay grown in Spanish Fork Canyon in central Utah. Baker et al. (23) first reported houndstongue poisoning in cattle. Six of nine Holstein steers died when fed chopped forage from a tall wheatgrass pasture infested with houndstongue in Trenton, UT. PA concentrations were 0.51% in leaves and 1.18% in seeds.

Houndstongue is also a noxious weed. Its velcro-like burs cling to socks and pant legs and are annoying to humans, irritating to dogs, and greatly reduce the value of wool. Houndstongue invades bare or abused areas on rangelands and is very common along roadways. It is one of the primary noxious weeds controlled by the Forest Service in the Intermountain and Rocky Mountain Regions.

Houndstongue is native to Europe but was introduced to North America and has become widely distributed. It is a biennial which establishes deep black roots and forms a large rosette the first year. The large dull-green leaves are long and slender and resemble a dog's tongue. The second year, a single stalk (3-12 dm) is produced. It is leafy to the top, but the stem leaves are smaller, narrow, and softly hairy. Several branched racemes arise from upper axils and produce dull reddish-purple flowers. Four attached nutlets form and their surface is covered with short stout prickles. These velcro-like seeds are the object of other common names, such as dog fur, sheep lice, woolmat, and beggartick.

Houndstongue can be controlled by 2,4-D or dicamba (2 kg/ha), picloram (0.25 kg/ha) when actively growing, or by EPTC or benfin in newly planted alfalfa. In established alfalfa, hexazinone applied in the fall or early spring when alfalfa is dormant is effective for control (24).

Fiddleneck or tarweed (*Amsinckia intermedia*) is an annual weed of waste areas and grain fields of California and the Pacific Northwest. It extends through the creosotebrush communities of Nevada, southern Utah, Arizona, New Mexico and Texas. Up until the 1940s, poisoning of horses, swine, and cattle from fiddleneck seeds in grain and grain screenings was common throughout the Pacific Northwest (3). The widespread use of herbicides to control weeds in grain has essentially eliminated poisoning problems. PA content averaged 0.24% in its leaves (19).

Comfrey (*Symphytum officinale*) is a deep-rooted perennial forb with large coarse prickly leaves. It is cultivated throughout the world and has widely reputed medicinal properties. It also contains at least eight PA.

436

Despite its wide use, it should be considered especially dangerous to children because of their higher susceptibility (3).

Paterson's curse or salvation Jane (*Echium plantagineum*) is one of the most important poisonous plant problems in Australia and New Zealand. *Heliotropium europaean* is another major poisonous plant of Australia. Species of *Echium* and *Heliotropium* (*E. vulgare, H. convolvulaceum,* and *H. curassavicum*) also grow in the U.S., but toxicity to livestock has not been documented. Borage (*Borago officinalis*) is cultivated in gardens and has escaped to waste places in the western U.S.

Summary

Pyrrolizidine alkaloids occur in a large number of plants throughout the world. They are toxic to both livestock and humans. They are primarily hepatotoxic, but can also affect the heart, lungs, and visceral organs. They can also be carcinogenic and mutagenic. PA-containing plants are generally not palatable to livestock but cause problems when the quantity or quality of desirable forages are low or when they contaminate hay or grain. Most of the plants are increaser or invader weeds. They inhabit disturbed sites and degraded range and pastures. They decline with improved range and pasture management. They can be controlled with herbicides registered for use on rangeland.

References

1. Huxtable RJ. In: *Toxicants of Plant Origin, Vol. 1 Alkaloids* (PR Cheeke, Ed.), pp. 41-86. CRC Press, Boca Raton, FL, 1989.
2. Smith LW and CCJ Culvenor. J. Nat. Prod. 44:129 (1981).
3. Cheeke PR and LR Shull. AVI Publishing Co., Westport, CN, 1985.
4. Sharrow SH, DN Ueckert, and AE Johnson. In: *The Ecology and Economic Impact of Poisonous Plants on Livestock Production* (LF James, MH Ralphs, and DB Nielsen, Eds.), pp. 181-196. Westview Press, Boulder, CO, 1988.
5. Johnson AE and RJ Molyneux. Amer. J. Vet. Res. 45:26-31 (1984).
6. Kingsbury JM. *Poisonous Plants of the United States and Canada*. Prentice-Hall, Englewood Cliffs, NJ, 1964.
7. Johnson AE, RJ Molyneux, and LD Stewart. Amer. J. Vet. Res. 46:577-582 (1985).
8. Johnson AE. In: *Symp. Proc. Pyrrolizidine (Senecio) Alkaloids: Toxicity, Metabolism, and Poisonous Plant Control Measures* (PR Cheeke, Ed), pp. 129-134. Oregon State Univ., Corvallis, 1979.
9. Cheeke PR. In: *Toxicants of Plant Origin, Vol. 1 Alkaloids* (PR Cheeke, Ed), pp. 1-22. CRC Press, Boca Raton, FL, 1989.

10. Snyder SP. Oregon Agr. Res. No. 255, 1972.

11. Isaacson DL and BJ Schrumpf. In: *Symp. Proc. Pyrrolizidine (Senecio) Alkaloids: Toxicity, Metabolism, and Poisonous Plant Control Measures* (PR Cheeke, Ed), pp. 163-169. Oregon State Univ., Corvallis, 1979.

12. Mathews FP. Texas Agric. Exp. Sta. Bull. 500, 1933.

13. Norris JJ. PhD Diss., Texas A&M Univ., College Station, 1951.

14. Dollahite JW. Southwest. Vet. 25:223-226 (1972).

15. Reagor JC. In: *Proc. Trans-Pecos Poison Plant Symp.* (A McGinty, Ed.), pp. 1a-1f. Texas Agr. Ext. Serv., Fort Stockton, TX, 1981.

16. Barkley TM. North America Floral Series II, Vol. 10:50-139 (1978).

17. Molyneux RJ and AE Johnson. J. Nat. Prod. 47:1030-1032 (1984).

18. Johnson AE, RJ Molyneux, and MH Ralphs. Rangelands 6:261-264 (1989).

19. Johnson AE, RJ Molyneux, and GB Merrill. J. Agric. Food Chem. 33:50-55 (1985).

20. Becker RB, WM Neal, Arnold PT Dix, and AL Shealy. J. Agric. Res. 50:911-922 (1935).

21. Knight AP, CV Kimberling, FR Stermitz, and MR Roby. J. Amer. Vet. Med. Assoc. 185:647-650 (1984).

22. Bagley CV, AE Johnson, KB Creer, and JO Evans. *Utah Farmer Stockman*, May 2, 1985, p 5.

23. Baker DC, RA Smart, MH Ralphs, and RJ Molyneux. J. Amer. Vet. Med. Assoc. 194:929-930 (1989).

24. Dewey SA. *Utah Weed Control Handbook*. Utah State Univ. Ext. Ser. EC 301 (1985).

44

Continuing the Forum
for Cooperation

Notes from the Special Concurrent Session

Noxious Weed Management Through Cooperation

A highlight of the 1990 Noxious Range Weeds National Conference was "Continuing the Forum for Cooperation," one of several concurrent sessions. Deen Boe (Forest Service), chairman of the concurrent session, stated that the forum for continuing cooperation was a follow-up to the 1989 Conference on Noxious Weeds in Billings, Montana. Boe said, "[this] meeting was billed as 'a forum for cooperation' and ... will build upon the need for strengthening intergovernmental coordination and cooperation to improve the effectiveness of weed management programs."

Reports were made on (1) accomplishments since the Billings meeting, including required action items such as incorporating the control and management of noxious weeds into the NEPA process; (2) development of noxious weed information and education materials; (3) the Department of Agriculture's policy on noxious weeds; (4) coordination issues and needs between federal and state agencies and with Canada; and (5) State noxious weed management issues. Establishment also was explored for a Noxious Weed Management Council among the state and federal agencies and Canada in the Greater Northwest Area.

The Logan forum concluded that the goals and objectives of management and control of noxious weeds and undesirable plants can best be achieved through improved awareness of problems and issues, commitment, and increased knowledge of what is needed to control and manage noxious weeds. Participants at the Logan meeting focused attention on the development of an action plan for the future, establishment of a Noxious Weed Management Council, and furthering communication.

Intergovernmental Noxious Weed Management Council

To strengthen cooperation and coordination of noxious weed management, the forum created an "Intergovernmental Noxious Weed Management Council." Federal and state agencies, the Canadian province of Alberta, and others are being considered as participants in the group. The charter and boundaries are still to be worked out, but the states involved include Colorado, Idaho, Montana, Nebraska, North Dakota, Oregon, South Dakota, Utah, Wyoming, and Washington. Barbara Mullin, State Weed Coordinator, Montana Department of Agriculture, will chair the group.

While still in the formative stages, the council will function in a coordinating capacity, conduct symposiums for exchange of information, improve awareness, and further education on noxious weed biology. Objectives and goals are flexibility and ability to move fast, center on public land agency needs, enhance interagency cooperation, and coordinate intergovernmental actions to create/improve working relationships and program efficiencies. Themes which will be considered, in descending order of priority, are: (1) information and education about noxious weeds; (2) improved funding for federal and state agencies; (3) standards for seed quality, quarantine, and weed testing; (4) noxious weed legislation (federal and state); (5) national noxious weed data base; (6) research; and (7) establishment of rapid response capability. The council will also use other forums to disseminate information, including meetings of the Society for Range Management and the Weed Science Society. These professional society meetings can provide opportunities for communication and exchange of information through symposia sessions.

The council will be organized at the regional (field) level. The executive committee will consist of the chairman of each working subcommittee. An ad hoc group was assigned to review potential coordination with existing groups, including the Interagency Noxious Weed Advisory Council (INWAC) and Great Plains Agricultural Council (GPC-14). This committee includes Barbara Mullin (Chairman), George Hittle (INWAC), Russ Lorenz (GPC-14), Buck Waters (Bureau of Land Management), and Curt Johnson (Forest Service).

Activity Reports

Reports were heard from representatives of federal, state, and county governments, universities, the province of Alberta, Canada, and others. Topics covered the proposed establishment of a Noxious Weed Initial Strike Force (to be patterned after federal wildland fire control which focuses on fast response and use of experienced crews); USDA certified seed policy; establishment of a national data base; economics of noxious weed management; funding of federal agencies, including long-term funding of research; compliance with the NEPA process; and states' issues and needs.

International Cooperation

The Logan meeting clearly established that noxious weeds are an international issue. Walter Yarish (Alberta) reported that coordination for management and control of noxious weeds is a high priority with other provinces and the United States, especially counties along the International Boundary. Yarish stated that they use an immediate action plan approach which includes spraying, digging, and destruction of noxious weeds. All weeds that are treated are scheduled in a five-year follow-up schedule and are continually monitored. Biological control is being used for leafy spurge. Cooperative work between the U.S. and Canada is badly needed.

Information and Education Materials

Control and management of noxious weeds must involve positive information and education materials for use in the private sector if it is to be successful. Jim Olivarez (Forest Service) reported that information and education materials are being presented to the private sector through postcard-type handbooks. Each of these booklets contains high quality color photos of noxious weeds and includes a message that states that control of any weed depends a great deal on early detection of its presence, and that the booklet was assembled to aid in that process.

Walter Yarish reviewed several handbooks, including use of pesticides, use and care of equipment, guides to crop protection, and planning considerations for vegetation management along roadways and rights-of-way. Posters and letter stuffing materials were also displayed. Materials from Canada are available upon request.

Noxious Weed Legislation

An overview of impending national noxious weed legislation through the 1990 Farm Bill was presented to forum participants. It was reported that the focus is on required coordination/cooperation between federal and state government, using an integrated management approach. Discussion revealed diverse views on the need for regional commissions, lack of funding identified in the impending legislation, and use of funds on federal lands. some state (e.g. Wyoming) stated they will not cost-share on federal lands, and will be looking for funding sources. Concerns were also expressed about cost-sharing on private lands.

Noxious Weed Data System

In the area of a national data system for noxious weed coverage, uncertainty about mapping reliability still exists. Mapping through a national data base is effective for some programs, but noxious weed coverage is variable. Mapping is focused on occurrence and weed density and would be GIS-compatible. Oversight control for noxious weed inclusion in any national data base would be through existing state survey coordinators. The current funding level is $4 million annually for the national data base. Utility of the data base is constrained at this funding level. States must be concerned about funding sources. In addition, information is only held for two years.

Economics of Noxious Weed Management

Ed Frandsen (Forest Service) reviewed the economics of noxious weed management. It was reported that noxious weeds and undesirable weeds have been a recurring problem since livestock producers first grazed cattle and sheep on the nation's rangelands. In spite of the problem, little research has been conducted on the economic magnitude of noxious weeds. Frandsen said noxious weeds have the potential of infesting all susceptible land and water in the United States, and can adversely affect food production, wilderness values, wildlife habitat, visual quality, forage production, reforestation, recreational opportunities, and land values. While many tax dollars are spent annually to manage, control, and eradicate noxious weeds and undesirable plants, the general public is either unaware or apathetic to this issue. Frandsen also reported that weed control on forested lands is generally economical. Tangible benefits have accrued due to wood production through control of noxious weeds. The full report follows as Chapter 45.

45

Economics of Noxious Weeds and Poisonous Plants[1]

Ed Frandsen and Deen Boe

Introduction

Control of undesirable plants, especially noxious weeds and/or poisonous plants, on rangelands requires knowledge of the alternative control measures and their related costs. Economic analysis provides information for decision making and for establishment of investment priorities. Economic efficiency analysis also provides information for comparing the effectiveness/efficiency of alternative control methods and provides input in the development of impact analyses required by Federal and state laws, such as the National Environmental Policy Act (NEPA) of 1969.

Review of the literature discloses that: (1) very little research has been focuses on the economics of noxious weed control; (2) what research that has been conducted is largely oriented to the impacts incurred by the western states' livestock industry; (3) while many tax dollars are spent each year to eradicate or control weeds, the general public is either unaware or apathetic to this issue; and (4) a significant data base is needed to identify and define the multiple resource relationships involved and the joint costs and benefits of noxious weed management. Bingham (1) states that "noxious weeds, by some accounts [are] one of the worst threats the West has faced. Most, but not all [undesirable plants or noxious weeds], are exotic species blooming and booming without natural adversaries." Millions of dollars are spent each year to combat

---

[1]Presented in the Symposium "Continuing the Forum for Cooperation."

weeds in the western states. For example, "Montana taxpayers spend over $5 million a year to kill, [control, manage or eradicate] them. In every western state but Arizona, New Mexico, and Colorado, authorities can enter private land to kill them and make the landowner pay. Soon a number of bills will come before Congress for a nationwide weed control law. Behind them is fear of wholesale loss of rangeland, crops, wildlife habitat, and soil stability."

Certain weeds grow at a phenomenal rate. For example, in Montana, "spotted knapweed alone infests 4.5 million acres in the 900 million acre state now. At the present 10% annual expansion rate, ...knapweed could eventually fill its potential habitat—fully half of Montana." Leafy spurge, another problem weed, "already [covers] one million acres in North Dakota, and is expanding at 25% a year. It has ...staked out a 500,000 acre claim in Montana."

The threat of noxious weed growth has been characterized as being similar to the "miracle of compound growth" that savings banks used to promote so heavily. In 25 years, a 100-acre patch of weed which expands 10% a year will have grown to 1000 acres. Suppression costs have grown even faster, given economic inflation. Given these circumstances, the impact of a spreading weed on agriculture, wildlife habitat, and other plants could be disastrous.

Dr. Brian Sindelar, range scientist, Montana State University, states that weeds are a "social disease," and goes on to say that "the whole scope of humanity's immensely complex relationship to nature is encompassed in the word weeds. Weeds have a life of their own. They cross boundaries. They mock our sense of property rights. They develop resistance. They are often propagated by the very activities we find necessary. They cost us money. Direct intervention has side effects we don't begin to understand. And one person's weed may benefit his neighbor."

Problems of definition have constrained legislation of weed control laws in some states. To get around the listing of certain species, some of which are beneficial to wildlife, the bills in Congress avoid a potential problem by not naming species. Rather, they focus on undesirable plants and define these as "any plants that are of little economic, aesthetic, or nutritional value; or are classified as exotic or noxious plants." An exotic plant is either "not a regular member of the native or natural community, of little economic value, or colonizes disturbed ground." A "noxious" plant is "aggressive, difficult to manage, detrimental, destructive, or poisonous; a carrier of insects or disease; parasitic; or directly or indirectly detrimental to the management of a desired ecosystem." Noxious weeds are declared through legislation, which differentiates them from other weeds. Most states, especially in the West, have identified weed species that are considered noxious.

While only estimated, the economic impacts of noxious weeds and poisonous plants on the western livestock industry are significant. But what about the impacts on other natural resource values, such as wildlife habitat, outdoor recreation use, soil productivity, and watershed productivity? From an economic perspective, the problem is an issue of joint product relationships, involving joint costs and benefits or joint costs and separable benefits. By definition, a joint cost is one that contributes to the production of more than one type of output. Little work has been done to define the integrated resource relationships involved with weed control of forested rangelands.

This paper has multiple objectives as follows: (1) To highlight the management interrelationships of natural resource disciplines involved in weed control; (2) Document the costs and benefits of weed control on rangelands; (3) To document the impacts of noxious weeds on the western range livestock industry. Demonstration examples are used to display efficiency and break-even points; (4) To overview accepted management practices for use on transitory (forested) rangelands; and (5) To identify needed research, especially in the area of economics of noxious weed control and management.

Joint Costs and Multiple Resource Values

In economic analysis, a joint cost contributes to the production of more than one type of output. For example, herbicide treatment of noxious weeds may serve more that one project purpose such as increasing forage production, improving wildlife habitat, increasing water yield, and improving water quality. Strict application of joint costs, especially where functional resource programs are involved, requires separating (separable costs) costs among the project beneficiaries. Literature review reveals that despite the level of impacts, very little "hard data" or accepted methodology exists in the area of joint cost/benefit analysis.

Other onsite and offsite impacts occur from the lack of controlling undesirable plants. At issue are problems with integrating multiple resource of land management, the need for data to document joint product relationships, and allocating costs and benefits among multiple beneficiaries.

Economic analysis of weed control requires several types of information for analysis of forested ranges including: treatment response relationships; physical and biological limitations; effectiveness of alternative weed control practices; and total costs for various practices. This information forms the framework for decision making on alternative treatment practices and the basis for quantifying benefits and costs. This section identifies the use of silvicultural practices and interrela-tionships involving range management, habitat management, etc. Case studies illustrate the application of economic efficiency of weed control on forested ranges.

Direct and indirect benefits result from weed control. Tangible benefits such as forage production and wood production on forested lands are easily quantified. Improved carrying capacity through reduction of competing, undesirable plants is a positive benefit, and easily measured. Stewart and Row (2) determined that benefits resulting from weed control on forested lands include "increased volume at harvest because of better stocking; height or diameter growth of trees; increased stand value because of improved species composition; higher value at the mill and lower felling, harvesting, and transportation costs because of larger stand diameter, and earlier return on investment from commercial thinning because of increased diameter growth." Other benefits involved in the joint production function, but more difficult to quantify, are the value of increased water yield to offsite or downstream users and improved habitat for some game and nongame wildlife species.

Non-market benefits involving environmental and visual resources may also accrue from weed control. These may include more rapid establishment of forest cover following timber harvest, including better visual appearance of clearcuts, improved visibility within established forest stands, and reduction of noxious and poisonous plants.

Estimating Costs of Weed Control and Other Treatments

Once tangible and indirect benefits of weed control are identified and quantified, it is time to determine the costs of weed control. On forested ranges, this should be done in conjunction with determining the costs of other practices, including silviculture, wildlife habitat improvement, range revegetation, etc. The direct costs of weed population management include materials, equipment, transportation, contracting costs and administration, monitoring, overhead, and standby time.

Treatment may also include various indirect costs such as mitigation of undesirable environmental effects and reduced production of other resource outputs. For example, increased fertilization to compensate for loss of nitrogen-fixing plants species, erosion control measures, or other activities may be necessary. All physical factors which affect treatment costs should be quantified and costs documented.

Weed Control and Timber Production

Weed control of forested lands, based on economic analysis, shows that it is generally feasible to control undesirable plants. Tangible benefits accrue to wood production through control of noxious weeds. Included are increased volume at harvest because of better stocking, height, and diameter growth of threes; increased stand value because of more species composition; and higher value at the mill.

Stewart and Row (2) studied the response of conifers and hardwoods to control of herbaceous weeds. Positive results showed "an increase in survival and height growth of small seedlings." Control of weed trees and shrubs that hinder conifer establishment and growth is a common silvicultural practice. From 78 published and 7 unpublished reports on the effects of natural levels of brush competition or brush control on conifer survival and growth, 79 reported a positive response in terms of short-term changes in tree survival, height growth, or diameter growth, five showed no response, and only one showed a negative response to reduced levels of competition from weeds or competing herbaceous growth.

In western Oregon, herbicide treatment of undesirable plants in a spruce hemlock forest type resulted in a positive present net value (PNV) of $43,424 per acre as opposed to a PNV of $37,958 when no herbicide was used, a difference of $5466 per acre. Similarly, economic analysis of establishing loblolly-shortleaf pine using intensive mechanical treatment versus aerial spraying with 2,4,5-T at the time of site preparation showed positive net benefits:

Mechanical treatment	B/C ratio: 4.99:1
Aerial spraying	B/C ratio: 5.61:1

The greater initial cost of intensive mechanical site preparation is partially offset by lower planting costs and by elimination of a release spray at age 4. A $58 per acre difference in site preparation cost results in only a $16 per acre difference in present net value between the two alternative forest management practices. In this case, both forest management practices involving control of undesirable plants or weeds are profitable, but the aerial spraying yields $0.54 greater return per dollar spent.

Poison Plant Control and Livestock Production

Infestation of noxious weeds and poisonous plants has been a recurring problem since livestock producers first grazed cattle and sheep on the vast areas of the nation's rangelands. Weed infestation not only disrupts harvesting of forage, but also causes serious economic impacts to livestock producers and dependent rural communities, especially in the western United States.

Economic impacts occur through direct and indirect processes. To the rangeland manager, weed infestation of sufficient size or magnitude reduces forage production. This impact occurs primarily through plant competition and the robbing of soil moisture needed by desirable forage species. to the livestock producer, noxious weeds and/or poisonous plants are principal cause of loss of rancher income. James et al. (3) report that "losses occur from deaths, abortions, birth defects, and pathologic effects on various organs that result in reproductive alterations, decreased performance, weight losses, wasting, and photosensitization."

Literature review discloses that very little has been done to quantify the economic impacts of noxious weeds and poisonous plants. The literature also reveals that economic losses from noxious plants are "difficult to quantify, hard to define, and almost impossible to get a handle on" (3). These observations are further attested to by the fact that even through acreage of noxious weed infestation on National Forest System lands is provided for in the Forest Service Range Automated Information System (FSRAMIS), the information is not collected by field units! At issue is the lack of inventory data and monitoring.

"Livestock poisoning by plants can usually be traced to problems of management, overall range condition, or both, rather that the mere presence of poisonous plants" (3). For example, lupine is more toxic to cattle than to sheep, yet thousands of sheep have died of lupine poisoning while few cattle have died. Conversely, halogeton (*Halogeton glomeratus*) is known to have about the same level of toxicity to both cattle and sheep, yet thousands of sheep have died but very few cattle. These differences can be attributed to management and use of rangelands and reflect the herding of sheep in large flocks or bands from place to place, as opposed to the free movement of cattle to independently search out feed and water.

Direct effects of noxious/poisonous plants on livestock and factors influencing these effects, though difficult to assess, are most obvious. Less obvious, and perhaps even more difficult to assess or quantify, are the indirect impacts. These include the costs of increased management, including fencing, herding, altered management, increased labor costs, supplemental feed, increased veterinary expenses, etc. While cost figures are difficult to obtain and appear tenuous, a best estimate of costs and benefits is required to define the problem of control-eradication program and to outline policy.

Economic Impacts of Poisonous Plants

Nielsen et al. (4), in a study of the economic effects of weed control, concluded that "although poisoning of livestock occurs throughout the U.S., it is more of a problem in the western states. Rangelands in the western United States comprise 75% of the land area. Historically, the forage on these lands can best be used by grazing animals, i.e., sheep, cattle, and big game wildlife. Foraging [of] native range plants is a low energy dependent use of these lands that requires little fossil fuel production costs, since the forage is renewed annually without tillage, fertilization, or mechanical harvesting. In essence, grazing of rangelands provides a cost-effective or least-cost way of converting solar energy into chemical energy (food), and therefore, must be used wisely."

In 1988, Nielsen estimated that total annual death loss was 3% for cattle and 8 to 10% for sheep (5). Assuming that one-third of the death loss in the western states is due to noxious weeds and/or poisonous plants, then the losses would be 280,560 cattle (1% of 28,056,000) and 264,215 sheep (3.5% of 7,549,000). Using average livestock prices of 1988, the economic losses would be $145,330,080 for cattle (280,056 head x $518) and $23,779,350 for sheep (264,215 head x $90). Total loss would be $169,109,430 (4).

In addition to death losses, reduction in reproduction efficiency also causes economic impacts. Losses in lamb and calf crops occur through reductions in calf and lamb crop percentages. Nielsen (5) reported that "the most meaningful lamb and calf crop percentage is the ratio of the number of offspring weaned to the total number of potential mothers." Death loss of calves and lambs prior to weaning is taken care of in the lamb and calf crops. Nielsen assumed that "poisonous plants decrease lamb and calf crops by 1%." Recognizing that this is income foregone or an economic opportunity cost, rancher income could be increased by the following amounts if poisonous plants were not a problem (4):

Calves:	220,430 head x $275/head =	$60,618,250
Lambs:	75,490 head x $60/head =	$4,529,400
TOTAL:		$65,147,650

Using 1989 prices, together with Nielsen's estimates of death loss for sheep and cattle (4) and the opportunity costs of income foregone from calves and lambs, the estimated economic loss is $340 million.

Reduced Animal Performance

Other economic impacts occur through reductions in animal performance, resulting from grazing poisonous plants. Smaller calves and/or lambs at weaning and time of sale result in lower sale revenues. Livestock that recover from poisoning often require longer feeding periods to reach marketable weights and translates into additional costs of livestock production.

Indirect Impacts

Indirect impacts, which are nearly impossible to quantify, are the adverse impacts which may occur through reputation. According to Nielsen (5), "animals marketed from known areas of poisonous plants that are known or suspected to have reduced performance potential through the finishing phase of production may cause lambs and calves from these areas to be discounted in the feeder markets."

Increased Livestock Production Costs and Land Treatment

Increases in livestock production costs and related management costs also occur. These costs include labor to find, treat, and/or remove affected animals; the cost of building and maintaining fences to keep livestock away from the plants; and underutilization of forage resources. Livestock stocked at lower rates than optimum levels or livestock kept off of range until late in the season in an attempt to reduce the consumption of the problem plants add costs to livestock producers that use rangelands in their livestock enterprise. Additional costs are incurred in controlling noxious and poisonous plants, including costs or spraying with herbicides, grubbing, plowing, pulling plants, burning, mowing, or other control programs.

Constraints on Land Use

Studies by James et al. (3) reveal that some areas of the western states "are almost forced to produce one type of animal because of poisonous plants. Larkspur-infested ranges may have been limited to sheep production for years because ranchers do not put cattle on them for fear of incurring heavy death losses." This is an economic loss to the rancher. Potential income reduction due to having to graze a type of livestock that can be grazed without fear of death loss or reductions in animal performance is an opportunity cost caused from not being able to make optimum use of the forage resource.

Poisonous Plants and Asset Values

Very few livestock producers operate without borrowing money, especially annual recurring operating capital. The increased uncertainty and risk resulting from noxious weeds and poisonous plants can have direct economic impacts on livestock producers' ability to borrow money, or the price they pay to borrow. If the potential losses from noxious or poisonous plants make ranch loans too risky, lenders may refuse to make a loan or increase interest rates and collateral requirements and shorten loan repayment periods. Each of these can add to the costs of livestock production.

Poisonous plants may also impact ranch real estate values. Nielsen (5) reports that "many ranchers are reluctant to permit research on their ranches concerning poisonous plants because they fear [that] calling attention to problems can reduce real estate values."

Real estate assets, especially ranch properties, are customarily valued through either the income approach (concepts and principles associated with capital budgeting) and/or the market or sales approach. The income approach capitalizes the earnings over a specified time period using a specified discount or interest rate. the theory of the income approach is that "the value of a property is the present worth of the net income it will produce during the remainder of its productive life" (6). Godfrey et al. (7) states that "many believe that this is the most valid approach to use when valuing resources because it reflects what the owner of the resources could afford to pay for the resource from the income to be received. This approach also has been used extensively to value Federal grazing privileges or permit values."

The sales of market approach also is used. This method assumes that "a prudent person will not pay more for property than the amount for which a comparable substitute property can be bought" (8). Adjustments are made between subject properties and comparable properties to bring values into comparability. If one parcel of land is infested with noxious weeds which detrimentally affects the returns or income obtained from using the area compared to another parcel that did not have noxious weeds or poisonous plants, some downward adjustment would be necessary to bring the parcels into full comparability. This adjustment would be made since the land containing noxious weeds would expect to be sold for less.

Economic Feasibility of Weed Control—Livestock Examples

Additional Forage for Livestock: Is It Worth It?

Tanaka and Conrad (9) are evaluating the economic feasibility of controlling whitetop (*Cardaria* spp.) on the sagebrush ecosystem of eastern Oregon. The project was designed to examine (1) the biological and ecological relationships of whitetop, crested wheatgrass, and big sagebrush and (2) the economic impact from whitetop competition on cattle ranches in Baker County, Oregon. The typical ranch in Baker County has 395 cows, a 94% calving rate (based on January 1 inventory), and an 18:1 cow-to-bull ratio. Twenty-four percent of the ranches hold calves over for sale as yearlings. The typical ranch owns 5900 acres, with 604 acres fallow or idle; 65 acres in grain crops; 512 acres of irrigated hay; 15 acres in dryland hay; 3224 acres of rangeland (sagebrush/grass understory); 1525 acres in crested wheatgrass; and 915 acres in irrigated pasture. The ranch also leases 68 acres of additional hayland and obtains 195 AUMs (animal unit month) from BLM administered rangelands, 174 AUMs from National Forest lands, and 262 AUMs from leased private rangelands.

The economic analysis is ongoing, but a demonstration example is provided. The objective is to determine the relationship between the desirable plant species (crested wheatgrass) and the undesirable species (whitetop and big sagebrush) and the resultant impact on the ranch's carrying capacity. The hypothesis of the research project is whitetop causes a reduction in the amount of spring forage and that forage availability in the spring is a limiting factor on ranch herd size. For example, if whitetop causes a percentage reduction in early spring forage, the ranch may have to reduce its *yearlong* herd size by an equal percentage to maintain a balanced carrying capacity. The economic impact of such a change can be reflected in what happens to the net ranch income and the resultant impacts on the ranch's income statement. By example, assuming that the amount of lost forage requires a herd size reduction—to stay within the carrying capacity—equivalent to $20,000 (a 15.8% reduction in sales). At the same time, annual variable costs will be reduced by $16,226, at the same percentage. The net income loss will be $3774 per year if this is allowed to occur. If whitetop is controlled so that these losses do not materialize, then this can be counted as the annual net benefit of the project. If the project requires treatment of 1000 acres, then the net annual benefit is $3.77 per acre per year.

Using this level of annual benefit, and an assumed real (inflation-free) interest rate of 8% and an effective life of five years for the weed control measure, the objective is to determine how much can be afforded or spent on a control project and break even. This calculation indicates that the $3.77 per acre received each year (over the 5 yr life of control) would be worth $15.46 today (present net value). So long as the control measure costs less than $15.46 per acre in this example (the economic threshold), it is economically feasible to treat the whitetop infestation. If it costs more than this, the ranch would be financially ahead by either reducing herd size or finding an alternative way of obtaining additional spring forage (e.g. leasing additional pasture or range) at a lower cost per acre.

The control of noxious weeds is affected by economics of scale. Nielsen (5), through the following hypothetical example, illustrates some points about the economic consideration of various alternative courses of action that should be considered. "A ten-acre infestation of a range weed that has the potential to spread over thousands of acres is located. Obviously, if the spread of the noxious weed plant can be controlled by the treatment of ten acres, one would do this." This is where the economics of this type of strategy needs to be considered. Assume there are 10,000 acres with the potential for being invaded by a noxious weed. Research in other areas indicates that the productivity of the range is decreased 50% when the invader plant moves in and takes over. The expected return to the land per animal unit month grazed is $8, and it requires six acres per AUM. Thus, the return is $1.33 per acre. After the rangeland is infested with the noxious weed plant, it requires 12 acres per AUM. This reduces the annual return to $0.67 per acre. The benefit from controlling this plant is the reduction in the range carrying capacity; if the invading plant is also poisonous to livestock, one would also have to include the value of animals not poisoned because the infestation was stopped. In the hypothetical example, the potential benefits of controlling a ten-acre infestation are $6700. Research shows that it costs about $30 per acre to control the noxious weed. The benefit/cost ratio (B/C) for weed control on the ten acres is $6700 ÷ 300 = 22.33, or 22 to 1. However, if one waits until 1000 acres are infested before treatment, the B/C ratio is 0.22 to 1 ($6700 ÷ 30,000), which says that treatment of the 1000 acres per year is not feasible. If a treatment has to be made each year to prevent the spread of the weed, the breakeven point is 223 acres. At this point, the B/C ratio is 1 to 1.

In the evaluation of the economics of weed control, the discounted net benefits should be equal to the discounted net costs of control. At this point, the B/C ratio equals 1 to 1, or breaks even. For example, suppose that a single treatment will stop the spread of the noxious plant for at least ten years. Control of the plant at this point will result in multi-year benefits. The initial control cost would be compared to the counted income stream of benefits from the 10,000 acres for the ten-year life of the control project. The present value of $0.67 per year for ten years discounted at 8% interest is $4.50 per acre. therefore, the present value of the potential benefits from 10,000 acres is $45,000. If prevention of infestation of 10,000 acres can be stopped by control of ten acres, the B/C ratio is 150 to 1. Control of the noxious weed plant would be economically feasible up to the point where 1500 acres were infested before anything was done. At 1500 acres, the B/C ratio would be 1 to 1.

These examples point out the economies of scale (size) that apply to the feasibility of noxious weed control. So long as the costs remain below the benefits to be derived, the project is economically feasible. Looking at the problem from the livestock producers' perspective, so long as the costs of control remain below the costs of acquiring substitute feed, assuming that the livestock herd remains the same, it is economically feasible to control noxious weeds.

Rangeland Forage and Wildlife Relationships

Rangelands provide habitat for big game wildlife along with domestic cattle. The purpose of this section is to identify the relationships between wildlife and rangeland forage and to estimate the value of additional forage that could be created through control of noxious weeds.

To grasp the relationship of wildlife and rangeland forage, it's necessary to understand that rangelands provide habitat (food, water, bedding ground, strutting grounds, etc.) for wildlife. It's also important to recognize that some degree of competition exists on most public rangelands between domestic livestock and big game animals, as well as between certain big game wildlife species. At issue is the allocation of land and forage between competing grazing ungulates.

Loomis et al. (10) state that "in terms of economic theory, determining the optimum mix of cattle and -wildlife species is relatively straight forward and can be derived in terms of the factor market for forage..., and that economic efficiency in forage allocation would require equating the value of marginal product of forage between cattle and each wildlife species."

In BLM's Final Environmental Impact Statement (EIS) for the Challis EIS Area (Idaho), it's stated that at least a 30% increase in deer numbers could be sustained with additional forage (11). The potential for increased carrying capacity for elk is estimated at about 20%. These estimates are consistent with Idaho State Department of Fish and Game's objectives for deer and elk herds in the Challis, Idaho, area (11).

Control or eradication of noxious weeds can produce increased forage for grazing and browsing. BLM's estimates of increased deer numbers resulting from increased forage can be used in conjunction with the value of an additional AUM of forage estimated by Loomis et al. (10). Loomis et al. (10) used the travel cost method to estimate marginal values of elk and deer in the Challis unit and calculate the marginal value of an AUM of forage to these species. Through this method, they were able to compare the marginal value of forage between wildlife and cattle as well as draw inferences about relative values of forage between different big game species.

To estimate the value of an AUM of forage to wildlife in the Challis unit, an understanding of the bioeconomic production relationships between harvest, big game populations, and habitat is required. According to Idaho Fish and Game (12), producing an additional 28 elk for harvest (huntable elk) annually would generally require a total increase of 378 elk in the Challis herd unit. Herd composition would be 19% bulls, 54% cows, and 27% calves. BLM indicates that each adult elk consumes between 0.4 and 0.67 AUMs of forage each month (11,13). For illustration purpose, Loomis et al. (10) used 0.54 AUMs per adult elk and half this amount for a calf, and derived a simple production function relationship of the number of elk available for harvest (herd unit 36) to quantity of forage [as follows]:

$$EH = 0.01322 \; AUM \quad \text{where } EH = \text{bull elk available for harvest}$$
$$AUM = AUM \text{ of forage consumed by elk}$$

Loomis et al. (10) estimated that a "25% increase in the number of elk harvested (28 more), generates ... an increase in net economic benefits of $14,075 annually. The derived marginal value of a harvested bull elk was $502. Combining the AUMs of forage consumed by elk with the marginal value of an elk ($502), the value of the AUM consumed by elk is $6.56 per AUM (value marginal product). This value ($6.65) is the maximum amount a hunter would bid for the increased forage to produce 25% more elk in hunt unit 36." Using similar methodology for deer in this herd unit, the marginal value product for deer is $6.33. Again, this is the maximum value a hunter would bid for the increased forage to produce 9.5 deer in order to have one more deer available for harvest. In comparison, similar methodology in herd unit 36B calculated marginal value products for elk of $8.25/AUM and for deer, $15.83/AUM. The large difference in forage value for deer in the two units relates to differences in marginal value per deer and to the higher marginal productivity of Unit 36B in producing deer. Specifically, it takes only an increase of 7.6 deer to produce one more available for harvest in Unit 36B as compared to 9.5 deer to produce one more for harvest in Unit 36 (12).

Noxious Weeds and Wildlife

Wolfe and Lance (14) and Jessup et al. (15) report that elk and antelope have been poisoned by locoweed and water hemlock on western U.S. rangelands. Laycock (16) states that "native wildlife species appear more resistant than livestock to poisoning from native plants, presumably because of the evolutionary process of selecting foods and developing preferences that would avoid harmful substances." Notwithstanding this, it can be concluded that the loss of forage to weed infestation over the long-term can have an adverse impact on big game population numbers. Over time, the loss of forage and desirable browse species to noxious weeds and poisonous plants could adversely impact elk, deer, and antelope populations. While it is difficult to quantify the impacts, the loss of big game would ultimately cause economic impacts.

Economics of Treating Noxious Weeds on National Forest System Lands

Economic effects of noxious weeds and/or poisonous plants are analyzed from a with and without relationship. To determine the economic feasibility of treating noxious weeds on National Forest System lands, three classes of information are required: (1) the existing situation; (2) the future with control of noxious weeds; and (3) the future without control. The "with" analysis focuses on the avoidance of future loss if investments in noxious weed or poisonous plant control are not made. The "without" scenario assumes that forage production declines without control or remains static.

Control of noxious weeds and/or poisonous plants on National Forest System lands is classified as project level work. In 1982, Forest Service Handbook (FSH) 2209.11, Range Project Effectiveness Analysis Handbook, was issued. The methodologies contained within the handbook are applicable for determining economic feasibility, especially where the analysis is focused on the potential loss of forage.

The relationship of with and without control of weeds would be evaluated through the following:

With Weed Control
[New (Added) AUMs - Current AUMs] x $/AUM

Without Weed Control
(Current AUMs - AUMs w/o Weed Control) x $/AUM

Equals TOTAL economic benefits of weed control, based on net AUMs derived including the loss of AUMs avoided.

Control of noxious weeds on grazing areas may increase carrying capacity above the existing obligation or avert reductions that might occur if the undesirable, competing weed plants are not controlled.

Costs of *with* and *without* land treatment are documented incrementally, during the analysis period as follows:

Nonstructural Range Improvement Costs
  Spraying
  Burning
  Grubbing
  Plowing

Total costs with weed control

Total costs without weed control

All costs, both Forest Service and rancher (permittee), must be converted to net present value. Net long-term costs equal the difference between costs with weed control as opposed to costs without control measures. Net annual costs equal the difference between annual costs with and without the weed control program.

To conduct an economic analysis of weed control on National Forest System lands, basic information is needed. Specifically, the acreage of weed infestation and type and location is needed. At this point in time, estimated acreage or coverage does not exist. Field inventory or estimates are necessary. From this data, costs and benefits of weed control can be derived that include operation and maintenance costs for the Agency and the Permittee (rancher). Comparisons are made between existing annual costs for weed control and long run costs with and without weed control.

Research Needs

This study, while limited in scope, points out that the economics of weed control have been primarily focused on the impacts incurred by the western range livestock industry. Little empirical data exist from which to make solid decisions. Information involving joint product relationships is required to fully understand the issue.

If national legislation is passed by the Congress, applied research should focus on inventory and analysis of the aerial extent of weed infestation or undesirable plant coverage, and the economic impacts involved. Research also should be focused on the integrated, joint product relationships that exist on rangelands.

Conclusions

Noxious weeds are a recurring problem and annually cause serious economic impacts to the western livestock industry. Loss of favorable forage production results primarily from competition from undesirable plants. Economic losses from undesirable plants (noxious weeds and/or poisonous plants) on agriculture, especially the range livestock industry, are significant. In 1989, annual losses in 17 western state are estimated to exceed $340 million. Losses vary by type and kind of livestock. Other economic losses may be incurred through the loss of harvestable wildlife. Forage production can be increased significantly through control and/or eradication of noxious plants. Even with low to moderate beef cattle prices, and with only 5 to 10 year treatment life or control, benefits may generally be expected to exceed the range improvement costs of controlling or eradicating noxious plants. If record 1979 beef cattle prices are used, benefits exceed costs by more than 30%.

Control of undesirable plants is also favorable to management of forest lands for timber production and harvest. Positive benefits are realized in reforestation of selected timber species.

As much as the problem is recognized, data problems of the past still prevail. Researchers state "it is impossible, or at least impractical, to get definitive answers from which to make empirical estimates of the total value of economic losses to the livestock industry caused by poisonous plants."

Research is needed in the economics of weed control. Specific areas of attention should be on joint product interrelationships and multiple resource benefits. Inventory of the aerial extent of noxious weed coverage also is needed, along with refinement of the impacts to agriculture—crops and livestock.

References

1. Bingham S. High Country News, Vol. 22, No. 5, pp. 18-19, March 12, 1990.
2. Stewart RE and C Row. USDA Forest Service Res. Paper SO-55, Southern Forest Experiment Station (1981).
3. James LF. In: *The Ecology and Economic Impact of Poisonous Plants on Livestock Production* (LF James, MH Ralphs, and DB Nielsen, Eds.), pp. 1-4. Westview Press, Boulder, CO, 1988.
4. Nielsen DR, NR Rimbey, and LF James. In: *The Ecology and Economic Impact of Poisonous Plants on Livestock Production* (LF James, MH Ralphs, and DB Nielsen, Eds.), pp. 5-14. Westview Press, Boulder, CO, 1988.
5. Nielsen DB. Unpublished paper.

458

6.   American Institute of Real Estate Appraisers. *The Appraisal of Real Estate*, p. 233. Chicago, 1962.
7.   Godfrey EB, DB Nielsen, and NR Rimbey. In: *The Ecology and Economic Impact of Poisonous Plants on Livestock Production* (LF James, MH Ralphs, and DB Nielsen, Eds.), pp. 17-25. Westview Press, Boulder, CO, 1988.
8.   American Institute of Real Estate Appraisers. 1983.
9.   Tanaka JA and V Conrad. Presented to SRM, Reno, NE, 1990.
10. Loomis J, D Donnelly, and C Sorg. Western Regional Research Publication W-133, Interim Report, pp. 157-170, 1988.
11. Bureau of Land Management. *Final Environmental Impact Statement: Proposed Domestic Livestock Grazing Program for the Challis Planning Unity.* Idaho State Office, Bureau of Land Management, 1977.
12. Tom Parker, Game Manager, Idaho Dept of Fish and Game, Salmon, ID. Personal Communication, Sept. 23, 1985.
13. Thomas JW. *Transactions of North American Wildlife and Natural Resources Conference*, pp. 455-468, 1984.
14. Wolfe GJ and WR Lance. J. Range Manage. 37:59-63 (1984).
15. Jessup DA, HC Boermans, and ND Kock. J. Amer. Vet. Med. Assoc. 189:1173-1175 (1986).
16. Laycock WA. J. Range Manage. 31:335-342 (1978).

Index

Printed and bound by CPI Group (UK) Ltd, Croydon, CR0 4YY

23/10/2024

01778240-0018